Results and Problems in Cell Differentiation

A Series of Topical Volumes in Developmental Biology

2

Origin and Continuity
of Cell Organelles

Edited by J. Reinert, Berlin, and H. Ursprung, Zürich

With contributions of

R. Baxter, Sittingbourne · C. E. Bracker, Yellow Springs
R. M. Brown, Jr., Chapel Hill · R. Buvat, Marseille
J. H. Campbell, Los Angeles · R. D. Campbell,
Irvine · F. A. L. Clowes, Oxford · M. Dauwalder, Austin
C. Fulton, Waltham · J. E. Kephart, Austin
A. P. Mahowald, Milwaukee · H. H. Mollenhauer, Lafayette
D. J. Morré, Lafayette · E. Schnepf, Heidelberg · W. Stubbe, Düsseldorf
L. G. Tilney, Philadelphia · W. G. Whaley, Austin

With 135 Figures

Springer-Verlag New York · Heidelberg · Berlin 1971

ISBN 0-387-05239-9 Springer-Verlag New York Heidelberg Berlin
ISBN 3-540-05239-9 Springer Verlag Berlin Heidelberg New York

© by Springer-Verlag Berlin · Heidelberg 1971. Library of Congress Catalog Card Number 75-132272. Printed in Germany. Typesetting, printing and bookbinding: Brühlsche Universitätsdruckerei Gießen.

Preface

The first volume of the series, on "The Stability of the Differentiated State" received many favorable reviews from the scientific community. Many readers seem to agree with us that publication of topical volumes is a worthwhile alternative to periodic compilations of rather unrelated, though up-to-date reviews.

Production of topical volumes is however, plagued with one great difficulty, that of "author synchronization". This difficulty explains the lag between volumes 1 and 2 of the series. Nevertheless we hope that the present volume will be appreciated as a valuable source of information on its central topic: How do cell organelles originate, and what mechanisms assure their continuity?

Tübingen, Berlin, Zürich,   W. BEERMANN, J. REINERT, H. URSPRUNG,
Heidelberg   H.-W. HAGENS

Contents

Origin and Continuity of Golgi Apparatus

by D. James Morré, H. H. Mollenhauer, and C. E. Bracker

Origin and Continuity of Cell Vacuoles

by Roger Buvat

Origin and Continuity of Polar Granules
by Anthony P. Mahowald

Centrioles
by Chandler Fulton

Origin and Continuity of Microtubules
by Lewis G. Tilney

Origin and Continuity of Desmosomes
by RICHARD D. CAMPBELL and JOHN H. CAMPBELL

On Relationships between Endosymbiosis and the Origin of Plastids and Mitochondria
by EBERHARD SCHNEPF and R. MALCOLM BROWN, JR.

Cell Organelles and the Differentiation of Somatic Plant Cells

by F. A. L. CLOWES

Contributors

Baxter, Robert, Dr., Shell Research Ltd., Woodstock Agricultural Research Centre, Sittingbourne, Kent (Great Britain)

Bracker, C. E., Dr., Charles F. Kettering Research Laboratory, Yellow Springs, Ohio (USA)

Brown Jr., R. Malcolm, Dr., Department of Botany, University of North Carolina, Chapel Hill, North Carolina (USA)

Buvat, Roger, Dr., Laboratoire de Cytologie Végétale, Faculté des Sciences, Université d'Aix-Marseille (France)

Campbell, John H., Dr., Department of Anatomy, School of Medicine, University of California, Los Angeles (USA)

Campbell, Richard D., Dr., Department of Developmental and Cell Biology, University of California, Irvine, California (USA)

Clowes, F. A. L., Dr., Botany School, South Parks Road, Oxford (Great Britain)

Dauwalder, Marianne, Dr., The Cell Research Institute, University of Texas, Austin, Texas (USA)

Fulton, Chandler, Dr., Department of Biology, Brandeis University, Waltham, Massachusetts (USA)

Kephart, Joyce E., Dr., The Cell Research Institute, University of Texas, Austin, Texas (USA)

Mahowald, Anthony P., Dr., Biology Department, Marquette University, Milwaukee, Wisconsin (USA)

Mollenhauer, H. H., Dr., Department of Biology, Purdue University, Lafayette, Indiana (USA)

Morré, D. James, Dr., Department of Botany and Plant Physiology, Purdue University, Lafayette, Indiana (USA)

Schnepf, Eberhard, Dr., Lehrstuhl für Zellenlehre der Universität, Heidelberg (Germany)

Stubbe, Wilfried, Dr., Botanisches Institut der Universität, Düsseldorf (Germany)

Tilney, Lewis G., Dr., Department of Biology, University of Pennsylvania, Philadelphia, Pennsylvania (USA)

Whaley, W. Gordon, Dr., The Cell Research Institute, University of Texas, Austin, Texas (USA)

Assembly, Continuity, and Exchanges in Certain Cytoplasmic Membrane Systems

W. Gordon Whaley, Marianne Dauwalder, and Joyce E. Kephart

The Cell Research Institute, University of Texas, Austin, Texas

Basic to the definition of a cell is a membrane separating the activities within it from the surrounding environment. In even the simplest organisms, the plasma membrane, or plasmalemma, bounding the cellular mass is characterized by definable structure and distinctive physiological properties. The structure, composition, and properties of membranes have all been subjects of extensive, interrelated studies. This work will deal with questions about their assembly, continuity, and exchange. Only brief consideration will be given to structure, composition, and properties to provide a basic understanding that will lend coherence to the other questions.

The eukaryotic cell is characterized by a substantial amount of intracellular compartmentalization by membranes. The membranes function both as selective barriers and reaction surfaces. That some sort of membrane transfer or flow exists in such cells has been apparent for some time (SCHNEPF, 1969). The focus here will be on exchanges among cellular membranes and the types of modifications that must occur concurrently with such exchanges. Partly because of lack of sufficient knowledge about the fundamental structure of membranes and its relation to specific functioning and about the precise molecular components of membranes, many of the basic questions cannot be answered with certainty. However, some observations can be made with assurance and the feasibility of some suggestions concerning interrelationships can be evaluated.

I. The Nature of the Membrane

Much knowledge about the physiological properties of biological membranes was accumulated in early studies of permeability, which also sparked a series of interpretations of the structure of membranes (GORTNER and GRENDEL, 1925, and others; see DANIELLI, 1967). Ultrastructural studies of recent years have accepted as one of their major challenges the testing of these interpretations and contributing to an understanding of the structure of the membrane. Although subject to certain limitations in meeting these challenges, these studies have strengthened the view that there are common elements of organization among all biological membranes. Combined with cytochemical and radioautographic investigations, they have served to broaden the concept of a membrane from that of essentially a permeability barrier to one which additionally includes its being a surface on which many reactions are carried out (see DALTON and HAGUENAU, 1968). Further, they have demonstrated that many membranes, or at least many membrane-associated materials, are characterized by specificity factors (see DAVIS and WARREN, 1967; SJÖSTRAND, 1968). This broadened concept of a biological membrane holds that many of the functional characteristics of the cell are associated with one or another of its membranes. Still more recent direct chemical analyses of purified membrane fractions (MADDY, 1966, 1967; BENEDETTI and EMMELOT, 1968; COOK, 1968a, b; MALHOTRA and VAN HARRE-VELD, 1968; ROTHFIELD and FINKELSTEIN, 1968; ROUSER et al., 1968; KORN, 1969) confirm postulates from the earlier work that the membranes are composed largely of lipids and proteins. They also make it evident that carbohydrates are frequent constituents. Information about the specific constituents of membranes associated with particular functions may well lead to much more satisfactory understanding of the differentiation of membranes and the transfers and transformations they undergo.

The advancement of techniques to the point where fractions of different cellular membranes can be separated and analysed has indicated that, despite the more or less common images, the makeup of membranes differs from species to species, from tissue to tissue, and from one cellular organelle to another in terms of the proportions of the various components and of the particular molecules in each class (Table 1, from KORN, 1969). As will be seen there are also clear indications of area differences in given membranes. Functional attributes of membranes relate to their specific molecular composition and architecture. It follows that if membrane transfer occurs, since it involves functional alteration, it must also be accompanied by changes in composition, structure, and specificity factors.

Table 1. *Protein and lipid content of membranes. From* KORN (1969). *The abbreviations are:* Cer, *cerebrosides;* DPG, *diphosphatidylglycerol;* GalDG, *galactosyldiglyceride;* PA, *phosphatidic acid;* PC, *phosphatidylcholine;* PE, *phosphatidylethanolamine;* PGaa, *amino acyl esters of phosphatidylglycerol;* Plas, *plasmalogen;* SL, *sulfolipid;* Sph, *sphingomyelin. For further explanation of sources, see* KORN (1969)

Membrane	Protein/Lipid	Cholesterol/Polar Lipid	Major Polar Lipids
	wt/wt	mole/mole	
Myelin	0.25	0.7—1.2	Cer, PE, PC
Plasma membranes			
Liver cell	1.0—1.4	0.3—0.5	PC, PE, PS, Sph
Ehrlich ascites	2.2		
Intestinal villi	4.6	0.5—1.2	
Erythrocyte ghost	1.5—4.0	0.9—1.0	Sph, PE, PC, PS
Endoplasmic reticulum	0.7—1.2	0.03—0.08	PC, PE, Sph
Mitochondrion			DPG, PC, PE, Plas
Outer membrane	1.2	0.03—0.09	
Inner membrane	3.6	0.02—0.04	
Retinal rods	1.5	0.13	PC, PE, PS
Chloroplast lamellae	0.8	0	GalDG, SL, PS
Bacteria			
Gram-positive	2.0—4.0	0	DPG, PG, PE, PGaa
Gram-negative		0	PE, PG, DPG, PA
PPLO	2.3	0	
Halophilic	1.8	0	Ether analogue PGP

The membrane images revealed by early electron microscopy studies (FINEAN, 1953; SJÖSTRAND and RHODIN, 1953; SJÖSTRAND and HANZON, 1954a; ROBERTSON, 1955) led ROBERTSON (1959) to postulate a fundamental structure which he termed the "unit membrane". The unit membrane concept relates, in part, to the pauci-molecular theory of membrane structure set forth by DAVSON and DANIELLI (see DAVSON and DANIELLI, 1943; see also DANIELLI, 1967). This theory was based largely on permeability studies and the properties of membranes revealed by them. The concept also relates, in part, to X-ray diffraction and other studies of the myelin sheath, many of them by SCHMITT and his co-workers (see ROBERTSON, 1959). Both the theoretical base of the unit membrane concept and the applicability of the results of the myelin studies to other cellular membranes have been questioned. Many investigators have called attention to the fact that uniform lipid bilayer structure does not adequately explain certain membrane phenomena, including the degree of differentiation that must exist between various cellular membranes or even spatially and/or ontogenetically within a given cellular membrane (see LUCY, 1968a). SJÖSTRAND (1968) has presented a summary of serious qualifications about the use of the myelin sheath as a model membrane. These questions have led several investigators to seek alternative interpretations of cellular membrane structure. Before considering these alternatives, a further description of the unit membrane concept is in order. This concept holds that cellular membranes are composed of double lipid layers and associated nonlipid material. (The concept of the unit membrane initially had reference to cell surface membrane. SJÖSTRAND had pointed out [SJÖSTRAND, 1953a, b] that there might be common principles of organization in both surface and intracytoplasmic membranes, although he also observed differences [SJÖSTRAND, 1956].)

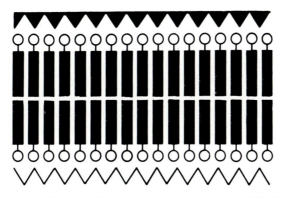

Fig. 1. Highly schematic diagram of the unit membrane pattern. The lipid polar groups are indicated by circles; the nonpolar carbon chains by bars and nonlipid monolayers by zig-zag lines. The upper zig-zag lines are partially blocked in to indicate that there is a chemical asymmetry in the membrane surface, the outside being chemically different from the inside in some important way. This may be due to the presence of a high concentration of mucopolysaccharide or mucoprotein on the outside surface with a dominance of protein on the inside. (This diagram is based upon ROBERTSON's interpretation of the myelin sheath which is an outer bounding membrane.) Diagram and legend from ROBERTSON (1965)

A recent (1965) version of ROBERTSON's diagram of the unit membrane is shown in Fig. 1. This version differs from the original diagram in that it takes into account the frequently emphasized asymmetry of cellular membranes and proposes that the nonlipid layer on one side may be protein and that on the opposite side partly polysaccharide.

Most of the alternatives to the unit membrane concept hold that the molecular components of membranes are arranged in subunits which are, in turn, associated to make up the membrane. The molecules of phospholipids in experimental phospholipid-water systems have been shown to be grouped in specific ways (LUZZATI and HUSSON, 1962; STOECKENIUS, 1962). Membrane breakdown studies have suggested that certain molecular components, perhaps including both phospholipids and proteins, are not degraded to the molecular level (HOKIN, 1968). These observations

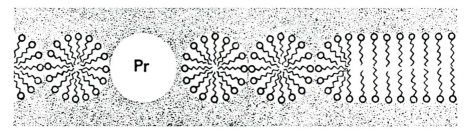

Fig. 2. Diagram of a cross-sectional view of a membrane in which globular micelles of lipid are in dynamic equilibrium with a bimolecular leaflet of lipid. A layer of protein and/or glycoprotein is shown on each side of the lipid layer. The structure of each lipid molecule is illutrated in a diagrammatic fashion: only a polar group (o) and a nonpolar moiety (⌇) are shown, and the lipid may be phospholipid or nonphospholipid. One globular micelle of lipid has been replaced by a globular protein molecule (pr) which may be a functional enzyme. Figure and legend from LUCY (1968b)

lend credence to the possibility of subunit structures in membranes. What appear to be subunits are detectable along the fracture planes of membranes prepared for electron microscopy studies by freeze etching (BRANTON, 1966). The subunit hypothesis is attractive for the possibilities it offers for explaining functional variation within membranes. LUCY (1964, 1968a, b) has suggested that membranes may be, in part, organized into subunits, and, in part, into molecular leaflets of the Danielli type (Fig. 2).

Some studies of phospholipid-water systems have shown different patterns of association of the phospholipid molecules and changes from one type of association to another as a result of modified conditions (STOECKENIUS, 1962; see Fig. 3). Presumably these different patterns of association and perhaps others could exist within a cellular membrane (see STOECKENIUS and ENGELMAN, 1969). The obvious functional differentiation in membranes may well relate to differences in structure in different parts of the membrane.

For further considerations of alternatives to the unit membrane concept, the reader is referred to KORN (1966), LUCY and GLAUERT (1967), GLAUERT and LUCY (1968), CHAPMAN (1968), SJÖSTRAND (1969) and STOECKENIUS and ENGELMAN (1969). Various models representing hypothetical interpretations of subunits of different character have now been adduced by a number of investigators.

Membrane dynamics is a complex subject which must deal with synthesis of membrane components, assembly, changes in area organization, continuous turnover of molecular constituents, and transfer. Some of the considerations have been summarized by SIEKEVITZ et al. (1967). No attempt will be made here to deal with questions of synthesis, but brief mention must be made of the turnover of membrane constituents because this represents transfer at one level. OMURA, SIEKEVITZ, and PALADE (1968) have shown continuing turnover of membrane proteins and lipids and demonstrated differences in the rates of this turnover between the proteins and the lipids and between different categories of lipids. HOKIN (1968) has reviewed evidence from a series of secretion stimulation studies which suggests that in membrane breakdown certain components are broken down to their building blocks

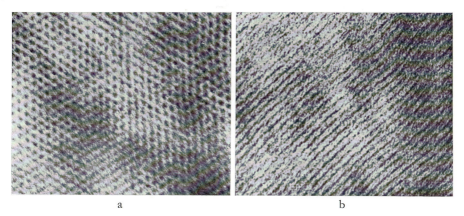

a   b

Fig. 3a and b. Electronmicrographs of phospholipids in water: (a) hexagonal phase; (b) lamellar phase. From STOECKENIUS (1962). Magnification (a) 532,000 $\times$; (b) 430,000 $\times$

whereas others are not. His results have led him to suggest that in the recirculation of membrane components, some may participate at the molecular level whereas others participate as part of subunits containing several phospholipids and proteins (see Fig. 19). This turnover, whether partly at the subunit level or not, must be considered in reference to the changing composition and specific functional state of the membrane. LUZZATI (1968) and LUZZATI et al. (1969) have postulated that any given portion of a membrane may be in a transient state. Such transient states relate to changes in functional characteristics of membranes, including those attendant upon transfer of membrane segments.

II. The Assembly of Membranes

A discussion of membrane assembly at this time must revolve around an unanswered question: are all membranes developed from already existing membrane or may at least some of them be formed by the assembling of components under the influence of conditions existing at a particular site in the cell?

There are no critical data to guide a choice of alternatives nor grounds for supposing that one or the other always pertains. Whatever the method of membrane assembly, there must be pools in the cell with which the molecular components of the membranes may be exchanged.

LUZZATI and HUSSON (1962) and LUZZATI et al. (1969) have developed a series of interpretations of the molecular arrangements in several different phases of dispersed phospholipids of the sort commonly found in membrane systems. STOECKENIUS (1962) was able to take electron micrographs that showed good agreement with the X-ray diffraction data concerning both hexagonal and lamellar phases of phospholipids and by altering conditions was able to bring about formation of the lamellar phase. Maddy (1967) has been able to fractionate membranes, separate their lipid and protein components, and reconstitute them structurally. Such *in vitro* experiments as these demonstrate that membrane-like arrangements of phospholipids and proteins can be formed in the absence of pre-existing membranes, but they do not prove that this is how membranes are assembled in the cell. Maddy's work (1967, 1969) suggests that much of the protein associated with the membrane lipids may have a structural function while a lesser amount of it is metabolically active, though he emphasizes the undesirability of attempting to draw a sharp line between structural and other protein. He concludes, in agreement with earlier ideas (see Danielli, 1967), that the protein adds stability to the membrane and then postulates that the mutability of membranes may relate primarily to the lipid fraction.

The comparability and differences between artificial "membranes" and membranes of living cells has been considered at some length by several investigators (see Tosteson, 1969; Tria and Scanu, 1969). In general, the observations tend to support the idea that the artificial systems may have some structural comparability with portions of functional cellular membranes. But the differences are numerous, once again indicating the complex character of biological membranes.

Direct studies of membrane composition and structure are beset by technical difficulties. The adaptation of electron microscopic studies to biological membranes brought the hope that some of the questions could be resolved by visualization of membrane structure. However, none of the techniques available for preparing cells for study is without the possibility of greatly modifying structure. Nonetheless it is possible to add to the earlier knowledge of the composition and general properties of membranes some observations concerning the nature of the differentiations involved and the changes associated with changes in cellular activity including various transfers and transformations.

III. The Growth and Transfer of Membranes

The movement of membrane components in the cell is readily detectable microscopically only in terms of organized membranes. There are detectable transfers of organized membranes that link together the nuclear envelope, the endoplasmic reticulum, the Golgi apparatus, the plasma membrane, the lysosomal system, and perhaps some other vesicles and vacuoles. This provides the rationale for considering these membranes part of an exchange system. There is little evidence that membranes of the plastids and the mitochondria are linked to this system at this level. Hence, they, and some other membranes about which evidence is not adequate, will be omitted from consideration here. It must be recognized, however, that visible transfers of membrane must constitute only a part of the total movement of membrane components and that exchanges between all of them and cellular pools supported by synthesis must occur.

A. The Nuclear Envelope

In all eukaryotic cells the nucleus is separated from the cytoplasm by a double-membrane envelope. This double-membrane structure encloses the so-called perinuclear space in which certain enzymes have been demonstrated cytochemically and a number of reactions presumably are carried out. This envelope is characterized by pores (Fig. 4). Some materials pass between nucleus and cytoplasm through the pores but materials pass through the membranes of the envelope as well. The relative

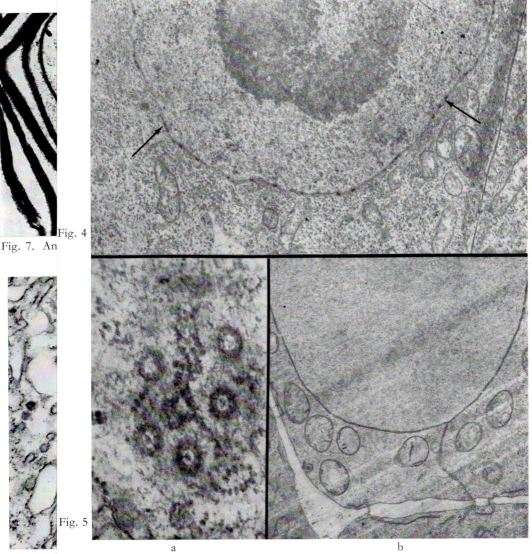

a   b

Fig. 4. Pores (see arrows) in a nuclear envelope demonstrated by treatment with $Zn(MnO_4)_2$. *Zea mays*. By H. H. MOLLENHAUER. Magnification 11,500 ×

Fig. 5a. Surface view of nuclear envelope showing pores and attached ribosomes. *Zea mays*. By MARIANNE DAUWALDER. Magnification 80,000 ×

Fig. 5b. Micrograph showing continuity of nuclear envelope with endoplasmic reticulum. Permanganate fixation fails to show ribosomes. *Zea mays*. By H. H. MOLLENHAUER. Magnification 9,500 ×

material surrounded by nuclear envelope membrane away from the envelope. This membrane could be transferred to the endoplasmic reticulum or to the plasma membrane. There are somewhat stronger suggestions of its transfer to the Golgi apparatus (J. C. Weston, personal communication; Falk, 1967; Falk und Kleinig, 1968; Fig. 8).

In many types of cells, among them certain highly differentiated secretion cells and early developmental stage cells, the Golgi apparatus is located close to the nuclear surface. As such cells become differentiated there may be a movement of the Golgi apparatus toward the more peripheral regions of the cytoplasm (Trelstad, 1970).

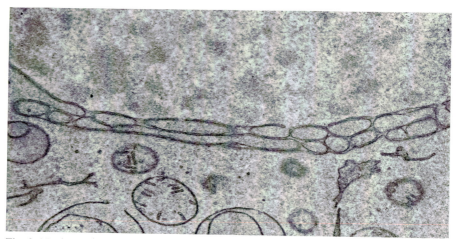

Fig. 9. Nuclear envelope which has proliferated in response to mechanical injury. *Zea mays*. Permanganate fixation fails to show ribosomes. By H. H. Mollenhauer. Magnification 17,800 ×

This positional relationship carries with it an implication that it may indicate more ready transfer of materials to the Golgi apparatus, possibly including nuclear membrane segments (Zamboni and Mastroianni, 1966a, b). Longo and Anderson (1969) have postulated that blebbing from the pronuclear envelope in early developmental stages in the rabbit may relate directly to development of the Golgi apparatus. There have been occasional interpretations of the formation of other organelles and inclusions from the nuclear envelope, but the evidence is very inconclusive.

Many types of cells have been shown to develop much more internal membrane when injured or exposed to toxic agents. Among such reactions, proliferation of what seems to be nuclear envelope is common (Fig. 9). The fate of such proliferated membrane has not been followed.

In some instances there are specialized transfers of nuclear envelope. Annulate lamellae result, in different patterns, from developments that begin with the transfer of nuclear envelope membrane outward into the cytoplasm (Kessel, 1968a; Fig. 10).

Carroll (1967) has shown separation of segments of the nuclear envelope and their subsequent transfer to constitute the delimiting membranes of developing ascospores in *Saccobolus*. In this instance there is a modification of segments of the nuclear envelope, transfer away from the initial sites, and ultimately fusion to form the spore membranes (Fig. 11). The modification in this instance appears to involve fusion of nuclear envelope and segments of the endoplasmic reticulum.

Material synthesized or assembled in the perinuclear space may be transferred in nuclear envelope-membrane enclosed vesicles to the cell surface and secreted in much the same fashion as secretory materials transferred from the Golgi apparatus. Bouck (1969) has shown that in certain species of algae where in flagella development the surface comes to be covered with microtubular shafts of specific architecture,

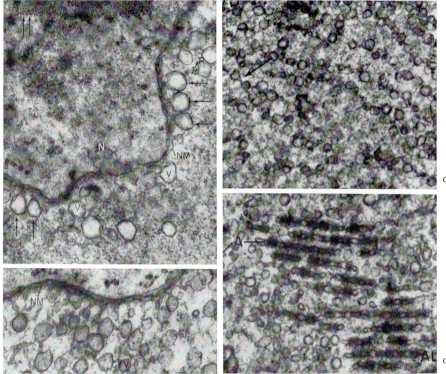

Fig. 10a—d. Development of annulate lamellae in oocytes of *Necturus*. (a) Blebbing of nuclear membrane (NM) to form vesicles (V) (see arrows); (b) continued blebbing forms numbers of vesicles which move out in the cytoplasm (CYV); (c) at a later stage these vesicles form chains which become organized into concentric layers; (d) fusions and other modifications result in annulate lamellae (AL). From Kessel (1963). For larger micrographs and more legend detail, see original. Magnification   (a)   27,000 ×;   (b)   27,000 ×;   (c)   18,000 ×;
(d) 24,000 ×

the presumptive stages of these shafts, or mastigonemes, can be found in expanded regions of the perinuclear space. The implication is that these structures are formed here (and probably in the endoplasmic reticulum as well), and then transported in membrane-bound vesicles to the cell surface where they are discharged and undergo subsequent arrangement and development in much the same manner as flagella scales secreted from the Golgi apparatus (Manton et al., 1965). There have been several observations that suggest that the assembling of secretory materials, most studied in the Golgi apparatus, can take place under certain conditions or in certain species in the endoplasmic reticulum or the nuclear envelope. These observations

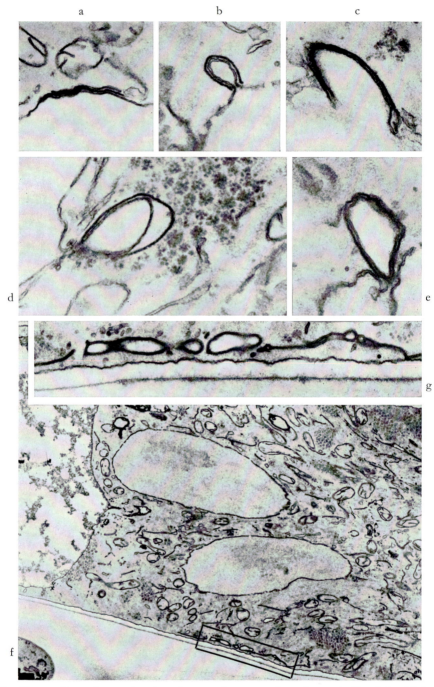

Fig. 11 a—g. Stages in transfer of nuclear envelope to surface of ascus in *Saccobolus*. (a) Area of presumptive blebbing on the surface of the fusion nucleus. ER cisternae are appressed to the nuclear envelope. (b) Nuclear envelope has evaginated; ER cisterna remains appressed, but still separate. (c) Stage in fusion of elements of the ER and nuclear envelope. (d) Detail of nuclear bleb and fusion with endoplasmic reticulum. (e) Cross section of nuclear bleb. (f) Membrane sac formation at the periphery of the ascus. (g) Detail of fusing membrane packets. By G. C. Carroll; partly from Carroll (1967). Magnification (a)-(e) 35,000 × ; (f) 9,000 × ; (g) 36,000 ×

tend to interrelate these systems, for a common feature seems to be transfer in membrane-enclosed vesicles with the membrane apparently being incorporated into the plasma membrane.

Brief consideration must be given to the breakdown of the nuclear envelope and its reconstitution in mitosis, for this development represents movement and extension of membrane. BAJER and MOLÈ-BAJER (1969) have summarized a good part of what is known about the breakdown and reconstitution of the nuclear envelope. They point out that in reconstitution some undetermined portion of the old envelope is apparently built into each new envelope, and that the endoplasmic reticulum seems to be involved. BAJER and MOLÈ-BAJER also noted that in their experimental material, *Haemanthus* endosperm, segments of nuclear envelope not incorporated into newly formed envelopes persist in identifiable form in the cytoplasm until the next division.

Transfers of membrane from the nuclear envelope are detectable in the development of the annulate lamellae, they are suggested in the blebbing off of vesicles that may move to the Golgi apparatus, and they are indicated in a few instances in which secretory products may move to the plasma membrane. If the endoplasmic reticulum is a derivative of the nuclear envelope, then a mass transfer of membrane might be involved with the separation of the endoplasmic reticulum from the nuclear envelope. Flow of membrane components into nuclear envelope must necessarily be great enough to compensate for molecular turnover. Whether actual membrane moves into the nuclear envelope has not been established, but the specialized structure of the envelope would make this seem unlikely.

B. The Endoplasmic Reticulum

The amount of endoplasmic reticulum (Fig. 12) varies with type of cell, stage of development, and level of activity, both in total and in terms of proportions between rough and smooth. There are now extensive data indicating that protein storage and possibly modification, lipid synthesis, and at least some carbohydrate synthesis take place in the endoplasmic reticulum. Membrane components may thus be largely of endoplasmic reticulum origin. Both normal changes in the amount of endoplasmic reticulum accompanying development and induced changes in the amount resulting from the introduction of toxic agents or injury have been followed in considerable detail. Much of this work is beyond the scope of this chapter, but some of it bears on the question of sites of assembly of membrane components and extension of membranes. Most of the induction experiments and comparisons of them with normal development changes have been carried out on liver cells. DALLNER, SIEKEVITZ, and PALADE (1966) have shown that in rapidly differentiating early stage cells formation of rough endoplasmic reticulum precedes that of smooth. These investigators postulated that membrane phospholipids and proteins were assembled in the rough endoplasmic reticulum and subsequently in some manner transferred to smooth endoplasmic reticulum. TZUR and SHAPIRO (1964) had shown, by labelling experiments, that the amount of phospholipid synthesis was regulated by the amount of protein available for binding. DALLNER, SIEKEVITZ, and PALADE (1966) concluded that the TZUR and SHAPIRO findings and their own work suggested that the amount of new membrane formed was determined by the amount of membrane protein available in the rough endoplasmic reticulum where the linkage of phospholipids and proteins apparently takes place.

Orrenius, Ericsson, and Ernster (1965) showed the administration of phenobarbital to bring about substantially increased enzyme activities related to detoxification and a concurrent increase in the extent of endoplasmic reticulum. That this membrane increase is a result of outgrowths and budding off from existing membranes was demonstrated by Orrenius and Ericsson (1966). They also showed the induced

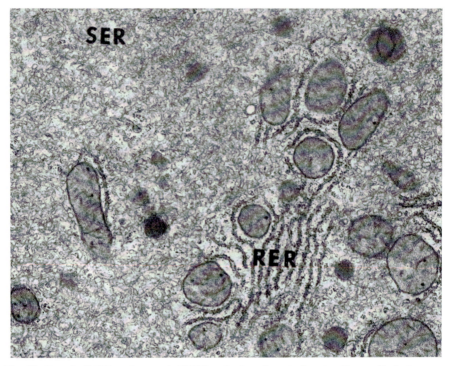

Fig. 12. Rough (RER) and smooth (SER) endoplasmic reticulum in a liver cell of a hamster. This micrograph was taken after administration of progesterone which increases the amount of smooth endoplasmic reticulum. From Emans and Jones (1968). Magnification 18,400 ×

enzyme activities to regress much sooner than the membranes disappeared, thus establishing a distinction between these and other membrane proteins. Stäubli, Hess, and Weibel (1969) are in agreement with Dallner, Siekevitz, and Palade (1966) that in normal development the increase in rough endoplasmic reticulum consistently precedes that of smooth endoplasmic reticulum. They have observed, however, that in the instance of the phenobarbital induction of membrane extension, rough endoplasmic reticulum and smooth endoplasmic reticulum increase simultaneously. These investigators suggest that in normal development smooth endoplasmic reticulum results from detachment of ribosomes from rough endoplasmic reticulum. They note that in the case of phenobarbital treatment metabolic processes are substantially modified and they discuss the possibility that in these instances the smooth endoplasmic reticulum might be budded off from the rough form and/or that various modifications in the normal transformations might apply. Kuriyama et al. (1969) have shown changes in the rate of degradation of the protein components of rat liver microsomal membranes at different stages after phenobarbital treatment. Reduced rates of membrane breakdown might well account for at least some membrane increase.

Mechanical injury not involving specific toxicity reactions will induce additional endoplasmic reticulum (WHALEY, KEPHART, and MOLLENHAUER, 1964). It is not known whether such instances involve increased synthesis of membrane components.

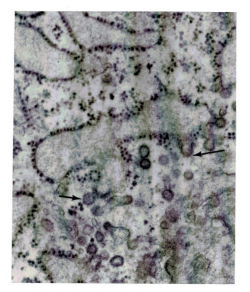

Fig. 13. Blebbing from rough endoplasmic reticulum (see arrows). *Helix aspersa* nuchal cell. By MARIANNE DAUWALDER. Magnification 39,000 ×

DALLNER, BERGSTRAND, and NILSSON (1968) and DALLMAN et al. (1969) have in ferred from studies of heterogeneous populations of microsomes that sectors of th endoplasmic reticulum vary with respect to enzymatic activities. This suggests functional differentiation in what may appear to be uniform membranes. This is consistent with some suggestions from morphological and cytochemical studies and with the general postulate that different activities characterize different regions of the endoplasmic reticulum. Some aspects of this differentiation may well relate to modifying the membrane for transfer. There is convincing electron micrographic evidence that such transfer takes place. Blebbing from one membrane of the endoplasmic reticulum is frequently observed (Fig. 13). The principal question appears to be not whether membrane is transferred from the endoplasmic reticulum, but where it goes. This question is difficult to answer from morphological observations because such membrane seems to be separated in the form of small vesicles, and there are usually many small vesicles in the cytoplasm, probably of different origins. One possibility would be direct transfer to the plasma membrane. Ross (1969), (see also Ross and BENDITT, 1965) has suggested such a possibility in connection with the secretion of the precursor of collagen from fibroblasts. The assumption in this case, as in other instances of secretion, would be that the transferred membrane is incorporated into the plasma membrane.

The extensive smooth endoplasmic reticulum in certain testicular cells (CHRI-STENSEN and FAWCETT, 1961; CHRISTENSEN, 1965) has been shown to be implicated in steroid biosynthesis. The manner in which the steroid hormones are transferred out of the cell in such instances is not known, but the observation that the Golgi apparatus appears not to be involved makes it necessary to consider the possibility that they may be transferred in membrane-bound vesicles from the endoplasmic reticulum. The morphology of these cells makes definite interpretations difficult. There may be continuities between this smooth endoplasmic reticulum and the plasma membrane, either persistent or transitory. Such continuities have been observed in some other types of cells.

Various gastric acid-producing cells and the chloride cells of certain marine organisms are characterized by extensive, smooth tubular systems concerning which some affinities with the plasma membrane are known. In gastric acid-secreting amphibian cells SEDAR (1969) has postulated that, during secretion, tubules and/or vesicles of this system may be incorporated into the plasma membrane which becomes greatly increased. He has interpreted findings of exogenous peroxidase in the tubular elements of the cell to indicate that, at times at least, the plasma membrane and the tubular elements are continuous. If membranes of the tubular system are actually incorporated into the plasma membrane, this is an instance of membrane transfer. RITCH and PHILPOTT (1969) have shown extensive continuities between the plasma membrane and the internal tubular system in gill cells. The problem of interpretation in these systems is complicated by lack of knowledge concerning the comparability of these smooth tubular systems in these specialized cell types and the smooth endoplasmic reticulum in other cell types.

FORTE, LIMLOMWONGSE, and FORTE (1969) in a developmental study of HCl-secreting cells in the tadpole have noted the extensive development of the Golgi apparatus at certain stages and consequently the diminution in the extent of this organelle concomitant with the buildup of the smooth tubular system. This observation has led them to the supposition that the Golgi apparatus may be the source of the membranes of this system. If these smooth membrane tubular systems are derived from the Golgi apparatus, they may not equate to smooth endoplasmic reticulum as it is usually defined. There is more evidence suggesting flow of membrane from the endoplasmic reticulum to the Golgi apparatus than in the reverse direction.

In some instances secretory products of the smooth endoplasmic reticulum are sulfated and there has been a suggestion (see LONG and JONES, 1967) that the Golgi apparatus may be the site of sulfation in this as it is in numerous other instances. On the whole, however, one can only note that the mechanisms of secretion from smooth endoplasmic reticulum, and particularly the possibility of membrane transfer, are not yet well understood.

There are significant implications for the transport of endoplasmic reticulum membrane to the Golgi apparatus in radioautographic studies having to do with the assembly and transport of various secretory products. For the most part, these have had to do with products assembled in the rough endoplasmic reticulum. The sequence of sites at which label can be detected has given rise to the suggestion that there is a mass transport of labelled protein from the endoplasmic reticulum to the Golgi apparatus (see PALADE, 1966), quite possibly packaged in membrane derived from the endoplasmic reticulum and subsequently incorporated into the membranes of the Golgi apparatus. There is no doubt that such proteinaceous material moves from one organelle to the other. There is as yet not adequate proof that its transport involves transfer of membrane. JAMIESON and PALADE (1967a, b; 1968a, b) have made detailed studies of such transport of material between the endoplasmic reticulum and the condensing vacuoles of the pancreatic exocrine cell. They have noted the consistent association of small vesicles with the peripheral region of the Golgi apparatus and the appearance of label in these vesicles at one stage in the sequence. The possible involvement of these vesicles in membrane transfer is considered further in the section on the Golgi apparatus.

The endoplasmic reticulum is a highly differentiated membrane system in association with which phospholipids and proteins can be synthesized. It is thus a source of membrane components. In normal development, smooth endoplasmic reticulum appears to be derived from rough endoplasmic reticulum, and in many instances the two types are continuous. There are some suggestions of transfer of segments from both the rough and smooth endoplasmic reticulum to the cell surface. At least transient continuities between the endoplasmic reticulum and the plasma membrane have been suggested. The movement of proteins from the endoplasmic reticulum to the Golgi apparatus has frequently been assumed to involve transfer of membrane and micrographic evidence of the blebbing off of membrane segments is convincing. Direct membrane transfer into the Golgi apparatus has not been adequately demonstrated. While in some instances there may be contribution of membrane from the nuclear envelope to the endoplasmic reticulum, the increase of endoplasmic reticulum both developmentally and experimentally without any convincing evidence of transfer of membrane segments into it would seem to attest to its capacity to assemble membrane.

C. The Golgi Apparatus

Inasmuch as the greatest amount of information about membrane extension and transfer has become available from studies of secretion, it is not surprising that the Golgi apparatus occupies a central position in any consideration of these processes. This organelle is characteristically a stack of smooth, membrane-bound cisternae (Fig. 14). In this discussion one face of the stack is designated as the proximal face, the opposite as the distal face.

There is no general agreement on what terms should be used to designate the opposite faces of the Golgi stack. GRASSÉ (1957) originally applied the term proximal to that face of the parabasal body in zooflagellates (now equated to the Golgi apparatus) which was consistently found in close proximity to the parabasal filament. He termed the opposite face distal and thus established an implication of differences between the two faces. It now appears that the characteristics and activities at the opposite faces may be quite different. By one interpretation cisternal envelopes are formed at one face and modified along the axis to become secretory. This has led to the designation of what is here termed the proximal face as the forming or immature face and what is here described as the distal face as the mature or secreting face. Until more specific information is available regarding the activities at the opposite faces, it seems best not to apply functional or developmental terms, but to rely on simple designations that do no more than emphasize differences. For this reason, proximal and distal will be used here, even though GRASSÉ's initial use of these terms implies a topographical relationship that appears to be more the exception than the rule.

That formation of cisternae takes place in the Golgi apparatus is attested to by the increase in number of cisternae in normal differentiation and by the fact that, under certain experimental conditions, the number of cisternae can be markedly increased (Hall and Witkus, 1964; Kephart, Dauwalder, and Whaley, 1966). Neutra and Leblond's (1966a) interpretation of the functioning of the Golgi

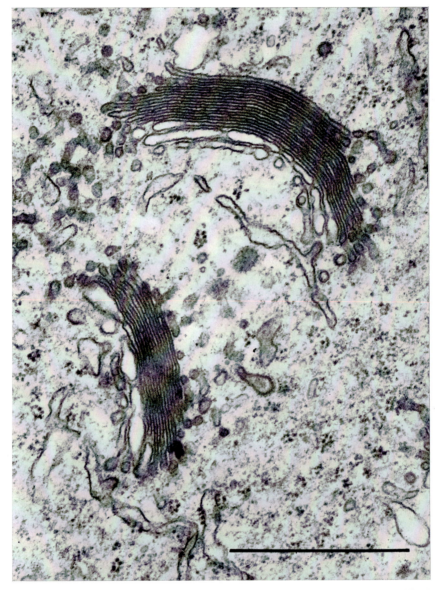

Fig. 14. Two Golgi "stacks". *Helix aspersa*. By Marianne Dauwalder. Magnification
49,000 ×

apparatus in colonic goblet cells is that cisternae at the distal face are completely transformed into secretory vesicles and the number of cisternae in the stack is kept relatively constant by the formation of new cisternae at the proximal face. Without attempting to answer the question of how the new cisternae are formed, NEUTRA and LEBLOND (1966a, b) have postulated that once formed each cisterna is successively displaced across the axis from one face to the other by a more recently formed cisterna. This suggestion is based upon radioautographic studies in which label was seen first in the cisternae near the proximal face, at later stages more or less step-wise

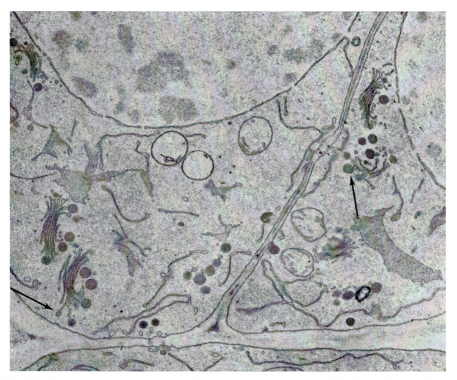

Fig. 15. Golgi apparatus of a form which produces secretory vesicles from the periphery of the cisternae (see arrows). *Zea mays.* By H. H. MOLLENHAUER. Magnification 17,800 ×

across the axis, and ultimately in mucigen granules successively displaced toward the apex of the cell. There are many instances in which vesicle production seems to be more or less confined to the edges of cisternae toward the distal face (Fig. 15). There seem to be several patterns of detachment of membrane from the Golgi cisternae, perhaps characteristic of particular sorts of secretory processes, and there are also suggestions that membrane destined for transfer to different sites may be separated in different ways. The existence of different patterns of separation raises some questions about the possible step-wise displacement of cisternae.

It has already been noted that there is frequently blebbing from the endoplasmic reticulum (and sometimes the nuclear envelope) adjacent to the Golgi apparatus and that small vesicles which could result from this blebbing are seen between the two organelles (Fig. 16). This observation has led to considerable speculation that new cisternae might result from transfer of segments of membrane from the endoplasmic reticulum to the Golgi apparatus. This hypothesis has been advanced with various degrees of directness by Sjöstrand and Hanzon (1954b), Essner and Novikoff (1962), Sjöstrand (1968) and a number of other investigators, mostly in the form of an assumption that cisternal envelopes are produced by fusion of small endoplasmic reticulum-derived vesicles. Proof by radioautographic means that there is a transfer of materials from the endoplasmic reticulum to the Golgi apparatus (see Palade, 1966) has lent a measure of support to this hypothesis. The assumption is that vesicles derived from the endoplasmic reticulum in this manner fuse at the proximal face of the Golgi apparatus to constitute new cisternae.

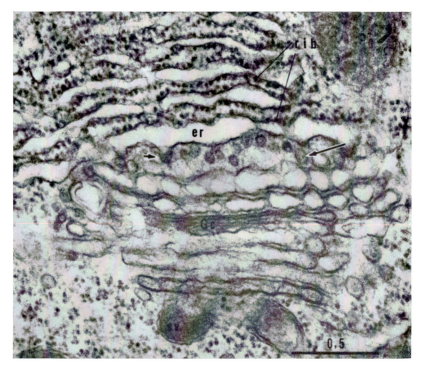

Fig. 16. Golgi cisternae (Gc) adjacent to endoplasmic reticulum which shows blebbing from a smooth profile (arrow) and vesicles between this blebbing surface and the proximal cisternae of the Golgi apparatus. Mealybug testis cell. From Whaley (1966). By F. R. Turner. Magnification 48,000 ×

Hirsch (1965) has proposed a related hypothesis suggesting that "buds" from the endoplasmic reticulum are transferred to the "Golgi field" where they undergo modification under the influence of this field to become cisternae, a single "bud" perhaps forming a cisterna. This does not necessarily assume fusion.

Such concepts of membrane transfer have many advocates, despite the absence of critical proof. If such transfer does take place, it is necessary to assume a change in the characteristics of the membrane concurrent with the transfer. The membranous components of the Golgi stack are quite different from those of the endoplasmic reticulum. The Golgi apparatus membranes are smooth, in part characterized by different enzyme activities, and generally have different functions than the endoplasmic reticulum. Differences in Golgi cisternae along the axis in the stack suggest that however cisternal membranes originate, they are subject to considerable developmental modification within the stack. Before considering this matter and some postulates of how membranes are further transferred, it seems desirable to give attention to an alternate possibility for cisternal envelope formation.

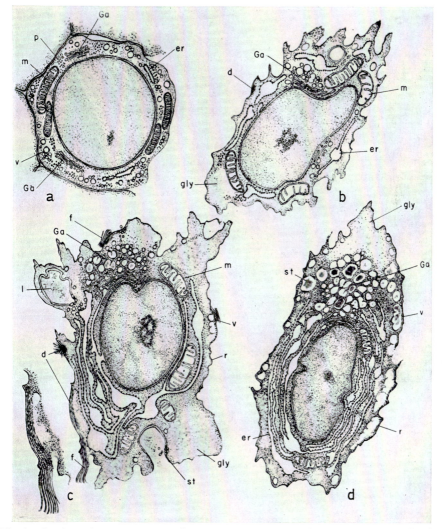

Fig. 17. A schematic representation of successive stages in development of the chondroblast showing notable increase in the membranes of the Golgi apparatus (Ga) accompanying the development of secretory functioning (from a to d). From GODMAN and PORTER (1960). For references to other cellular components, see the original

There are many instances in which Golgi cisternal envelopes appear to be constituted without any obvious blebbing from other membrane systems and without any apparent involvement of small vesicles. These instances seem to indicate the possibility of the membranes arising by some other means than the transfer of already formed membrane. The *in vitro* experiments on the formation of phospholipid lamellar structures and their associations with proteins raise at least the theoretical possibility that "new" membrane may be formed within the cell. If such new membrane formation does take place, the Golgi apparatus might be a principal site of the activity.

Whatever may be the manner of the assembly of cisternal envelopes, once they are formed they are subject to substantial increases in extent. Numerous developmental studies have shown relatively small Golgi cisternae to become much more extensive with the onset of secretory functioning. This change is nicely illustrated in the Godman and Porter (1960) study of the development of the secreting chondroblast (Fig. 17). Flickinger (1969a) has shown a comparable developmental increase in Golgi apparatus membrane in rat epididymis and added the observation that the rate of this increase varies at different stages and that the increase is accompanied by morphological differentiation of the cisternae.

After the cell has entered a secreting phase there is a continuing membrane increase providing the membrane that encloses the secretory vesicles (see Fig. 15). In some but not all instances, secretory vesicles separated from the Golgi apparatus also continue to increase in size with a concomitant increase of membrane. Membrane increase in the Golgi apparatus could be brought about by the transfer of membrane segments from elsewhere or by drawing of membrane components from pools. It may be that the formation of cisternal envelopes involves one process, and their extension, the other. There are almost always small vesicles in the vicinity of the Golgi stack but whether they are in transit toward the stack or away from it, whether they have been derived from some other cell component or from the stack itself, or whether they represent "shuttle" vesicles, as Jamieson and Palade (1966) have postulated, has not been determined. Again, much membrane transfer at a level that cannot be visualized may be involved (Fawcett, 1962). If both mechanisms for membrane increase apply in growth of Golgi apparatus membrane, the balance may depend on the stage of functional activity.

Questions about transfer of membrane from the Golgi apparatus elsewhere can be answered with more certainty. One such transfer is from the Golgi apparatus to the plasma membrane. It has been proposed that small vesicles (sometimes described as "empty") may be involved in direct transfer of Golgi apparatus-assembled and differentiated membrane to the plasma membrane (see Arnold, 1967). Another suggestion of this is seen in the laying down of the new plasma membrane in cell plate formation in plant cells (Whaley, Dauwalder, and Kephart, 1966; Fig. 18). The assembling of plasma membrane in the Golgi apparatus and its subsequent transfer is also indicated by some studies of morphologically specialized membrane. Hicks (1966) in a study of superficial squamous cells of rat transitional epithelium has shown regions of Golgi cisternal membranes to be asymmetrically thickened in a distinctive pattern, vesicles in the cytoplasm showing comparable patterns of thickening, and regions of the plasma membrane showing the same thickening. A somewhat similar relationship of Golgi apparatus activity to the production of more uniformly thickened plasma membrane has been suggested by Falk (1969) from studies of an alga.

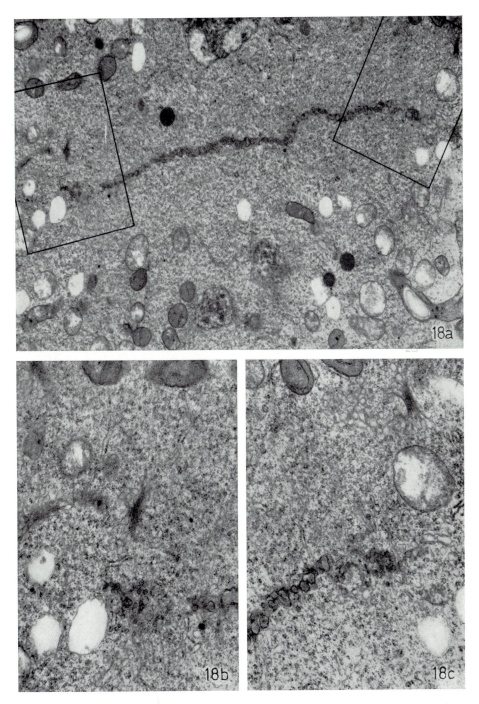

Fig. 18. (a) Telophase in a root apex of *Zea mays* showing the fusion stage in plate formation. Note Golgi apparatus at periphery of plate. (b and c) enlargements of areas indicated in (a). From WHALEY, DAUWALDER, and KEPHART (1966). Magnification (a) 7,800 × ; (b), (c) 19,800 ×

Various cytochemical methods have demonstrated that the protein-polysaccharides of cell coats are assembled in the Golgi apparatus and then transferred to the cell surface in Golgi-derived vesicles (RAMBOURG, HERNANDEZ, and LEBLOND, 1969). Incorporation of Golgi apparatus-derived membrane into the plasma membrane when the coat material is exteriorized seems to be involved in these cases. REVEL and ITO (1967) have postulated that this may resemble other instances of secretion except that the exteriorized Golgi apparatus-assembled material remains attached to the cell membrane to give the cell surface distinctive characteristics. BONNEVILLE and WEINSTOCK (1970) have assigned to the Golgi apparatus a principal function in determining the complex macromolecular architecture of the cell surface, including the membrane, as have several other investigators.

The transfer of membrane in association with classically defined secretory products is firmly established. PALADE postulated in 1959 that Golgi apparatus-derived membrane might become part of the plasma membrane during the process in which zymogen granules are transferred to the outside of the pancreatic exocrine cell. Radioautographic studies indicating the movement of secretory products from the Golgi apparatus to and through the plasma membrane lend support to this idea, even though they have not been concerned with membrane labels (see LEBLOND and WARREN, 1965). The manner in which secretory products are discharged through the plasma membrane (discussed under the section on plasma membrane) also supports the idea, and further confirmation comes from dimensional comparability between the plasma membrane and the membranes toward the distal face of the Golgi apparatus.

HELMINEN and ERICSSON (1968) have followed the transfer of milk proteins in lactating mammary glands of rats from the endoplasmic reticulum to the Golgi apparatus and then to the cell surface. They found the membrane bordering the milk proteins to undergo a change in thickness of the order of 30 to 40 Å during transport from the endoplasmic reticulum to the cell surface and suggested that this change takes place within the Golgi apparatus. GROVE, BRACKER, and MORRÉ (1968) have shown the cisternal membranes on the proximal face of the Golgi apparatus in a fungus to be comparable to those of the endoplasmic reticulum and the membranes of successive cisternae toward the distal face to be modified in thickness until they equate to the plasma membrane (cf. MORRÉ, MOLLENHAUER, and BRACKER, this volume, Fig. 5, p. 89). They have also noted an equivalence of the proximal face membranes and membranes of the nuclear envelope.

In considering changes in the character of membranes across the Golgi stack, some attention must be given to aspects of differentiation other than simple changes in dimensions. DALLNER and ERNSTER (1968) have pointed out that three types of proteins may have to be considered in relation to membranes: hypothetical structural proteins, enzymes, and loosely associated secretory proteins. There is no direct evidence concerning structural proteins in the Golgi apparatus. There have been a number of demonstrations of enzymatic differences between cisternae at one face of the apparatus and those at the other (NOVIKOFF, 1965; NOVIKOFF, ROHEIM, and QUINTANA, 1966; NOVIKOFF, ALBALA, and BIEMPICA, 1968; DAUWALDER, WHALEY, and KEPHART, 1969). Such enzymatic differences undoubtedly relate to the build-up of secretory products and perhaps their modification. No specific changes in the lipid components have yet been identified, though it seems likely that such changes also are part of the differentiation pattern. All that is known about the carbohydrates in the Golgi apparatus pertains to the secretory products rather than directly to the membranes.

SJÖSTRAND (1968) has interpreted such changes as those that have been demonstrated in membrane-associated enzymes to suggest that the principal function of the Golgi apparatus is the building and differentiation of membranes and that the assembling of the secretory products should be viewed as a function of the synthesized and assembled membrane. If membrane, whether associated with secretory products or not, is transferred from the Golgi apparatus to the plasma membrane, some of the characteristics of the plasma membrane must be determined by activities in the Golgi apparatus. The recognition of Golgi apparatus membrane by plasma membrane lends plausibility to such a transfer but the incorporation of Golgi apparatus-derived membrane in the plasma membrane must be accompanied by transformation which gives the transferred membrane characteristics of the plasma membrane. That such transformation takes place is supported by observations on contacts between zymogen granules and the plasma membrane.

ICHIKAWA (1965), in a study of the effects of secretin and pancreozymin on the secretion of zymogen from exocrine cells of the canine pancreas, noted that zymogen granules not in contact with the plasma membrane do not coalesce, no matter how tightly packed, but that a granule membrane will fuse with that of another granule if one is in contact with the plasma membrane. AMSTERDAM, OHAD, and SCHRAMM (1969) made this same observation and pointed out that phase-contrast microscopy yields observations supporting this sequential fusion of zymogen granules. Based on the same supposition of transformation of granule membrane into plasma membrane, AMSTERDAM and his co-workers interpreted their findings to indicate deep penetration of plasma membrane (due to granule membrane fusions with it) into the cytoplasm to provide for more rapid secretion than would be provided by individual granule secretion. They pointed out that the fusion of zymogen granule membranes only after one of them had become part of the plasma membrane must imply some highly specific biochemical alteration, and they suggested that cyclic 3'5'-adenosine monophosphate (AMP) might be involved in this process, noting that cyclic 3'5'-AMP had been shown to play a part in membrane modifications in other systems.

PALADE (1959) noted that massive incorporation of zymogen granule membranes into the plasma membrane would greatly increase it and postulated that there must be some compensation for this increase. He suggested that segments of plasma membrane might form small endocytotic vesicles and possibly be transferred back to the Golgi region. AMSTERDAM, OHAD, and SCHRAMM (1969), studying secretion from parotid glands of the rat stimulated by isoprenaline, observed numerous vesicles of this sort in the apical region of the cell at the time the lumen surface was being reduced following secretion. HAND (1970), however, studying secretion from von Ebner's gland of the rat stimulated to secrete by fasting-refeeding or pilocarpine, failed to find such vesicles, though he noted that the secretion in this instance was less massive than in the Amsterdam study. Such recycling of segments of membrane would be difficult to follow with certainty inasmuch as the nuclear envelope, the endoplasmic reticulum, the Golgi apparatus, and the plasma membrane may all be sources of small vesicles which sometimes occur in the cytoplasm in substantial numbers. Perhaps the only conclusion that can be drawn is that the plasma membrane is seen in several instances to invaginate and form vesicles of various sorts (for an example see Palade and Bruns, 1968) and hence such direct retransfer of membrane is possible. Fawcett (1962) has made another suggestion by which such compensation might take place: membrane components might be withdrawn from the plasma membrane as molecules and then reassembled into visible membranes in the Golgi

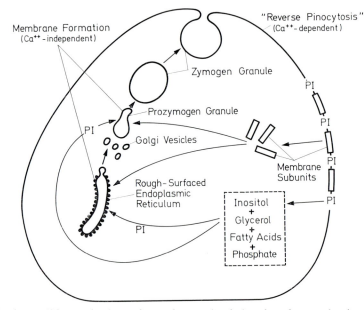

Fig. 19. A possible mechanism of membrane circulation based on subunit relocation. According to this mechanism the membrane added to the plasmalemma by coalescence of zymogen granule membrane is released back into the cytoplasm by breakdown into subunits. The breakdown into subunits could be brought about by the hydrolysis of phosphatidylinositol which might link the subunits together. Membrane which was lost from the rough-surfaced endoplasmic reticulum and from the smooth-surfaced Golgi membranes in the formation of zymogen granules is replaced by reassembly of subunits in these structures. This is brought about by resynthesis of phosphatidylinositol in these intracellular membranes, which joins together the subunits. Figure and legend from Hokin (1968)

region. Hokin (1968), emphasizing membrane breakdown to different levels of molecular organization, has made a somewhat comparable suggestion, although he notes that the endoplasmic reticulum may also be implicated (Fig. 19).

All membrane transferred from the Golgi apparatus is by no means directly incorporated into the plasma membrane. Much of it becomes membrane bounding several categories of small vacuoles and vesicles. Some of these may also derive membrane from the plasma membrane and other intracytoplasmic membrane systems (see vacuoles and vesicles).

The unanswered question of how the Golgi apparatus replicates, or how it is formed if it does not replicate, involves either a question of membrane movement or actual membrane assembly from molecular components. A number of investigators have suggested that replication takes place by a division of the Golgi stack (see Whaley, 1966). Others have suggested a transformation of other types of cellular components into Golgi apparatus (see WHALEY, 1966; and WARD and WARD, 1968). FLICKINGER (1968a, 1969b) has observed that the Golgi apparatus disappears in enucleated amoebae and reappears in such cells after they have been renucleated. Actinomycin-D inhibition of nuclear RNA synthesis reduced the size of the Golgi apparatus, but did not cause their disappearance (FLICKINGER, 1968b). In FLICKINGER's several experiments modification of the Golgi apparatus was accompanied

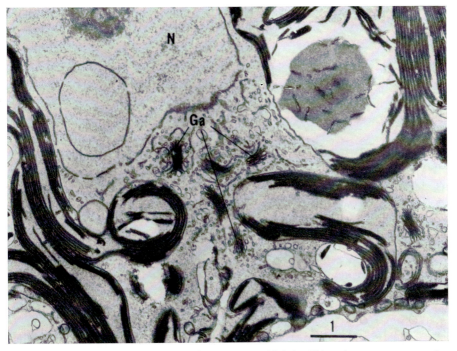

Fig. 20. Increase in the number of Golgi apparatus (Ga) in *Chlamydomonas* accompanying colchicine-induced polyploidy. Nucleus (N). A comparable section of a control cell would show only one or two Golgi apparatus. From WHALEY (1966). By PATRICIA WALNE. Magnification 12,700 ×

by modification of the endoplasmic reticulum. That the presence and number of Golgi apparatus is in some manner directly related to nuclear activity is also implicit in the buildup of the number of Golgi apparatus concurrent with development of colchicine-induced polyploidy in certain instances (WALNE, 1967; Fig. 20).

In some slime molds the plasmodial stages appear to be without Golgi apparatus. In at least one instance (R. W. SHEETZ, personal communication) there is a suggestion of the possibility of membrane transfer from the nuclear envelope to the plasma membrane at this stage. However, when spore-forming stages are entered, Golgi apparatus become apparent (GOODMAN and RUSCH, 1970) and membrane seems to be transferred from them to the cell surface as part of a secretory process. Such instances as these, which suggest the appearance of the Golgi apparatus at a specific developmental stage, accord with many observations of the appearance of the centrioles in reproductive stages in certain lower plants.

Drawing any firm conclusions about the apparent absence of the Golgi apparatus is beset by several difficulties. One is the absence of any easy means of distinguishing between a single Golgi cisterna and any other smooth membrane component. Another lies in the identification of the Golgi apparatus in cells in different physiological states. Cells of unhydrated plant embryos sometimes seem to be without typical Golgi apparatus, which subsequently become apparent upon hydration of the cells. But the problems involved in fixation and microscopy leave an uncertainty as to whether, in such instances, there is a development or a modification of the apparatus (Yoo, 1970). The evidence is simply too inadequate to serve as a basis for deciding whether the Golgi apparatus replicates or is formed in some other manner.

All the evidence indicates that the Golgi apparatus functions in the assembly of membrane from segments transferred from elsewhere or from molecular components or both. It is a center of membrane differentiation, both structurally and with respect to particular enzyme activities. It is a site from which membrane is transferred into the plasma membrane or to certain other membrane-bound organelles, notably the lysosomal system (Whaley, 1968). Its activities are obviously correlated with the activities of other membranes in the cell so as to provide a balance appropriate to the particular functions of the cell at a given time, with some sort of recycling of membrane components being involved.

D. The Plasma Membrane

The plasma membrane and the various cell coat or other extraprotoplastic materials associated with it seem clearly to be products of intracellular metabolism. The possibility seems to exist for flow of material from the nuclear envelope and the endoplasmic reticulum to the plasma membrane without mediation by the Golgi apparatus. However, it seems likely that much plasma membrane is, in one way or another, transferred from the Golgi apparatus. In some instances, such as in the rapidly developing pollen tube, major extension of the plasma membrane parallels intense Golgi apparatus activity and it could largely result from transfer of membrane segments (Fig. 21). A notable example of plasma membrane increase related to intracytoplasmic activity can be seen in the extension of the membrane making up the retinal rods and cones where folded plasma membrane associated with pigments comes to be a predominant part of the cellular structure (Young and Droz, 1968). The substantial increase in the plasma membrane of the pancreatic exocrine cell which accompanies secretion of zymogen but is then compensated by some sort of mechanism has already been discussed. Plasma membrane extension is, however, by no means always accompanied by conspicuous secretory activity. It may well be that even if segments of membrane are transferred during secretion, the mechanism of plasma membrane extension under other circumstances is different. Brandt, Reuben, and Grundfest (1968) have postulated that the increase and diminution of the extent of muscle fiber membrane involves reversible incorporation of cytoplasmic molecules into the membrane.

Despite the fact that plasma membrane may be more or less directly derived from intracellular membranes, it has a number of unique characteristics relating to its position and function, at least many of which are acquired at the cell surface itself. Enzymatic differences between plasma membranes and other cellular membranes have been discussed by Novikoff et al. (1962), de-Thé (1968) and many other investigators. Additional differences are set forth in Table I. It is known that the plasma membrane is characterized by definite surface specificity related to genetic type and even to organ (see DAVIS and WARREN, 1967; NEVILLE, 1968). These specificity factors are often primary determinants in cell associations. Major investigations relating the cell surface to cell aggregation have been summarized by HUMPHREYS (1967). NEVILLE (1968) has proposed that every cell type may have a characteristic surface protein that determines specificity. There is, however, abundant evidence that in some instances this specificity involves more than simply proteins—notably proteins variously conjugated with polysaccharides. MOSCONA (1968) has shown that when the cell coat is removed from sponge cells of like type, they will not reaggregate until either they have formed additional cell coat or the proper cell

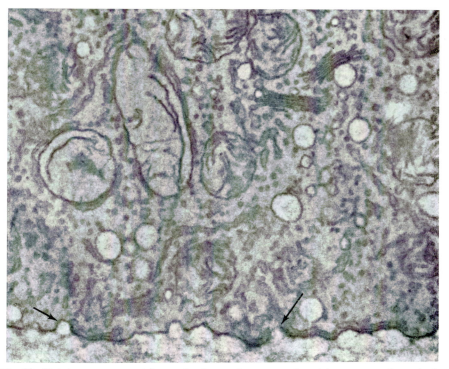

Fig. 21. Golgi apparatus vesicle production and passage of vesicle contents through the plasma membrane (arrows) during pollen tube germination in *Parkinsonia aculeata*. By D. LARSON. Magnification 46,700 ×

coat factor has been added to the medium. In this case the specificity has been demonstrated to be a glycoprotein with some associated lipid (MOSCONA, 1968). In the higher organisms cell coats are generally known to be largely mucopolysaccharide or glycoprotein (RAMBOURG, HERNANDEZ, and LEBLOND, 1969). The bulk of evidence indicates that these mucopolysaccharides or glycoproteins are assembled in the Golgi apparatus and transported to the cell surface, with incorporation of the transport vesicle membrane into the plasma membrane. That coat material apparently does not constitute a structural part of the membrane is suggested by the fact that the cell maintains its integrity when the coat material is removed by enzymatic digestion (MOSCONA, 1968; OVERTON, 1968, 1969). It has already been observed that such transfer of membrane must be accompanied by transformation.

There undoubtedly are structural and compositional changes in the plasma membrane associated with variations in function. Some such differences are readily identifiable morphologically (HICKS, 1966); some are indicated by cytochemical enzyme

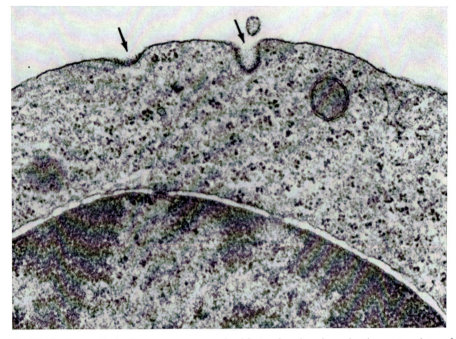

Fig. 22. Endocytosis in bone marrow erythroblasts showing invaginations at points of ferritin adsorption (arrows). From FAWCETT (1965). Magnification 39,200 ×

activity localizations (DE-THÉ, 1968); some can be made detectable by the adsorption of microscopically identifiable tracer molecules, for example, ferritin (FAWCETT, 1965). Additional techniques would undoubtedly reveal the plasma membrane to be a structural and functional mosaic subject to continuing transient differentiation. Evidence of this is to be seen in such phenomena as pinocytosis and phagocytosis. BRANDT and FREEMAN (1967) have shown both structural and electrical resistance changes in the plasma membrane of the giant amoeba *Chaos chaos* accompanying the induction of pinocytosis. PALADE and BRUNS (1968) have shown image changes in the membranes of certain endothelial cells — some of them apparently associated with the invaginations of the membrane to form vesicles and others associated with the fusion of plasmalemma vesicles with the membrane. FAWCETT (1965) has shown sites of ferritin adsorption to be involved in endocytotic activity (Fig. 22).

KORN and his associates (WEISMAN and KORN, 1967; KORN and WEISMAN, 1967; WETZEL and KORN, 1969) have demonstrated that in amoebae phagocytotic vesicles will carry latex beads into the protoplast and that such vesicles may subsequently fuse with each other and perhaps with digestive vacuoles. WETZEL and KORN were able to isolate membrane fractions from preparations of such induced phagocytotic vesicles. These membranes proved to have dimensions comparable to plasma membranes. ULSAMER, SMITH, and KORN (1969) undertook to determine whether such phagocytosis involved compositional changes in the segments of plasma membrane transferred. They found some suggestions of such changes, as had KARNOVSKY and his collaborators (see ULSAMER, SMITH, and KORN, 1969), but did not find that phagocytosis stimulated incorporation of labelled precursors into the lipids of the membranes as might have been expected in view of the apparent increase of membrane involved. They noted, however, that overall analyses might not demonstrate localized effects.

Although direct transfer from the nuclear envelope or the endoplasmic reticulum cannot be excluded, it would appear that a large part of the plasma membrane and its associated surface materials are assembled in and transferred as vesicles from the Golgi apparatus. In pinocytosis and phagocytosis membrane is transferred back into the cytoplasm in visible segments as the bounding membrane of various vacuoles and vesicles. There is also apparently movement of membrane components in both directions that is not microscopically detectable. Secretory function does not always bring about increase in the cell surface and such activities as pinocytosis and phagocytosis are not always accompanied by transfer of visible membrane to the plasma membrane. Hence there must be a balance between the detectable and the nondetectable movement of components, which, subject to change with cellular activity, would seem, in part, to be governed by the functions of the lysosomal system.

E. Vacuoles and Vesicles

Some of the most clearly defined instances of membrane transfer in the cell involve the formation of various types of vacuoles and vesicles. The plasma membrane, the Golgi apparatus, probably the endoplasmic reticulum, and quite possibly the nuclear envelope may all play some part in vacuole or vesicle formation. Most is probably known about the formation and functioning of the lysosomes, which are organelles that contain a number of hydrolases, among which acid phosphatase is commonly used as a marker.

Secondary lysosomes can be equated, in many instances, to endocytic vacuoles formed by invaginations of the plasma membrane. Proof of this formation rests in the long recognized morphological picture of infolding of the plasma membrane, separation of vesicles from it, and their transport into the cytoplasm. Further proof resides in the frequent demonstration in such vesicles of substances with some of the same cytochemical activities as the cell coat (see RAMBOURG, HERNANDEZ, and LEBLOND, 1969). The commonly accepted assumption for the development of intracellular digestive activity in these structures; i. e., for conversion of some of them into functional lysosomes, is that enzymes are transmitted to them in smaller, primary lysosomes derived from the Golgi apparatus (DE DUVE and WATTIAUX, 1966; NOVIKOFF, ROHEIM, and QUINTANA, 1966; FRIEND and FARQUHAR, 1967). This involves the fusion of membrane transported inward from the plasma membrane with membrane coming directly from the Golgi apparatus. It will be recalled that Golgi apparatus-derived membrane and plasma membrane appear to be comparable in dimensions. Recognition between them leading to fusion has already been mentioned.

Novikoff and his associates (Holtzman, Novikoff, and Villaverde, 1967) have described another sort of membrane-bound entity with acid hydrolase activity under the name of GERL – the intent was to designate some interrelationship of the Golgi apparatus, the endoplasmic reticulum and the lysosomes. In Novikoff's GERL the acid phosphatase activity appears in a cisternal-type element in association with the distal face of the Golgi apparatus (Fig. 23; Novikoff, 1967). GERL might

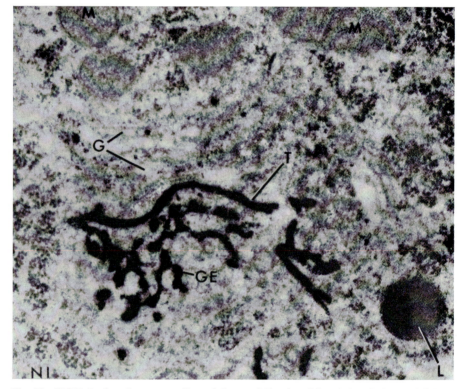

Fig. 23. GERL in dorsal root ganglion of the rat. Mitochondrion (M). Nissl material (NI). Golgi apparatus (G). Lysosome (L). Tubular portion of GERL (T). Fenestrated region (GE). From Novikoff (1967). For details of preparation, see original. Magnification 52,250 ×

represent a separated and somewhat modified cisterna or portion of a cisterna of the Golgi apparatus, or it might represent a profile of smooth endoplasmic reticulum in which enzyme development has been mediated by the Golgi apparatus. Whether GERL membranes are transferred from the Golgi apparatus or are modified regions of endoplasmic reticulum, the development of their specific enzymatic characteristics appears to be a Golgi apparatus function. GERL membranes may represent sites from which enzyme-loaded primary lysosomes are transferred to the secondary lysosomes, though this same transfer seems frequently to take place directly from the more distal cisternae of the Golgi apparatus.

The concept of the functioning of the lysosomes in intracellular digestion holds two other conditions of significance with respect to membrane components. Lysosomes may engulf cellular components, including membrane-bound organelles such as the mitochondria, and the digestive processes may bring about a breakdown of the membranes of these components, some of which may be returned to general cellular pools for re-utilization. Not all the products of this digestion are discharged into cellular pools. There may ultimately be transport of the lysosomes containing debris to the cell surface where a fusion of membrane with the plasma membrane facilitates a discharge to the outside of the protoplast (see DE DUVE and WATTIAUX, 1966).

Multivesicular bodies have been noted by several investigators to contain inner vesicles of the same order of magnitude as the small vesicles frequently seen associated with the Golgi apparatus (Fig. 24). Certain of the staining reactions of multivesicular bodies have been interpreted by DE DUVE and WATTIAUX (1966) and others to warrant their classification among lysosomes. FRIEND (1969) has studied the staining reactions of the small vesicles and the matrix of the multivesicular bodies and determined that the vesicles react similarly to other small vesicles surrounding the Golgi apparatus. FRIEND has postulated that these small vesicles are derived from Golgi cisternae which are not characterized by lysosomal enzyme activity. He postulates that clusters of these small vesicles somehow become enveloped by segments of smooth membrane, and that fusions of Golgi-derived vesicles containing acid hydrolases with this system then impart lysosomal activity to the matrix surrounding the small inner vesicles. HOLTZMAN, NOVIKOFF, and VILLAVERDE (1967) have interpreted the structure of multivesicular bodies present in neurons as small vesicles surrounded by segments of GERL. Both HOLTZMAN and DOMINITZ (1968) and FRIEND (1969) have shown exogenous peroxidase to be incorporated into the matrix of the multivesicular bodies. This incorporation of peroxidase, which is a marker for endocytosis, indicates that the outer membrane of the multivesicular body may be immediately derived, in part, from the plasma membrane. Multivesicular bodies may then result from transfer to a single organelle of segments of plasma membrane and two different derivatives of membrane from the Golgi apparatus. It has been suggested that these organelles may be specialized forms of lysosomes with particular functions in relation to membrane turnover.

Lysosomal activity has been shown to be a normal function in many types of animal cells; it is especially notable in certain disease conditions and in degenerating cells (see DINGLE and FELL, 1969). The possible lysosomal activity of plant cells has been much less investigated. Some of the marker enzymes of lysosomes have been identified in certain storage bodies and vacuoles in plant cells (see MATILE, 1969). If plant cell vacuoles are a site of digestive activity in the same sense as the lysosomes of animal cells, they may function similarly in the redistribution of membrane components.

The origin of the plant cell vacuolar membrane has never been satisfactorily explained. The endoplasmic reticulum, the Golgi apparatus, or both, development only from pre-existing vacuoles or origin by the modification of other organelles have all been proposed (for a brief discussion see MESQUITA, 1969). Whatever their origin, the membranes of vacuoles are sometimes subject to redistribution, as when a large vacuole is divided in the course of division of a highly differentiated cell (SINNOTT and BLOCH, 1940). In certain instances higher plant cell vacuole membranes appear to undergo substantial extension into the vacuole where they have specialized functions in connection with crystal development (ARNOTT, 1966).

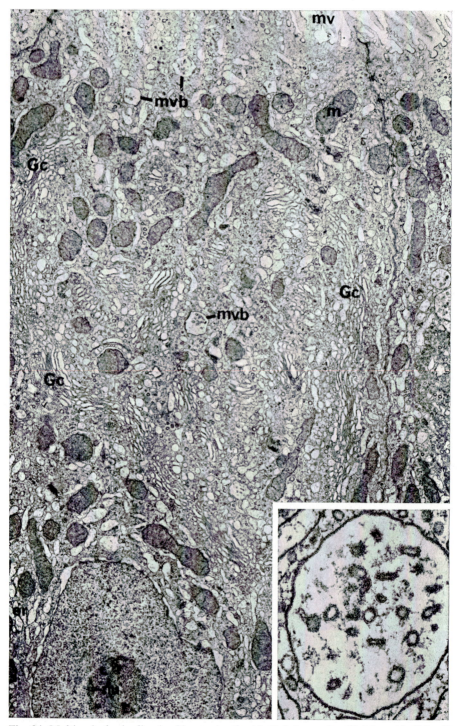

Fig. 24. Multivesicular bodies (mvb) associated with Golgi complex (Gc) in epithelial cells of the cauda epididymis of the rat. Mitochondrion (m). Endoplasmic reticulum (er). Nucleolus (nu). Microvilli (mv). Insert: enlargement of a multivesicular body. From Friend (1969). Magnification 12,200 ×; Insert: 70,000 ×

In certain types of cells endocytotic processes produce numbers of vesicles which appear to function in the intake of various materials. An example is the engulfment and inward movement of yolk and other proteins by the invagination of the cell

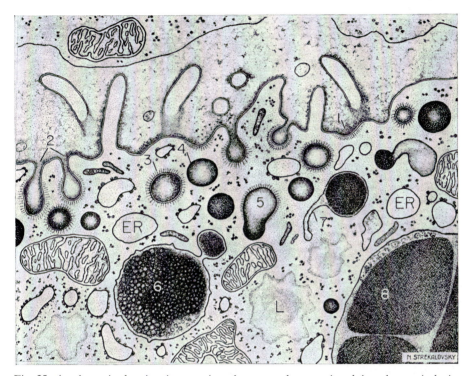

Fig. 25. A schematic drawing interpreting changes and events involving the cortical pits and coated vesicles of the mosquito oocyte. At (1) is shown the first stage of invagination into the oocyte of the protein-coated plasma membrane from the intercellular space. The fully developed pit (2), by pinching off, forms the coated vesicle (3). These vesicles lose their bristles to form dense spheres of similar size (4), which then fuse with other dense spheres (5). Often a flattened empty sac is attached to the droplet (7). This sac may be the membrane remnant of a vesicle or perhaps some element of the Golgi complex that has recently fused with the droplet. The larger droplets (6) coalesce to form the large crystalline proteid yolk bodies (8) of the oocyte. Other conspicuous and characteristic inclusions and organelles of the oocyte cytoplasm are mitochondria, vesicles of the rough-surfaced endoplasmic reticulum (ER), lipid (L) and ribosomes. At the top of the drawing, microvilli project into the intercellular space fronting on the follicular epithelium. Note the absence of adhering material on the membranes of the follicular epithelial cells. Figure and legend from ROTH and PORTER (1964)

surface to surround masses of material which are then cut off in membrane-bound vesicle form (ROTH and PORTER, 1962, 1964; Fig. 25). Thyroglobulin is transferred back into the cell by a somewhat different process but one which also involves activity of the plasma membrane (SELJELID, 1967). There are many other instances in which endocytosis involving transfer of plasma membrane is important in the uptake of material. In such transfer coated vesicles are frequently seen in association with the plasma membrane and the Golgi apparatus. ROTH and PORTER (1962, 1964) hypothesized that these vesicles have a specialized function and suggested it might be protein uptake. Fawcett (1965) called attention to the fact that there is some modification of the cell surface involved in the formation of coated vesicles and he also postulated that proteins, perhaps among other classes of substances, are taken into the cell by this activity. It has subsequently been shown that polysaccharides, probably as components of glycoproteins, are also taken into the cell in coated vesicles (Favard-Séréno, 1969).

Friend and Farquhar (1967) and Garant and Nalbandian (1968) have distinguished two classes of coated vesicles: relatively large vesicles seen more frequently in association with the plasma membrane and smaller ones seen, for the most part, in the vicinity of the Golgi apparatus. Both concluded that the larger coated vesicles transport materials from the cell inward and the smaller ones transport lytic enzymes from the Golgi apparatus to the lysosomes.

Droller and Roth (1966) have ascribed the buildup of yolk in the development of egg cells of the guppy in part to the activity of plasma membrane-derived coated vesicles and in part to accumulation of yolk in the Golgi apparatus which become extensively proliferated at this stage. They assume that membranes of both types of yolk-containing vesicles fuse to form yolk droplets. They have also noted the accumulation of a flocculent substance in the endoplasmic reticulum at this stage and the likely separation of this material from the endoplasmic reticulum in smooth, membrane-bound vesicles which may subsequently fuse. They look upon the terminal stage in yolk droplet formation as being the fusion of yolk-containing, membrane-bound vesicles from all of these sources: the plasma membrane, the Golgi apparatus, and the endoplasmic reticulum. To this picture Ulrich (1969) has added the possible fusion of yolk-engulfing plasma membrane with the rough endoplasmic reticulum. Yolk is a complex material containing protein, carbohydrates and lipids (see Kessel, 1968b). The relative importance played by any one of the organelles or mechanisms of the cell in its ultimate constitution must also vary from one organism to another. Yolk production seems, in any event, to involve what Anderson (1968) has described as the conjoined efforts of several organelles, some of which are involved in the incorporation of exogenous material. The fusion of the various yolk-containing vesicles represents a clear instance of the fusion of membranes from multiple sources and the combining of material assembled in connection with the activities of these membranes.

Various types of coated or spiny vesicles, or vesicles with otherwise modified surfaces, are involved in other cellular activities. They are frequently characteristic of particular stages of development, as in the case of root cell vesicles associated with slime production (see Newcomb, 1967). Neither their origin nor their direction of transfer, if they are transport vesicles, has been sufficiently identified to permit meaningful discussion.

Microbodies, including the peroxisomes, which are distinguished by the presence of specific oxidases have been identified in a number of different types of cells, both animal and plant (Novikoff and Biempica, 1966; Beard and Novikoff, 1969; Frederick and Newcomb, 1969). The rather consistent association of microbodies with profiles of the endoplasmic reticulum (Novikoff and Shin, 1964; see also Hruban and Rechcigl, 1969) has given rise to the suggestion that the membrane of these organelles is derived from the endoplasmic reticulum.

Organelles such as those of the lysosomal system and the microbodies with their specific enzymatic characteristics are now being related to specific cellular functions. Most cells are characterized, in one stage of development or another, by different types of small vesicles which remain to be defined chemically and to be assigned particular functions. Functional activities of some vacuoles are also still to be defined.

Such evidence as there is about the origin of membranes of vacuoles and vesicles suggests that they might be transferred from any of the sources mentioned. They might also, in some instances at least, be newly formed from pools of components in the cytoplasm. In any event, their components are subject to return to cellular pools for recirculation. The lysosomal system with such specialized components as the multivesicular bodies may well be the functional regulator in what appears to be a transfer and recirculation system of great diversity.

IV. Concluding Remarks

The manner in which molecules are assembled into membranes is unclear. Membrane growth is an established fact and there is considerable experimental evidence to suggest the possibility of membrane formation. Assembled membrane may be transferred from one site to another in the cell. Numerous such transfers, known to be associated with certain specific cellular activities, are detectable microscopically. Examples include transfers of membrane from the nuclear envelope in the formation of annulate lamellae; transfers from the endoplasmic reticulum, particularly notable during movement of protein in secretory cells; transfers of Golgi apparatus-derived secretory-vesicle membrane to the plasma membrane; transfers of membrane of hydrolytic enzyme-containing vesicles from the Golgi apparatus to charge the lysosomes; and transfers from the plasma membrane back into the cytoplasm in formation of various types of endocytotic vacuoles and vesicles. Other examples are seen in the fusion of membranes from several sources to form yolk droplets and probably multivesicular bodies.

Most such instances of detectable transfer are not balanced by detectable transfer in the opposite direction. This fact and what is known of molecule turnover in membranes would seem to make it essential to assume substantial membrane component movement at a level that is not microscopically detectable. It would thus appear that consideration must be given to an exchange system in which some of the exchanges are in the form of membrane segments and others are not.

Each of the cellular membranes is a highly differentiated structure which has particular characteristics related to its function and which probably is subject to localized transient modifications as well. Transfers between functionally different membranes depend upon two considerations: recognition between membranes, which makes fusion possible, and modification of the incorporated segments to relate them functionally to the new site. Such modifications require substantial turnover of molecular components in exchange with cellular pools and may involve synthesis of specific components *in situ*.

The complex picture of synthesis and recycling of "membrane" in detectable and nondetectable phases must obviously depend upon the presence of regulatory systems to maintain the balances. Among these are the Golgi apparatus functioning in the assembly and differentiation of visible membranes and the lysosomal system functioning in membrane breakdown. The Golgi apparatus contribution of enzymes to the lysosomal system would seem to interrelate the two.

Both because they localize activities and substrates and because they function as reaction surfaces, membranes are key units of biological structure. This makes understanding of their assembly, continuity, and exchange critical to the interpretation of cellular function. While processes of membrane organization, functioning, and breakdown cannot be considered apart from other cellular processes, cellular specialization and efficiency are reflected quite directly in the degree of organization of the cellular membranes.

Acknowledgement

The authors wish to acknowledge the assistance of Dr. Audrey N. Slate and Mrs. Elaine Kaufman. Further they wish to acknowledge that many of the conclusions drawn here have come from research supported by NSF Research Grant Nr. GB 17778, and a grant from the Faith Foundation.

References

Amsterdam, A., Ohad, I., Schramm, M.: Dynamic changes in the ultrastructure of the acinar cell of the rat parotid gland during the secretory cycle. J. Cell Biol. **41**, 753—773 (1969).

Anderson, E.: Cortical alveoli formation and vitellogenesis during oocyte differentiation in the pipefish, *Syngnathus fuscus*, and the killifish, *Fundulus heteroclitus*. J. Morph. **125**, 23—60 (1968).

Arnold, J. M.: Organellogenesis of the cephalopod iridophore: cytomembranes in development. J. Ultrastruct. Res. **20**, 410—421 (1967).

Arnott, H. J.: Studies of calcification in plants. In: Third European Symposium on Calcified Tissues (Ed.: H. Fleisch, H. J. J. Blackwood and M. Owen), pp. 152—157. Berlin-Heidelberg-New York: Springer 1966.

Bajer, A., Molè-Bajer, J.: Formation of spindle fibers, kinetochore orientation, and behavior of the nuclear envelope during mitosis in endosperm. Fine structural and *in vitro* studies. Chromosoma (Berl.) **27**, 448—484 (1969).

Beard, M. E., Novikoff, A. B.: Distribution of peroxisomes (microbodies) in the nephron of the rat. A cytochemical study. J. Cell Biol. **42**, 501—518 (1969).

Behnke, O., Moe, H.: An electron microscope study of mature and differentiating Paneth cells in the rat, especially of their endoplasmic reticulum and lysosomes. J. Cell Biol. **22**, 633—652 (1964).

Benedetti, E. L., Emmelot, P.: Structure and function of plasma membranes isolated from liver. In: The Membranes (Ed.: A. J. Dalton and F. Haguenau), pp. 33—120. New York: Academic Press 1968.

Bonneville, M. A., Weinstock, M.: Brush border development in the intestinal absorptive cells of *Xenopus* during metamorphosis. J. Cell Biol. **44**, 151—171 (1970).

Bouck, G. B.: Chromatophore development, pits, and other fine structure in the red alga, *Lomentaria baileyana* (Harv.) Farlow. J. Cell Biol. **12**, 553—569 (1962).

— Fine structure and organelle associations in brown algae. J. Cell Biol. **26**, 523—528 (1965).

— Extracellular microtubules. The origin, structure, and attachment of flagellar hairs in *Fucus* and *Ascophyllum* antherozoids. J. Cell Biol. **40**, 446—460 (1969).

Brandt, P. W., Freeman, A. R.: Plasma membrane: substructural changes correlated with electrical resistance and pinocytosis. Science **155**, 582—585 (1967).

— Reuben, J. P., Grundfest, H.: Correlated morphological and physiological studies on isolated single muscle fibers. II. The properties of the crayfish transverse tubular system: localization of the sites of reversible swelling. J. Cell Biol. **38**, 115—129 (1968).

Branton, D.: Fracture faces of frozen membranes. Proc. nat. Acad. Sci. (Wash.) **55**, 1048—1056 (1966).

CARROLL, G.: The ultrastructure of ascospore delimitation in *Saccobolus kerverni*. J. Cell Biol. **33**, 218—224 (1967).

CHAPMAN, D. (editor): Biological Membranes. Physical Fact and Function. London and New York: Academic Press 1968.

CHRISTENSEN, A. K.: The fine structure of testicular interstitial cells in guinea pigs. J. Cell Biol. **26**, 911—936 (1965).

— FAWCETT, D. W.: The normal fine structure of opossum testicular interstitial cells. J. biophys. biochem. Cytol. **9**, 653—670 (1961).

COOK, G. M. W.: Chemistry of membranes. Brit. med. Bull. **24**, 118—123 (1968a).

— Glycoproteins in membranes. Biol. Rev. **43**, 363—391 (1968b).

DALLMAN, P. R., DALLNER, G., BERGSTRAND, A., ERNSTER, L.: Heterogeneous distribution of enzymes in submicrosomal membrane fragments. J. Cell Biol. **41**, 357—377 (1969).

DALLNER, G., ERNSTER, L.: Subfractionation and composition of microsomal membranes: a review. J. Histochem. Cytochem. **16**, 611—632 (1968).

— BERGSTRAND, A., NILSSON, R.: Heterogeneity of rough-surfaced liver microsomal membranes of adult, phenobarbital-treated, and newborn rats. J. Cell Biol. **38**, 257—276 (1968).

— SIEKEVITZ, P., PALADE, G. E.: Biogenesis of endoplasmic reticulum membranes. I. Structural and chemical differentiation in developing rat hepatocyte. J. Cell Biol. **30**, 73—96 (1966).

DALTON, A. J., HAGUENAU, F. (editors): The Membranes. New York and London: Academic Press 1968.

DANIELLI, J. F.: The formation, physical stability, and physiological control of pauci-molecular membranes. In: Formation and Fate of Cell Organelles (Ed.: K. WARREN), pp. 239—253. New York: Academic Press 1967.

DAUWALDER, M., WHALEY, W. G., KEPHART, J. E.: Phosphatases and differentiation of the Golgi apparatus. J. Cell Sci. **4**, 455—497 (1969).

DAVIS, B. D., WARREN, L. (editors): The Specificity of Cell Surfaces. Englewood Cliffs, N. J.: Prentice-Hall, Inc. 1967.

DAVSON, H., DANIELLI, J. F.: The Permeability of Natural Membranes, London-New York: Cambridge Univ. Press 1943.

DE DUVE, C., WATTIAUX, R.: Functions of lysosomes. Ann. Rev. Physiol. **28**, 435—492 (1966).

DE-THÉ, G.: Ultrastructural cytochemistry of the cellular membranes. In: The Membranes (Ed.: A. J. DALTON and F. HAGUENAU), pp. 121—150. New York: Academic Press 1968.

DINGLE, J. T., FELL, H. B. (editors): Lysosomes in Biology and Pathology, parts 1 and 2, Amsterdam-London: North-Holland Publishing Co. 1969.

DROLLER, M. J., ROTH, J. F.: An electron microscope study of yolk formation in *Lebistes reticulata* guppyi. J. Cell Biol. **28**, 209—232 (1966).

EMANS, J. B., JONES, A. L.: Hypertrophy of liver cell smooth surfaced reticulum following progesterone administration. J. Histochem. Cytochem. **16**, 561—571 (1968).

ESSNER, E., NOVIKOFF, A. B.: Cytological studies on two functional hepatomas. Inter-relations of endoplasmic reticulum, Golgi apparatus, and lysosomes. J. Cell Biol. **15**, 289—312 (1962).

FALK, H.: Zum Feinbau von *Botrydium granulatum* Grev. (*Xanthophyceae*). Arch. Mikrobiol. **58**, 212—227 (1967).

— Fusiform vesicles in plant cells. J. Cell Biol. **43**, 167—174 (1969).

— KLEINIG, H.: Feinbau und Carotinoide von *Tribonema* (*Xanthophyceae*). Arch. Mikrobiol. **61**, 347—362 (1968).

FAVARD-SÉRÉNO, C.: Capture de polysaccharides par micropinocytose dans l'ovocyte du grillon en vitellogenèse. J. Microscopie **8**, 401—414 (1969).

FAWCETT, D. W.: Physiologically significant specializations of the cell surface. Circulation **26**, 1105—1125 (1962).

— Surface specializations of absorbing cells. J. Histochem. Cytochem. **13**, 75—91 (1965).

FINEAN, J. B.: Further observations on the structure of myelin. Exptl. Cell Res. **5**, 202—215 (1953).

Flickinger, C. J.: The effects of enucleation on the cytoplasmic membranes of *Amoeba proteus*. J. Cell Biol. **37**, 300—315 (1968a).
— Cytoplasmic alterations in amebae treated with actinomycin D. A comparison with the effects of surgical enucleation. Exp. Cell Res. **53**, 241—251 (1968b).
— The pattern of growth of the Golgi complex during the fetal and postnatal development of the rat epididymis. J. Ultrastruct. Res. **27**, 344—360 (1969a).
— The development of Golgi complexes and their dependence upon the nucleus in amebae. J. Cell Biol. **43**, 250—262 (1969b).
Forte, G. M., Limlomwongse, L., Forte, J. G.: The development of intracellular membranes concomitant with the appearance of HCl secretion in oxyntic cells of the metamorphosing bullfrog tadpole. J. Cell Sci. **4**, 709—727 (1969).
Franke, W. W.: Isolated nuclear membranes. J. Cell Biol. **31**, 619—623 (1966).
— Zur Feinstruktur isolierter Kernmembranen aus tierischen Zellen. Z. Zellforsch. **80**, 585—593 (1967).
— Scheer, U.: The ultrastructure of the nuclear envelope of amphibian oocytes: a reinvestigation. I. The mature oocyte. J. Ultrastruct. Res. **30**, 288—316 (1970).
Frederick, S. E., Newcomb, E. H.: Cytochemical localization of catalase in leaf microbodies (peroxisomes). J. Cell Biol. **43**, 343—353 (1969).
Friend, D. S.: Cytochemical staining of multivesicular body and Golgi vesicles. J. Cell Biol. **41**, 269—279 (1969).
— Farquhar, M. G.: Functions of coated vesicles during protein absorption in the rat vas deferens. J. Cell Biol. **35**, 357—376 (1967).
Garant, P. R., Nalbandian, J.: Observations on the ultrastructure of ameloblasts with special reference to the Golgi complex and related components. J. Ultrastruct. Res. **23**, 427—443 (1968).
Gibbs, S. P.: Nuclear envelope-chloroplast relationships in algae. J. Cell Biol. **14**, 433—444 (1962).
Glauert, A. M., Lucy, J. A.: Globular micelles and the organization of membrane lipids. In: The Membranes (Ed.: A. J. Dalton and F. Haguenau), pp. 1—32. New York-London: Academic Press 1968.
Godman, G. C., Porter, K. R.: Chondrogenesis, studied with the electron microscope. J. biophys. biochem. Cytol. **8**, 719—760 (1960).
Goodman, E. M., Rusch, H. P.: Ultrastructural changes during spherule formation in *Physarum polycephalum*. J. Ultrastruct. Res. **30**, 172—183 (1970).
Gortner, E., Grendel, F.: On bimolecular layers of lipoids on the chromocytes of the blood. J. exp. Med. **41**, 439—443 (1925).
Grassé, P.-P.: Ultrastructure, polarité et reproduction de l'appareil de Golgi. C. R. Acad. Sci. (Paris) **245**, 1278—1281 (1957).
Grove, S. N., Bracker, C. E., Morré, D. J.: Cytomembrane differentiation in the endoplasmic reticulum-Golgi apparatus-vesicle complex. Science **161**, 171—173 (1968).
Hall, W. T., Witkus, E. R.: Some effects on the ultrastructure of the root meristem of *Allium cepa* by 6 aza uracil. Exp. Cell Res. **36**, 494—501 (1964).
Hand, A. R.: The fine structure of von Ebner's gland of the rat. J. Cell Biol. **44**, 340—353 (1970).
Helminen, H. J., Ericsson, J. L. E.: Studies on mammary gland involution. I. On the ultrastructure of the lactating mammary gland. J. Ultrastruct. Res. **25**, 193—213 (1968).
Hicks, R. M.: The function of the Golgi complex in transitional epithelium. Synthesis of the thick cell membrane. J. Cell Biol. **30**, 623—643 (1966).
Hirsch, G. C.: The "Golgi apparatus" or the lamellar-vacuolar field in the electron microscope. In: Intracellular Membraneous Structure (Ed.: S. Seno and E. V. Cowdry), pp. 197—206. Okayama: Chugoku Press Ltd. 1965.
Hokin, L. E.: Dynamic aspects of phospholipids during protein secretion. Int. Rev. Cytol. **23**, 187—208 (1968).

HOLTZMAN, E., DOMINITZ, R.: Cytochemical studies of lysosomes, Golgi apparatus and endo-
plasmic reticulum in secretion and protein uptake by adrenal medulla cells of the rat.
J. Histochem. Cytochem. **16**, 320—336 (1968).
— NOVIKOFF, A. B., VILLAVERDE, H.: Lysosomes and GERL in normal and chromatolytic
neurons of the rat ganglion nodosum. J. Cell Biol. **33**, 419—435 (1967).
HRUBAN, Z., RECHCIGL, M., JR.: Microbodies and related particles. Int. Rev. Cytol. Suppl. 1,
1—251 (1969).
HUMPHREYS, T.: The cell surface and specific cell aggregation. In: The Specificity of Cell
Surfaces (Ed.: B. D. DAVIS and L. WARREN), pp. 195—210. Englewood Cliffs, N. J.:
Prentice-Hall, Inc. 1967.
ICHIKAWA, A.: Fine structural changes in response to hormonal stimulation of the perfused
canine pancreas. J. Cell Biol. **24**, 369—385 (1965).
JAMIESON, J. D., PALADE, G. E.: Role of the Golgi complex in the intracellular transport
of secretory proteins. Proc. nat. Acad. Sci. (Wash.) **55**, 424—431 (1966).
— — Intracellular transport of secretory proteins in the pancreatic exocrine cell. I. Role of
the peripheral elements of the Golgi complex. J. Cell Biol. **34**, 577—596 (1967a).
— — Intracellular transport of secretory proteins in the pancreatic exocrine cell. II. Trans-
port to condensing vacuoles and zymogen granules. J. Cell Biol. **34**, 597—615 (1967b).
— — Intracellular transport of secretory proteins in the pancreatic exocrine cell. III.
Dissociation of intracellular transport from protein synthesis. J. Cell Biol. **39**, 580—588
(1968a).
— — Intracellular transport of secretory proteins in the pancreatic exocrine cell. IV.
Metabolic requirements. J. Cell Biol. **39**, 589—603 (1968b).
KEPHART, J. E., DAUWALDER, M., WHALEY, W. G.: Ultrastructural responses of cells to
deleterious conditions. J. Cell Biol. **31**, 59A—60A (1966).
KESSEL, R. G.: Electron microscope studies on the origin of annulate lamellae in oocytes
of *Necturus*. J. Cell Biol. **19**, 391—414 (1963).
— Annulate lamellae. J. Ultrastruct. Res. Suppl. **10**, 5—82 (1968a).
— Electron microscope studies on developing oocytes of a coelenterate medusa with special
reference to vitellogenesis. J. Morph. **126**, 211—248 (1968b).
KORN, E. D.: Structure of biological membranes. Science **153**, 1491—1498 (1966).
— Cell membranes: structure and synthesis. Ann. Rev. Biochem. **38**, 263—288 (1969).
— WEISMAN, R. A.: Phagocytosis of latex beads by *Acanthamoeba*. II. Electron microscopic
study of the initial events. J. Cell Biol. **34**, 219—226 (1967).
KURIYAMA, Y., OMURA, T., SIEKEVITZ, P., PALADE, G. E.: Effects of phenobarbital on the
synthesis and degradation of the protein components of rat liver microsomal membranes.
J. biol. Chem. **244**, 2017—2026 (1969).
LANG, N. J.: Electron microscopy of the Volvocaceae and Astrephomenaceae. Amer. J.
Botany **50**, 280—300 (1963).
LEBLOND, C. P., WARREN, K. B. (editors): The Use of Radioautography in Investigating
Protein Synthesis. Symp. Internat. Soc. Cell Biol., Vol. 4. New York-London: Academic
Press 1965.
LONG, J. A., JONES, A. L.: The fine structure of the zona glomerulosa and the zona fasci-
culata of the adrenal cortex of the opossum. Amer. J. Anat. **120**, 463—487 (1967).
LONGO, F. J., ANDERSON, E.: Cytological events leading to the formation of the two-cell
stage in the rabbit: association of the maternally and paternally derived genomes. J. Ul-
trastruct. Res. **29**, 86—118 (1969).
LUCY, J. A.: Globular lipid micelles and cell membranes. J. theor. Biol. **7**, 360—373 (1964).
— Theoretical and experimental models for biological membranes. In: Biological Mem-
branes. Physical Fact and Function (Ed.: D. CHAPMAN), pp. 233—288. London-New
York: Academic Press 1968a.
— Ultrastructure of membranes: micellar organization. Brit. med. Bull. **24**, 127—129
(1968b).
— GLAUERT, A. M.: Assembly of macromolecular lipid structures *in vitro*. In: Formation
and Fate of Cell Organelles (Ed.: K. WARREN), pp. 19—37. New York-London:
Academic Press 1967.

LUZZATI, V.: X-ray diffraction studies of lipid-water systems. In: Biological Membranes. Physical Fact and Function (Ed.: D. CHAPMAN), pp. 71—123. London-New York: Academic Press 1968.
— HUSSON, F.: The structure of the liquid-crystalline phases of lipid-water systems. J. Cell Biol. **12**, 207—219 (1962).
— GULIK-KRZYWICKI,T., TARDIEU, A., RIVAS, E., REISS-HUSSON, F.: Lipids and membranes. In: The Molecular Basis of Membrane Function (Ed.: D. C. TOSTESON), pp. 79—93. Englewood Cliffs, N. J.: Prentice-Hall, Inc. 1969.
MADDY, A. H.: The chemical organization of the plasma membrane of animal cells. Int. Rev. Cytol. **20**, 1—65 (1966).
— The organization of protein in the plasma membrane. In: The Formation and Fate of Cell Organelles (Ed.: K. WARREN), pp. 255—273. New York-London: Academic Press 1967.
— Some problems relating to the chemical composition of membranes. In: The Molecular Basis of Membrane Function (Ed.: D. C. TOSTESON), pp. 95—108. Englewood Cliffs, N. J.: Prentice-Hall, Inc. 1969.
MALHOTRA, S. K., VAN HARREVELD, A.: Molecular organization of membranes of cells and cellular organelles. In: The Biological Basis of Medicine (Ed.: E. E. BITTAR and N. BITTAR), vol. 1, pp. 3—68. London-New York: Academic Press 1968.
MANTON, I., RAYNS, D. G., ETTL, H., PARKE, M.: Further observations on green flagellates with scaly flagella: the genus *Heteromastix* Korshikov. J. mar. biol. Ass. U. K. **45**, 241—255 (1965).
MATILE, PH.: Plant lysosomes. In: Lysosomes in Biology and Pathology, part 1 (Ed.: J. T. DINGLE and H. B. FELL), pp. 406—430. Amsterdam-London: North-Holland Publishing Co. 1969.
MESQUITA, J. F.: Electron microscope study of the origin and development of the vacuoles in the root tip cells of *Lupinus albus* L. J. Ultrastruct. Res. **26**, 242—251 (1969).
MOSCONA, A. A.: Cell aggregation: properties of specific cell-ligands and their role in the formation of multicellular systems. Devel. Biol. **18**, 250—277 (1968).
NEUTRA, M., LEBLOND, C. P.: Synthesis of the carbohydrate of mucus in the Golgi complex as shown by electron microscope radioautography of goblet cells from rats injected with glucose-H³. J. Cell Biol. **30**, 119—136 (1966a).
— — Radioautographic comparison of the uptake of galactose-H³ and glucose-H³ in the Golgi region of various cells secreting glycoproteins or mucopolysaccharides. J. Cell Biol. **30**, 137—150 (1966b).
NEVILLE, D. M., JR.: Isolation of an organ specific protein antigen from cell-surface membrane of rat liver. Biochim. biophys. Acta (Amst.) **154**, 540—552 (1968).
NEWCOMB, E. H.: A spiny vesicle in slime-producing cells of the bean root. J. Cell Biol. **35**, C17—C22 (1967).
NOVIKOFF, A. B.: Enzymic activities and functional interrelations of cytomembranes. In: Intracellular Membraneous Structure (Ed.: S. SENO and E. V. COWDRY), pp. 277—290. Okayama: Chugoku Press Ltd. 1965.
— Enzyme localization and ultrastructure of neurons. In: The Neuron (Ed.: H. HYDÉN), pp. 255—318. Amsterdam-London-New York: Elsevier Publ. Co. 1967.
— BIEMPICA, L.: Cytochemical and electron microscopic examination of Morris 5123 and Reuber H-35 hepatomas after several years of transplantation. In: Biological and Biochemical Evaluation of Malignancy in Experimental Hepatomas. Gann Monogr. **1**, 65—87 (1966).
— SHIN, W.-Y.: The endoplasmic reticulum in the Golgi zone and its relations to microbodies, Golgi apparatus and autophagic vacuoles in rat liver cells. J. Microscopie **3**, 187—206 (1964).
— ALBALA, A., BIEMPICA, L.: Ultrastructural and cytochemical observations on B-16 and Harding-Passey mouse melanomas. The origin of premelanosomes and compound melanosomes. J. Histochem. Cytochem. **16**, 299—319 (1968).
— ROHEIM, P. S., QUINTANA, N.: Changes in rat liver cells induced by orotic acid feeding. Lab. Invest. **15**, 27—49 (1966).

NOVIKOFF, A. B., ESSNER, E., GOLDFISCHER, S., HEUS, M.: Nucleosidephosphatase activities of cytomembranes. In: The Interpretation of Ultrastructure (Ed.: R. J. C. HARRIS), pp. 149—192. New York: Academic Press 1962.

OMURA, I., SIEKEVITZ, P., PALADE, G. E.: Turnover of constituents of the endoplasmic reticulum membranes of rat hepatocytes. J. biol. Chem. **242**, 2389—2396 (1968).

ORRENIUS, S., ERICSSON J. L. E.: Enzyme-membrane relationship in phenobarbital induction of synthesis of drug-metabolizing enzyme system and proliferation of endoplasmic membranes. J. Cell Biol. **28**, 181—198 (1966).

—— —— ERNSTER, L.: Phenobarbital-induced synthesis of the microsomal drug-metabolizing enzyme system and its relationship to the proliferation of endoplasmic membranes. A morphological and biochemical study. J. Cell Biol. **25** (3/1), 627—639 (1965).

OVERTON, J.: Localized lanthanum staining of the intestinal brush border. J. Cell Biol. **38**, 447—452 (1968).

— A fibrillar intercellular material between reaggregating embryonic chick cells. J. Cell Biol. **40**, 136—143 (1969).

PALADE, G. E.: Functional changes in the structure of cell components. In: Subcellular Particles (Ed.: T. HAYASHI), pp. 64—83. New York: Ronald Press 1959.

— Structure and function at the cellular level. J. Amer. med. Ass. **198**, 815—825 (1966).

— BRUNS, R. R.: Structural modulations of plasmalemmal vesicles. J. Cell Biol. **37**, 633—649 (1968).

PARKS, H. F.: Unusual formations of ergastoplasm in parotid acinous cells of mice. J. Cell Biol. **14**, 221—234 (1962).

PORTER, K. R.: The endoplasmic reticulum: some current interpretations of its form and functions. In: Biological Structure and Function (Ed.: T. W. GOODWIN and O. LINDBERG), vol. 1, pp. 127—155. London-New York: Academic Press 1961.

RAMBOURG, A., HERNANDEZ, W., LEBLOND, C. P.: Detection of complex carbohydrates in the Golgi apparatus of rat cells. J. Cell Biol. **40**, 395—414 (1969).

REVEL, J.-P., ITO, S.: The surface components of cells. In: The Specificity of Cell Surfaces (Ed.: B. D. DAVIS and L. WARREN), pp. 211—234. Englewood Cliffs, N. J.: Prentice-Hall, Inc. 1967.

RITCH, R., PHILPOTT, C. W.: Repeating particles associated with an electrolyte-transport membrane. Exp. Cell Res. **55**, 17—24 (1969).

ROBERTSON, J. D.: Recent electron microscope observations on the ultrastructure of the crayfish median-to-motor giant synapse. Exp. Cell Res. **8**, 226—229 (1955).

— The ultrastructure of cell membranes and their derivatives. Biochem. Soc. Symp. **16**, 3—43 (1959).

— Current problems of unit membrane structure and substructure. In: Intracellular Membraneous Structure (Ed.: S. SENO and E. V. COWDRY), pp. 379—433. Okayama: Chugoku Press Ltd. 1965.

ROSS, R.: Wound healing. Sci. Amer. **220** (No. 6), 40—50 (1969).

— BENDITT, E. P.: Wound healing and collagen formation. V. Quantitative electron microscope radioautographic observations of proline-H³ utilization by fibroblasts. J. Cell Biol. **27**, 83—106 (1965).

ROTH, T. F., PORTER, K. R.: Specialized sites on the cell surface for protein uptake. In: Electron Microscopy: Fifth International Congress for Electron Microscopy (Ed.: S. S. BREESE, JR.), vol. 2, LL4. New York-London: Academic Press 1962.

—— —— Yolk protein uptake in the oocyte of the mosquito *Aedes aegypti* L. J. Cell Biol. **20**, 313—332 (1964).

ROTHFIELD, L., FINKELSTEIN, A.: Membrane biochemistry. Ann. Rev. Biochem. **37**, 463—496 (1968).

ROUSER, G., NELSON, G. J., FLEISCHER, S., SIMON, G.: Lipid composition of animal cell membranes, organelles and organs. In: Biological Membranes. Physical Fact and Function (Ed.: D. CHAPMAN), pp. 5—69. London-New York: Academic Press 1968.

SCHNEPF, E.: Membranfluß und Membrantransformation. Ber. Dtsch. Bot. Ges **82**, 407—413 (1969).

SEDAR, A. W.: Uptake of peroxidase into the smooth-surfaced tubular system of the gastric acid-secreting cell. J. Cell Biol. **43**, 179—184 (1969).

Seljelid, R.: Endocytosis in thyroid follicle cells. III. An electron microscopic study of the cell surface and related structures. J. Ultrastruct. Res. **18**, 1—24 (1967).

Siekevitz, P., Palade, G. E., Dallner, G., Ohad, I., Omura, T.: The biogenesis of intracellular membranes. In: Organizational Biosynthesis (Ed.: H. J. Vogel, J. O. Lampen and V. Bryson), pp. 331—362. New York-London: Academic Press 1967.

Sinnott, E. W., Bloch, R.: Cytoplasmic behavior during division of vacuolate plant cells. Proc. nat. Acad. Sci. (Wash.) **26**, 223—227 (1940).

Sjöstrand, F. S.: Electron microscopy of mitochondria and cytoplasmic double membranes. Ultra-structure of rod-shaped mitochondria. Nature (Lond.) **171**, 30—31 (1953a).

— Electron microscopy of mitochondria and cytoplasmic double membranes. Systems of double membranes in the cytoplasm of certain tissue cells. Nature (Lond.) **171**, 31—32 (1953b).

— The ultrastructure of cells as revealed by the electron microscope. Int. Rev. Cytol. **5**, 455—533 (1956).

— Ultrastructure and function of cellular membranes. In: The Membranes (Ed.: A. J. Dalton and F. Haguenau), pp. 151—210. New York-London: Academic Press 1968.

— Morphological aspects of lipoprotein structures. In: Structural and Functional Aspects of Lipoproteins in Living Systems (Ed.: E. Tria and A. M. Scanu), pp. 73—128. London-New York: Academic Press 1969.

— Hanzon, V.: Membrane structures of cytoplasm and mitochondria in exocrine cells of mouse pancreas as revealed by high resolution electron microscopy. Exp. Cell Res. **7**, 393—414 (1954a).

— — Ultrastructure of Golgi apparatus of exocrine cells of mouse pancreas. Exp. Cell Res. **7**, 415—429 (1954b).

— Rhodin, J.: The ultrastructure of the proximal convoluted tubules of the mouse kidney as revealed by high resolution electron microscopy. Exp. Cell Res. **4**, 426—456 (1953).

Stäubli, W., Hess, R., Weibel, E. R.: Correlated morphometric and biochemical studies on the liver cell. II. Effects of phenobarbital on rat hepatocytes. J. Cell Biol. **42**, 92—112 (1969).

Stoeckenius, W.: Some electron microscopical observations on liquid-crystalline phases in lipid-water systems. J. Cell Biol. **12**, 221—229 (1962).

— Engelman, D. M.: Current models for the structure of biological membranes. J. Cell Biol. **42**, 613—646 (1969).

Tosteson, D. E. (editor): The Molecular Basis of Membrane Function. Englewood Cliffs, N. J.: Prentice-Hall, Inc. 1969.

Trelstad, R. L.: The Golgi apparatus in chick corneal epithelium: changes in intracellular position during development. J. Cell Biol. **45**, 34—42 (1970).

Tria, E., Scanu, A. M. (editors): Structural and Functional Aspects of Lipoproteins in Living Systems. London-New York: Academic Press 1969.

Tzur, R., Shapiro, B.: Dependence of microsomal lipid synthesis on added protein. J. Lipid Res. **5**, 542—547 (1964).

Ulrich, E.: Étude des ultrastructures au cours de l'ovogenèse d'un poisson téléosteen, le danio, *Brachydanio rerio* (Hamilton-Buchanan). J. Microscopie **8**, 447—478 (1969).

Ulsamer, A. G., Smith, F. R., Korn, E. D.: Lipids of *Acanthamoeba castellanii*. Composition and effects of phagocytosis on incorporation of radioactive precursors. J. Cell Biol. **43**, 105—114 (1969).

Walne, P. L.: The effects of colchicine on cellular organization in *Chlamydomonas*. II. Ultrastructure. Amer. J. Botany **54**, 564—577 (1967).

Ward, R. T., Ward, E.: The multiplication of Golgi bodies in the oocytes of *Rana pipiens*. J. Microscopie **7**, 1007—1020 (1968).

Watson, M. L.: The nuclear envelope. Its structure and relation to cytoplasmic membranes. J. biophys. biochem. Cytol. **1**, 257—270 (1955).

Weisman, R. A., Korn, E. D.: Phagocytosis of latex beads by *Acanthamoeba*. I. Biochemical properties. Biochemistry **6**, 485—497 (1967).

Wetzel, M. G., Korn, E. D.: Phagocytosis of latex beads by *Acanthamoeba castellanii* (Neff). III. Isolation of the phagocytic vesicles and their membranes. J. Cell Biol. **43**, 90—104 (1969).

WHALEY, W. G.: Proposals concerning replication of the Golgi apparatus. In: Organisation der Zelle. III. Probleme der biologischen Reduplikation. 3. Wissenschaftliche Konferenz der Gesellschaft Deutscher Naturforscher und Ärzte (Ed.: P. SITTE), S. 340—371. Berlin-Heidelberg-New York: Springer 1966.
— The Golgi apparatus. In: The Biological Basis of Medicine (Ed.: E. E. BITTAR and N. BITTAR), vol. 1, pp. 179—208. London-New York: Academic Press 1968.
— DAUWALDER, M., KEPHART, J. E.: The Golgi apparatus and an early stage in cell plate formation. J. Ultrastruct. Res. 15, 169—180 (1966).
— KEPHART, J. E., MOLLENHAUER, H. H.: The dynamics of cytoplasmic membranes during development. In: Cellular Membranes in Development (Ed.: M. LOCKE), pp. 135—173. New York-London: Academic Press 1964.
YOO, B. Y.: Ultrastructural changes in cells of pea embryo radicles during germination. J. Cell Biol. 45, 158—171 (1970).
YOUNG, R. W., DROZ, B.: The renewal of protein in retinal rods and cones. J. Cell Biol. 39, 169—184 (1968).
ZAMBONI, L., MASTROIANNI, L., JR.: Electron microscopic studies on rabbit ova. I. The follicular oocyte. J. Ultrastruct. Res. 14, 95—117 (1966a).
— — Electron microscopic studies on rabbit ova. II. The penetrated tubal ovum. J. Ultrastruct. Res. 14, 118—132 (1966b).

Origin and Continuity of Mitochondria

Robert Baxter

Shell Research Ltd., Woodstock Agricultural Research Centre, Sittingbourne, Kent

I. Introduction

Although mitochondria have been studied for nearly a century, and their role as the "power houses" of the aerobic cell has been recognised for twenty years, the mode of replication of these very important subcellular organelles is by no means clearly understood. Research effort in the area of mitochondrial biogenesis is at present intense and is increasing rapidly along the interconnected lines of biochemistry, cytology, genetics and molecular biology. Numerous review articles have recently appeared on the subject (e.g. WORK, COOTE, and ASHWELL, 1968; ROODYN, 1968; LLOYD, 1969; WAGNER, 1969) as well as at least two books (ROODYN and WILKIE, 1968; and SLATER, TAGER, PAPA, and QUAGLIARIELLO, 1968) and the reader is referred to these for detail and discussion in depth of the several aspects of the subject. The intention in this chapter is to trace the outline of our knowledge of mitochondrial genesis and continuity and to indicate how evidence from recent experiments is causing a continual modification of our understanding of how mitochondria are formed, and how they multiply.

II. Mitochondrial Biogenesis: the Machinery

In the early years of mitochondrial study, long before the advent of the electron microscope, the visual similarity between mitochondria and bacteria was recognised. Mitochondria were thought by some (e.g. ALTMANN, 1890) to be semi-autonomous elementary structures, capable, as are bacteria, of self-replication within the cell. (For a discussion of the early ideas concerning mitochondria, see NOVIKOFF, 1961).

The idea that mitochondria may have evolved, through endosymbionts, from invading bacteria has been raised and considered afresh in recent years (e.g. SAGAN, 1967). The renewed interest in this concept followed the discovery that mitochondria contained genetic material, in the form of a specific DNA, together with the components of a protein synthesis system which differed from the microsomal system but possessed many properties in common with that of bacteria. Furthermore mitochondria appeared to possess inheritable systems which were independent of the nuclear system (see GIBOR and GRANICK, 1964).

Mitochondrial DNA has now been prepared from numerous species of plants and animals (e.g. NASS, NASS, and AFZELIUS, 1965) and is probably universal. It differs from nuclear DNA in possessing a greater buoyant density during caesium chloride density gradient centrifugation (see ROODYN and WILKIE, 1968). Mitochondrial DNA also renatures more readily after heating than does nuclear DNA (BORST and RUTTENBERG, 1966) and ease of renaturation reflects a more homogeneous base composition and a more limited genetic capability. It was thought that mitochondrial DNA, like bacterial DNA, consisted of closed "circular" molecules (NASS, 1966), and AVERS (1967) has suggested that the respiratory deficiency of "petite" yeast mutants may be a consequence of impaired transcription of mitochondrial DNA, which, in this mutant, is predominantly of an "open" form. However, SUYAMA and MIURA (1968) have now reported the absence of circular mitochondrial DNA forms in *Tetrahymena* and WOLSTENHOLME and GROSS (1968) found French bean mitochondrial DNA to be linear rather than circular.

Mitochondrial DNA does not appear to have histones associated with it as has nuclear DNA of higher organisms. In this respect mitochondria are similar to bacteria.

Mitochondria appear to contain ribosomes which are smaller in diameter than "cytoplasmic" ribosomes (e.g. SWIFT, 1965) and yeast mitochondria contain RNA species of 23S and 16S (WINTERSBERGER, 1966a) which would correspond to a 70S ribosome of bacterial type than rather the 80S ribosome of the cytoplasm. Ribosome-like particles with sedimentation values close to 70S have been isolated from yeast (KÜNTZEL and NOLL, 1967) and maize (WILSON, HANSON, and MOLLENHAUER, 1968) but values of 81S (RIFKIN, WOOD, and LUCK, 1967) and 55S (O'BRIEN and KALF, 1967a and b) have also been reported, and the extent of degradation suffered by the particles during isolation is not yet clear. Polysome-like aggregations of ribosomes have been observed in sections of yeast mitochondria (VIGNAIS, HUET and ANDRÉ, 1969).

High-molecular-weight RNA species associated with mitochondria which differ in sedimentation value from cytoplasmic ribosomal RNA have been reported in yeast (ROGERS, PRESTON, TITCHENER, and LINNANE, 1967), *Neurospora* (KÜNTZEL and NOLL, 1967) and HeLa cells (VESCO and PENMAN, 1968). *Neurospora* mitochondrial RNA has a different base ratio from that of the cytoplasmic ribosomes (RIFKIN, WOOD and LUCK, 1967) and hybridizes readily with mitochondrial DNA but not with nuclear DNA (WOOD and LUCK, 1969). High degrees of hybridizing ability have also been shown between mitochondrial RNA and DNA of yeast (WINTERSBERGER, 1967, and FUKUHARA, 1967a) and *Tetrahymena* (SUYAMA, 1967), suggesting that mitochondrial DNA codes for mitochondrial ribosomal RNA.

Mitochondrial ribosomes require a higher concentration of magnesium to maintain their integrity than do cytoplasmic ribosomes (RIFKIN, WOOD, and LUCK, 1967; KÜNTZEL, 1969a). Instability in low concentrations of magnesium is also characteristic of bacterial ribosomes, and furthermore the mitochondrial ribosome sub-units themselves show striking resemblances to bacterial ribosome sub-units (KÜNTZEL, 1969a).

A proportion of the mitochondrial RNA appears to be intimately associated with membranes (ROODYN, 1962; KROON, 1965). FUKUHARA (1967a) has reported a membrane-associated yeast RNA which is neither ribosomal nor transfer RNA, but which hybridizes with, and is therefore probably coded for by, mitochondrial DNA. It is suggested that some membrane-associated RNA may be synthesised by mitochondria and then exported to other cell membranes (ATTARDI, 1968).

The transfer RNA (tRNA) molecules corresponding to several amino acids have been found in *Neurospora* mitochondria, and at least three of these appear to be specific for mitochondria, as are the respective activating enzymes (Barnett and Brown, 1967; Barnett, Brown, and Epler, 1967). More recently Buck and Nass (1968) have shown a range of tRNA species which are found exclusively in the mitochondria of rat liver, and that cytoplasmic synthetases for a given amino acid will not acylate the "corresponding" mitochondrial tRNA (Buck and Nass, 1969). Furthermore, comparative and competitive hybridization experiments have shown rat liver mitochondrial leucyl-tRNA to be more closely related to mitochondrial DNA than is "cytoplasmic" leucyl-tRNA (Nass and Buck, 1969). This type of evidence indicates that there exists an amino-acid-activating machinery in mitochondria which is different in mechanism, and possibly origin, from that of the cytoplasm.

When the mechanism of protein chain initiation is considered, the similarity between the mitochondrial and bacterial system again becomes evident. Protein chain initiation in bacteria involves the mediation of N-formyl-methionyl transfer RNA. This material has now been detected in the mitochondria of yeast, rat liver (Smith and Marcker, 1968) and HeLa cells (Galper and Darnell, 1969) but is absent from the (extra-mitochondrial) cytoplasm.

The presence, in mitochondria, of enzymes concerned with the synthesis of nucleic acids is now well established. RNA polymerase activity has been demonstrated in mitochondria isolated from a range of tissues (e.g. Neubert, Helge, and Merker, 1965), and the enzyme appears to be DNA-dependent (Wintersberger, 1964; Luck and Reich, 1964) and concerned directly with mitochondrial amino acid incorporation (Kalf, 1964; Kroon, 1965).

Yeast mitochondria have been shown to contain DNA polymerase (Wintersberger, 1966b) and more recently Kalf and Ch'ih (1968) have succeeded in isolating the enzyme from rat liver mitochondria. Mitochondrial DNA polymerase appears to be involved in DNA replication rather than repair (Karol and Simpson, 1968) and possesses properties which are different from those of the nuclear enzyme. These include a differing requirement for metal ions (Meyer and Simpson, 1968) and sensitivity to mutagenic dyes (Meyer and Simpson, 1969). Yeast mitochondrial DNA polymerase appears to be smaller than its nuclear counterpart, and is active at different stages of the cell cycle (Iwashima and Rabinowitz, 1969). Visual evidence showing what appears to be rat liver mitochondrial DNA in the process of replicating has been presented by Kirschner, Wolstenholme, and Gross (1968) (Fig. 1).

The ability of isolated mitochondria to incorporate amino acids into protein by a process consistent with the known requirements for protein synthesis is now generally recognised (see Roodyn and Wilkie, 1967) and the claim that such incorporation was invariably due to contaminating bacteria or microsomes has been largely discounted (Roodyn, 1968; Work, Coote, and Ashwell, 1968). The mitochondrial amino acid incorporating system differs from the microsomal system in that it is insensitive to ribonuclease and does not require added "pH5 enzyme" or cell sap (e.g. Roodyn, Reis, and Work, 1961).

In contrast to the microsomal system, which is insensitive to D-*threo*-chloramphenicol (So and Davie, 1963), mitochondrial protein synthesis is inhibited by the drug (Wintersberger, 1965; Kroon, 1965; Wheeldon and Lehninger, 1966). The sensitivity of the mitochondrial system to chloramphenicol is particularly intriguing because the drug also inhibits protein synthesis in bacteria (see Gale, 1963) and chloroplasts (EISENSTADT and BRAWERMAN, 1964). Inhibition in bacteria seems to involve a binding of the drug to the ribosome (VAZQUEZ, 1963) and both chloroplasts and mitochondria are reported to contain ribosomes similar in size (70S) to those of bacteria (BOARDMAN, FRANCKI, and WILDMAN, 1965; KÜNTZEL and NOLL, 1967).

On the other hand, cycloheximide, a powerful inhibitor of microsomal protein synthesis (ENNIS and LUBIN, 1964) has little inhibitory effect upon protein synthesis

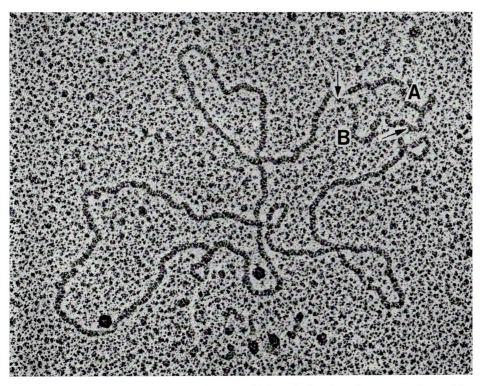

Fig. 1. A circular molecule of rat liver mitochondrial DNA showing what seems to be partial duplication. The molecule appears forked at two points (arrows). The two short segments, A and B, between the forks, are of equal length (0.6 μ) and the figure could represent an early stage in the replication of the DNA molecule. Magnification × 99,400. (By permission, from KIRSCHNER, R. H., WOLSTENHOLME, D. R., GROSS, N. J.: Proc. nat. Acad. Sci. (Wash.) **60**, 1466—1472 (1968).

by isolated mitochondria (BORST, KROON, and RUTTENBERG, 1967) or bacteria (ENNIS and LUBIN, 1964). Thus there exists a striking similarity between bacteria and mitochondria with respect to the sensitivities of their protein synthesising systems to these and other drugs (see BORST, KROON, and RUTTENBERG, 1967, and ROODYN and WILKIE, 1968).

Mitochondria, being membraneous structures, contain a high proportion of lipid as well as protein. Although some of the enzymes involved in the synthesis of some mitochondrial lipids are known to be extra-mitochondrial (WILGRAM and KENNEDY,

1963; Schneider, 1963), phospholipid precursors such as choline, serine and ethano-lamine have been shown to be incorporated by mitochondria isolated from rat liver (Kaiser and Bygrave, 1968; Kaiser, 1969) and locust (Bygrave and Kaiser, 1969). The transfer of monosaccharides from nucleoside diphosphate monosaccharide precursors into gluco-protein of the inner membranes has been demonstrated in isolated rat liver mitochondria (Bosmann and Martin, 1969).

III. Limitations of Mitochondrial Autonomy

Isolated mitochondria posses the ability to incorporate small-molecular-weight precursors into their fabric. Furthermore, mitochondria contain what appears to be a complete set of functional components for the synthesis of protein. This set of components differs in many ways from that of the microsomal protein synthesis system, and possesses some capability for generation. But by no means does it necessarily follow that mitochondria are autonomous bodies, replicating and re-generating independently of nuclear control and cytoplasmic involvment. On the contrary, several lines of evidence have indicated a heavy dependence of mitochon-dria on the rest of the cell.

The quantities of certain materials concerned with protein synthesis in mito-chondria appear to be very limited. For example, if the molecular weight of a typical mitochondrial DNA molecule is taken as 10 million, it could code for only about 5,000 amino acids or about 30 proteins of a molecular weight of 20,000 (Sinclair and Stevens, 1966). The electron transporting system alone must account for a mole-cular weight of nearly two million (Lehninger, 1964) and although mitochondria may contain more than one molecule of DNA (Borst, Kroon, and Ruttenberg, 1967) and mitochondrial DNA molecules from some species may be relatively large (Suyama and Miura, 1968; Wolstenholme and Gross, 1968) it seems extremely unlikely that mitochondrial DNA can carry enough information to code for a whole mitochondrion. This is especially true if the mitochondrial DNA is of more homo-genous base composition than nuclear DNA (Dubuy, Mattern, and Riley, 1966) and is therefore probably lacking in genetic capability.

Recent RNA-DNA hybridization experiments by Wintersberger (1967) have indicated that mitochondrial ribosomal RNA may constitute one of the few classes of mitochondrial macromolecules which is coded for by mitochondrial DNA.

The proteins of mitochondria do not all seem to be synthesised at the same rate. When isolated rat-liver mitochondria are incubated with radioactive amino acids, the proteins of the insoluble membrane-associated type ("structural proteins") label more intensely than the soluble "enzymatic" proteins (Roodyn, 1962; Wheeldon and Lehninger, 1966; Beattie, Basford, and Koritz, 1967), while the outer mitochondrial membrane does not label well under such conditions (Neupert, Brdiczka, and Bücher, 1967). Examination of the time-course of labelling with radioactive amino acids *in vivo* reveals a complex pattern (Fig. 2) in rat kidney mito-chondria. The insoluble types of protein label more intensely at first, but the soluble proteins become much more radioactive after a few hours. Eventually the intensity of labelling reaches an even level throughout the mitochondrion (Beattie, Basford, and Koritz, 1966). The lag period in the labelling of the soluble mitochondrial proteins could well indicate that they are synthesised at an extra-mitochondrial site and subsequently imported. Experiments with radiolabelled precursors and rat liver have indicated that cytochrome c is synthesised *in vivo* on the microsomes and then transferred to the mitochondria (GONZALES-CADAVID and CAMPBELL, 1967; PENNIALL and DAVIDIAN, 1968). A transfer from microsomes to mitochondria of

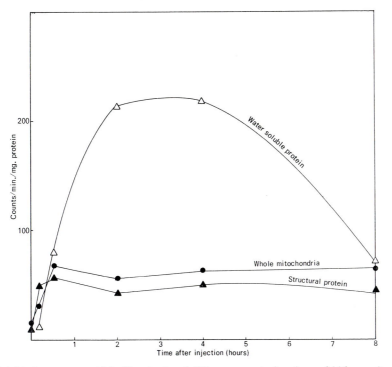

Fig. 2. The time-course of labelling *in vivo* of different protein fractions of kidney mitochondria from rats injected with a radioactive amino acid. (By permission, after ROODYN, D. B., WILKIE, D.: The biogenesis of mitochondria. London: Methuen 1968; from data of BEATTIE, D. S., BASFORD, R. E., KORITZ, S. B.: Biochemistry **5**, 926—930 (1966).)

protein (KADENBACH, 1967a) and of cytochrome c (KADENBACH, 1967b) has also been demonstrated *in vitro*. Evidence for the involvement of nuclear genes in the production of mitochondrial leucyl-tRNA synthetase (GROSS, McCOY and GILMORE, 1968) and malic dehydrogenase (LONGO and SCANDALIOS, 1969) adds weight to the idea of extamitochondrial synthesis of mitochondrial enzyme proteins.

It should be pointed out that our understanding of the nature of "structural protein" is undergoing modification. Originally considered to be a homogeneous protein concerned with the alignment of respiration carriers (CRIDDLE, BOCK, GREEN, and TISDALE, 1962) the fraction was later shown to consist of several proteins (e.g. HALDAR, FREEMAN and WORK, 1966) and more recently to contain appreciable amounts of denatured enzymic material (SCHATZ and SALTZGABER, 1969). It would be interesting to know if the proteins synthesised by isolated mitochondria could in fact be enzymes, possibly including adenosine triphosphatase.

Recent *in vivo* labelling experiments with rat liver (BEATTIE, 1969) have indicated that the protein and lipid of the mitochondrial outer membrane are synthesised before

4*

either the inner "structural" components or the soluble enzymes. The synthesis of outer membrane may be a prerequisite for the formation of the internal structures.

The mitochondrial outer envelope may well be synthesised by a process which differs from that of the internal membranes. Outer membranes of rat liver mitochondria appear to possess a turn-over rate which is similar to that of the microsomes, and much faster than that of mitochondrial inner membranes (Brunner and Neupert, 1968; Beattie, 1969). Further similarity between outer membrane and endoplasmic reticulum is reflected in their lipid composition (Parsons et al., 1967) and shown by their similar pattern of choline incorporation *in vivo* (Bygrave and Bücher, 1968). Furthermore, several proteins, including rotenone-insensitive NADH-cytochrome c reductase and cytochrome b_5, are common to mitochondrial outer membrane and the endoplasmic reticulum, but are not found in mitochondrial inner membranes (Sotto-casa et al., 1967). However, there are enzymes (e.g. monoamine oxidase) which are present in the mitochondrial outer membrane but largely absent from the endoplasmic reticulum, so the analogy cannot be taken too far (see Schnaitman and Greena-walt, 1968). The precise biosynthetic relationship between the endoplasmic reticulum and the mitochondrial outer envelope will await a deeper understanding of membrane organization in general.

Further evidence that mitochondria are not biosynthetically self-sufficient is apparent from work with yeast mutants. The cytoplasmic mutant giving rise to "petite" forms of *Saccharomyces* (Ephrussi, 1953) appears to be deficient in functional mitochondrial DNA (see Roodyn and Wilkie, 1968, for discussion of the cytoplasmic genetics of yeast and *Neurospora*). The quantity of DNA in "petite" mutant mitochondria appears about the same as that in wild-type mitochondria, but the base composition is very much simpler and presumably lacking in information (Meh-rotra and Mahler, 1968). Mitochondria are not absent from "petite" mutants, as would be expected if the whole mitochondrion were coded for by an intact mitochondrial DNA. They are, however, lacking in internal membrane structure and deficient of cytohaemin (Tuppy and Birkmayer, 1969).

Chloramphenicol is an inhibitor of both bacterial and mitochondrial protein synthesis, apparently because of its interference with the functioning of the 70s ribosome (e.g. Vazquez, 1963). Yeast grown in the presence of chloramphenicol contains mitochondria which are lacking in internal structure and cytochromes a, a_3, b and c_1. The production of cytochrome c, outer membrane and soluble enzymes is not affected by the drug (Clark-Walker and Linnane, 1967). Kleitke and Wollen-berger (1968) have also found that chloramphenicol fails to inhibit the hormone-induced formation of soluble mitochondrial enzymes in rats. The inference to be drawn from this kind of work is that proteins which are produced in the presence of chloramphenicol are probably not synthesised on the mitochondrial ribosomes. Conversely, those proteins whose production is inhibited by the drug are likely to be formed by a process involving the mitochondrial ribosomes.

Sebald et al. (1969) inhibited cytoplasmic ribosome-based protein synthesis in intact locusts using cycloheximide, a drug which does not affect protein synthesis by isolated mitochondria. The labelling of the mitochondria with radioactive amino acids was modified to a pattern very similar to that obtained using isolated mitochondria, indicating that the mitochondrial system makes the same limited numbers of proteins *in vivo* as *in vitro*.

The differentiation of the cytoplasmic protein synthesis site from the mitochondrial protein synthesis site by the use of chloramphenicol and cycloheximide has led KÜNTZEL (1969 b) to suggest that mitochondrial ribosomal proteins are not synthesised by the mitochondrial system (although mitochondrial ribosomal RNA may be; see WINTERSBERGER, 1967). Although the possible interdependence of the two protein synthetic systems has not been investigated, use of cycloheximide and chloramphenicol has prompted ASHWELL and WORK (1968) to suggest that the proteins produced under the influence of mitochondrial DNA in the intact cell constitute but a small percentage of the total proteins in the mitochondria.

In short, it appears that the capacity of mitochondria to code for, and synthesise, their own substance is very limited, and that the bulk of mitochondrial materials are brought in from cytoplasmic microsomal sites of construction. Very little is known about the movement of nucleic acids around the cell, however, and the use of nuclear information by the mitochondria, in the form, say, of imported messenger RNA, could be considerable. The assessment of such movements will await a more comprehensive grasp of the mobilities of substances in general around the cell, and greater knowledge of membrane structure and permeability.

IV. The Replication of Mitochondria

Our understanding of the process whereby mitochondria are produced is still very incomplete. In a review of the earlier observations in this field, LEHNINGER (1964) classified the various theories of possible routes of mitochondrial genesis into three main groups.

1. Formation from other membraneous structures in the cell.
2. Growth and division of pre-existing mitochondria.
3. *De novo* synthesis from submicroscopic precursors.

The formation of mitochondria by "pinching off" or "budding" from pre-existing cell structures has been suggested for a range of cell membranes, including those of the plasmalemma (ROBERTSON, 1959), endoplasmic reticulum, nuclear envelope, and Golgi bodies (see NOVIKOFF, 1961). But the support for such processes has relied heavily upon electron micrographs of cell sections, and such evidence, in the absence of supporting biochemical data, cannot be wholly conclusive. Part of the problem undoubtedly lies in our fragmentary knowledge of the structure and composition of, and differences between, cell membranes in general. Indeed, the similarities discussed earlier, between the outer mitochondrial membrane and the endoplasmic reticulum, could lend weight to the idea that at least this part of the mitochondrion is derived in some way from the endoplasmic reticulum.

The experiments of LUCK (1963), in which radioactive choline was supplied to a choline-requiring mutant of *Neurospora crassa*, represent the major biochemical evidence for the theory that mitochondria replicate by increase in size followed by division. By means of radio-autography at timed intervals after exposure to the radioactive choline, LUCK observed a random pattern of labelling in the mitochondria of daughter cells. If the mitochondria had been formed from pre-existing membraneous structures or *de novo* from primary building materials, the labelling pattern would have been non-random. Recent measurements of mitochondrial volume change in *Neurospora* grown under controlled conditions have indicated that mitochondria are produced, probably by division, in synchronous "fronts" in tissue of a specific physiological age (Hawley and Wagner, 1967). Experiments with yeast (Smith et al., 1968) and *Tetrahymena* (Charret and André, 1968) have shown that mitochondrial DNA is produced at different times in the cell cycle than is nuclear DNA. It is quite possible that this production of mitochondrial DNA is associated in time with mitochondrial division.

Electron microscopic evidence for mitochondrial division by fission, although plentiful, is difficult to assess. The danger of producing artefacts is very real because of the harsh chemical and physical agents brought to bear on the test material during processing. Interpretation is not made easier by the ability of mitochondria to undergo extreme changes in shape *in vivo* (e.g. Ritchie and Hazeltine, 1953) which may or may not be associated with mitochondrial fission. There are numerous reports of mitochondria connected to each other by narrow bridges of membrane, especially

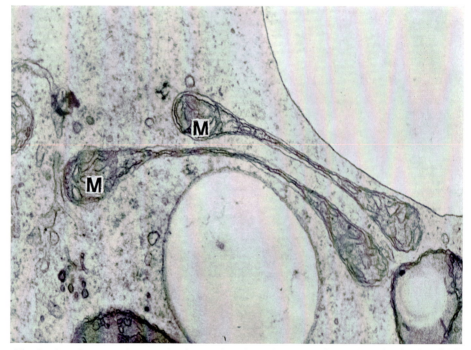

Fig. 3. An electron micrograph of a section through a pair of cup-shaped mitochondrial bodies (M) in developing callus tissue of fern. Each profile shows two masses of mitochondrial material separated by a narrow "bridge" consisting of two double membranes enclosing a strip of mitochondrial matrix. Fixed with glutaraldehyde and osmium tetroxide and stained with lead citrate. Magnification ×35,000. (By permission, from Bagshaw, V., Brown, R., Yeoman, M. M.: Ann. Botany (N.S.) **33**, 35—44 (1969).)

in rapidly metabolising tissue, and it is thought that such figures may represent mitochondria in an early stage of fission (Bahr and Zeitler, 1962; Claude, 1965; Diers, 1966). By observing serial sections of rat liver, Stempak (1967) was able to show that "dumb-bell-shaped" mitochondria can be sections of cup-shaped bodies. Such bodies have also been observed in rapidly-growing tissues of fern (Fig. 3) (Bagshaw,

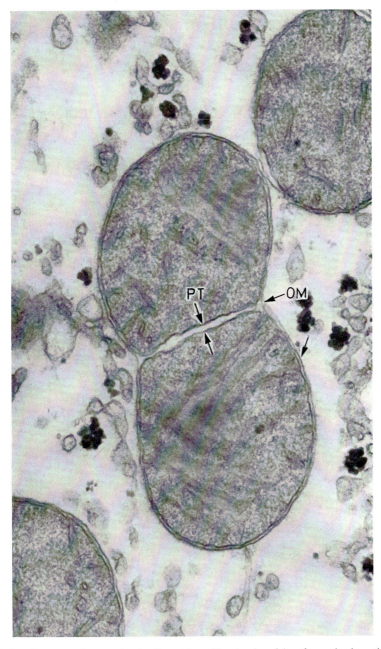

Fig. 4. An electron micrograph of a "partitioned" mitochondrion from the hepatic tissue of a rat which had been fed on a diet containing an azo-dye. The contents of the organelle appear to be divided into two cristae-containing segments separated by a partition (PT), while the whole is surrounded by a continuous outer membrane (OM). Fixed with osmium tetroxide and stained with lead hydroxide. Magnification × 53,400. (By permission, from LAFONTAINE, J. G., ALLARD, C.: J. Cell Biol. 22, 143—172 (1964).)

Brown, and Yeoman, 1969), and may represent the beginning stages of division (Hawley and Wagner, 1967).

An early stage in mitochondrial division may involve the separation of mitochondrial contents into two or more compartments. The presence of mitochondria with internal "partitions" has been reported for several cell types (see Tandler et al.,

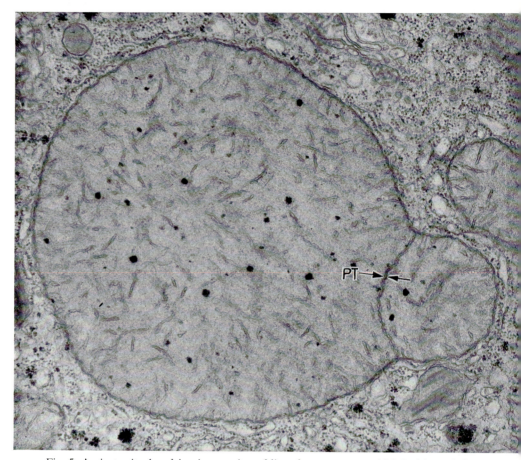

Fig. 5. A giant mitochondrion in a section of liver from a mouse which had been deficient of riboflavin for 6 weeks and allowed to recover for 1 day. A small segment of mitochondrial material, of about the size of a normal mitochondrion, appears to be separated by a partition (PT) from the bulk of the mitochondrial contents. Fixed in osmium tetroxide and stained with lead citrate. Magnification ×36,000. (By permission, from Tandler, B., Erlandson, R. A., Smith, A. L., Wynder, E. L.: J. Cell Biol. **41**, 477—493 (1969).)

1969) although the possibility that they are manifestations of mitochondrial fusion cannot easily be ruled out. Lafontaine and Allard (1964) have presented electron micrographs of rat liver mitochondria which exhibit what appear to be partitions dividing the inner membrane complex into two masses, the whole being surrounded by a continuous outer membrane (Fig. 4). Tandler et al. (1969) have demonstrated the partitioning of mitochondria in liver which was recovering from riboflavin deficiency (Figs. 5 and 6). In this tissue, giant mitochondria caused by the deficiency were reverting to normal size, and it seemed likely that, in this case at least, the partitions represented a stage in mitochondrial division rather than fusion.

The possibility of *de novo* synthesis of mitochondria arose with experiments in the early part of the century, when mitochondria-containing larvae were seen to develop from sea urchin egg cytoplasm which had apparently been freed of mitochondria by centrifugation (see NOVIKOFF, 1961). Using the greater resolving power of the electron microscope it was later shown that mitochondria could not be dislodged

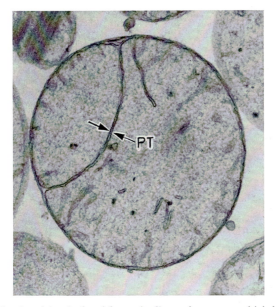

Fig. 6. A giant mitochondrion isolated from the liver of a mouse which had been deficient of riboflavin for 8 weeks and allowed to recover for 3.5 hours. The partition (PT) which divides the mitochondrial matrix appears to be continuous with the inner mitochondrial membrane. Stained with uranyl acetate and lead tartrate. Magnification × 47,000. (By permission, from TANDLER, B., ERLANDSON, R. A., SMITH, A. L., WYNDER, E. L.: J. Cell Biol. **41**, 477—493 (1969).)

by centrifugation of the egg (LANSING, HILLIER, and ROSENTHAL, 1952). In the earlier experiments, mitochondria had probably been present in the "centripetal end" of the egg cell after all, and these mitochondria could have served as precursors in subsequent mitochondrial production.

The absence in the cytoplasm of enzymic materials specifically characteristic of mitochondria prompted LEHNINGER (1964) to point out that *de novo* synthesis would involve an unlikely instantaneous integration of components into a functional organelle. In recent years, however, evidence for the presence of cytochrome c in rat liver microsomes (e.g. DAVIDIAN, PENNIALL and ELLIOTT, 1968), its synthesis there (GONZALES-CADAVID and CAMPBELL, 1967), and its transfer to mitochondrial fraction (PENNIALL and DAVIDIAN, 1968) has been reported. Transfer of preformed protein (KADENBACH, 1967a) and even ribosomes (GEORGATSOS and PAPASARANTOPOULOU,

1968) to mitochondria has been suggested, and it seems that quite large precursors of the mitochondrial fabric can be incorporated.

Facultative anaerobic yeast cells, grown under anaerobic conditions contain organelles, "promitochondria", which are lacking in functional electron-transporting enzymes and possibly in inner membrane structures (Wallace, Huang, and Linnane, 1968; Criddle and Schatz, 1969). When oxygen is supplied, the organelles become functionally respiring aerobically within a few hours. The adaption process does not appear to involve an increase in energy consumption or protein synthetic activity (Bartley and Tustanoff, 1966; Fukuhara, 1967b). Thus relatively minor modifications of high-molecular-weight pre-mitochondrial units may be involved during mitochondrial adaption, and the formation of mitochondria from large, membraneous structures has been suggested (Jayaraman et al., 1966).

Quite complex structures, such as flagella and viruses, can be reconstituted from their sub-units *in vitro* (see Roodyn and Wilkie, 1968) and separated complexes of the mitochondrial electron-transport system are capable of being recombined to form functional units (e.g. Hatefi, 1963) under conditions which also promote the formation of membrane structures (Green and Tzagoloff, 1966). The spontaneous formation of mitochondria by the coordinated integration of substructures is not out of the question, and such a process then begins to reapproach, in a sense, the concept of *de novo* synthesis.

But it is perhaps unwise to group the evidence to suit one or other of a limited number of clear-cut methods by which mitochondria could replicate. The actual situation is probably complex, and it may well be that different methods of replication take place in different tissues, and at different stages in development. One could imagine the early mitochondria being formed from membrane structures in the developing embryo. Concentrations of mitochondria around the nuclear membrane have been noted in embryonic tissues from several phyla (North and Pollak, 1961; Bell and Mühlethaler, 1962) and formation of mitochondria from this membrane could involve the transfer of nuclear genetic information essential for subsequent mitochondrial growth and multiplication by division. The multiplication of mitochondria could then proceed by the incorporation of large pre-fabricated molecules and associations of molecules, with division by fission when the mitochondria reached a critical size.

V. Discussion and Conclusion

The conclusion to be drawn from the evidence at hand seems to be that the mitochondrion is deeply integrated biochemically with the remainder of the cell it occupies. Much of its own fabric, including probably its outer envelope and many of the more "soluble" enzymatically active proteins, is synthesised at extra-mitochondrial sites, under the genetic control of nuclear DNA. The integration of the mitochondrion is not complete, however, and the organelle does contain a genetic capability, albeit a limited one. A peculiar bacteria-like protein synthesis mechanism enables the organelle to synthesise a small proportion of its own substance.

de Duve has suggested that mitochondria are relatively recent arrivals in the cell, and that their appearance enabled higher organisms to cope successfully with an atmophere which was becoming increasingly rich in oxygen through the activity of photosynthetic organisms. This is an attractive theory involving some fascinating speculation on the evolution of metabolic pathways (see DE DUVE and BAUDHUIN, 1966; MÜLLER, HOGG, and DE DUVE, 1968) and would fit with the "invasion" concept. Possible stages in the evolution of mitochondria have been suggested (SAGAN, 1967; ROODYN and WILKIE, 1968).

But if the small, specifically mitochondrial, biosynthetic capability really does represent the vestige of what was once the regenerative system of an invading symbiont, then a drastic loss of autonomy has occurred since the original bridgehead was established.

The processes whereby mitochondria are initiated and by which they replicate are not yet understood, although it seems likely that the incorporation of relatively large subunits into existing structures, followed eventually by fission, is involved. The formation of mitochondria from pre-existing membrane structures, at least in the early stages of embryogenesis, seems a possibility. A thorough knowledge of mitochondrial biosynthesis will await a greater understanding of the biosynthetic control mechanism operating in the cell, a deeper knowledge of the movement of molecules and complexes through the cell, and a greater understanding of the structure of the cell components, particularly the membranes, which modify such movements.

References

ALTMANN, R.: Die Elementarorganismen und ihre Beziehungen zu den Zellen, S. 145. Leipzig: Viet 1890.

ASHWELL, M. A., WORK, T. S.: Contrasting effects of cycloheximide on mitochondrial protein synthesis *in vivo* and *in vitro*. Biochem. biophys. Res. Commun. **32**, 1006—1012 (1968).

ATTARDI, G., ATTARDI, B.: Mitochondrial origin of membrane-associated heterogeneous RNA in Hela cells. Proc. nat. Acad. Sci. (Wash.) **61**, 261—268 (1968).

AVERS, C. J.: Heterogeneous length distribution of circular DNA filaments from yeast mitochondria. Proc. nat. Acad. Sci. (Wash.) **58**, 620—627 (1967).

BAGSHAW, V., BROWN, R., YEOMAN, M. M.: Changes in the mitochondrial complex accompanying callus growth. Ann. Botany (N. S.) **33**, 35—44 (1969).

BAHR, G. F., ZEITLER, E.: Study of mitochondria in rat liver. Quantitative electron microscopy. J. Cell Biol. **15**, 489—501 (1962).

BARTLEY, W., TUSTANOFF, E. F.: The effect of metabolic inhibitors on the development of respiration in anaerobically grown yeast. Biochem. J. **99**, 599—603 (1966).

BARNETT, W. E., BROWN, D. H.: Mitochondrial transfer ribonucleic acids. Proc. nat. Acad. Sci. (Wash.) **57**, 452—458 (1967).

— EPLER, J. L.: Mitochondrial-specific amino-acyl-RNA synthetases. Proc. nat. Acad. Sci. (Wash.) **57**, 1775—1781 (1967).

BEATTIE, D. S.: The biosynthesis of the protein and lipid components of the inner and outer membranes of rat liver mitochondria. Biochem. biophys. Res. Commun. **35**, 67—74 (1969).

— BASFORD, R. E., KORITZ, S. B.: Studies on the biosynthesis of mitochondrial protein components. Biochemistry **5**, 926—930 (1966).

— — — The inner membrane as the site of the *in vitro* incorporation of L-[^{14}C] leucine into mitochondrial protein. Biochemistry **6**, 3099—3110 (1967).

BELL, P. R., MÜHLETHALER, K.: The fine structure of the cells taking part in oogenesis in *Pteridium aquilinum* (L) Kuhn. J. Ultrastruct. Res. **7**, 452—466 (1962).

BOARDMAN, N. K., FRANCKI, R. I. B., WILDMAN, S. G.: Protein synthesis by cell-free extracts from tobacco leaves. 2. Association of activity with chloroplast ribosomes. Biochemistry **4**, 872—876 (1965).

BORST, P., RUTTENBERG, G. J. C. M.: Renaturation of mitochondrial DNA. Biochem. biophys. Acta (Amst.) **114**, 645—647 (1966).

BORST, P., KROON, A. M., RUTTENBERG, G. J. C. M.: Mitochondrial DNA and other forms of cytoplasmic DNA. In: Genetic elements properties and function (Ed.: D. SHUGAR). London-New York: Academic Press 1967.

BOSMANN, H. B., MARTIN, S. S.: Mitochondrial autonomy: incorporation of monosaccharides into glycoprotein by isolated mitochondria. Science **164**, 190—192 (1969).

BRUNNER, G., NEUPERT, W.: Turnover of outer and inner membrane proteins of rat liver mitochondria. FEBS Letters **1**, 153—155 (1968).

BUCK, C. A., NASS, M. M. K.: Differences between mitochondrial and cytoplasmic transfer RNA and aminoacyl transfer RNA synthetases from rat liver. Proc. nat. Acad. Sci. (Wash.) **60**, 1045—1052 (1968).

— — Studies on the mitochondrial tRNA from animal cells. 1. A comparison of mitochondrial and cyoplasmic tRNA and amino-acyl-tRNA synthetases. J. molec. Biol. **41**, 67—82 (1969).

BYGRAVE, F. L., BÜCHER, T.: Synthesis *in vivo* of the lecithin components of the inner and outer membranes of rat liver mitochondria. FEBS Letters **1**, 42—45 (1968).

— KAISER, W.: The magnesium-dependent incorporation of serine into the phospholipids of mitochondria isolated from the developing flight muscle of the African locust, *Locusta migratoria*. Europ. J. Biochem. **8**, 16—22 (1969).

CHARRET, R., ANDRÉ, J.: La synthèse de l'ADN mitochondrial chez *Tetrahymena pyriformis*. Etude radioautographique quantitative au microscope electronique. J. Cell Biol. **39**, 369—381 (1968).

CLARK-WALKER, G. D., LINNANE, A. W.: The biogenesis of mitochondria in *Saccharomyces cerevisiae*. A comparison between cytoplasmic respiratory-deficient mutant yeast and chloramphenicol-inhibited wild-type cells. J. Cell Biol. **34**, 1—14 (1967).

CLAUDE, A.: The morphology and significance of dumb-bell shaped mitochondria in early stages of regenerating liver. J. Cell Biol. **27**, 146 A (1965).

CRIDDLE, R. S., BOCK, R. M., GREEN, D. E., TISDALE, H.: Physical characteristics of proteins of the electron transfer system and interpretation of the structure of the mitochondria. Biochemistry **1**, 827—842 (1962).

— SCHATZ, G.: Promitochondria of anaerobically grown yeast. 1. Isolation and biochemical properties. Biochemistry **8**, 322—334 (1969).

DAVIDIAN, N., PENNIALL, R., ELLIOTT, W. B.: Origin of mitochondrial enzymes. 2. The subcellular distribution of cytochrome c in rat-liver tissue. FEBS letters **2**, 105—108 (1968).

DE DUVE, C., BAUDHUIN, P.: Peroxisomes (microbodies and related particles). Physiol. Rev. **46**, 323—357 (1966).

DIERS, L.: On the plastids, mitochondria, and other cell constituents during oögenesis of a plant. J. Cell Biol. **28**, 527—543 (1966).

DUBUY, H. G., MATTERN, C. F. T., RILEY, F. L.: Comparison of the DNAs obtained from brain nuclei and mitochondria of mice and from the nuclei and kinetoplasts of *Leishmania enriettii*. Biochem. biophys. Acta (Amst.) **123**, 298—305 (1966).

EISENSTADT, J. E., BRAWERMAN, G.: The protein-synthesizing systems from the cytoplasm of *Euglena gracilis*. J. molec. Biol. **10**, 392—402 (1964).

ENNIS, H. L., LUBIN, M.: Cycloheximide: Aspects of inhibition of protein synthesis in mammalian cells. Science **146**, 1474—1476 (1964).

EPHRUSSI, B.: Nucleo-cytoplasmic relations in microorganisms. Oxford: Clarendon Press 1953.

FUKUHARA, H.: Informational role of mitochondrial DNA studied by hybridization with different classes of RNA in yeast. Proc. nat. Acad. Sci. (Wash.) **58**, 1065—1072 (1967a)..

— Protein synthesis in non-growing yeast. The respiratory adaption system. Biochem biophys. Acta (Amst.) **134**, 143—164 (1967b).

GALE, E. F.: Mechanisms of antibiotic action. Pharmacol. Rev. **15**, 481—530 (1963).

GALPER, J. B., DARNELL, J. E.: The presence of N-formyl methionyl-tRNA in HeLa cell mitochondria. Biochem. biophys. Res. Commun. **34**, 205—214 (1969).

GEORGATSOS, J. G., PAPASARANTOPOULOU, N.: Evidence for the cytoplasmic origin of 78S ribosomes of mouse liver mitochondria. Arch. Biochem. **126**, 771—775 (1968).

GIBOR, A., GRANICK, S.: Plastids and mitochondria: inheritable systems. Science **145**, 890—897 (1964).

GONZALES-CADAVID, N. F., CAMPBELL, P. N.: The biosynthesis of cytochrome c. Sequence of incorporation *in vivo* of [¹⁴C] lysine into cytochrome c and total proteins of rat-liver subcellular fractions. Biochem. J. **105**, 443—450 (1967).

GREEN, D. E., TZAGOLOFF, A.: The mitochondrial electron transport chain. Arch. Biochem. **116**, 293—304 (1966).

GROSS, S. R., McCOY, M. G., GILMORE, E. B.: Evidence for the involvement of a nuclear gene in the production of the mitochondrial leucyl tRNA synthetase of *Neurospora*. Proc. nat. Acad. Sci. (Wash.) **61**, 253—260 (1968).

HALDAR, D., FREEMAN, K., WORK, T. S.: Biogenesis of mitochondria. Nature (Lond.) **211**, 9—12 (1966).

HATEFI, Y.: The pyridine nucleotide-cytochrome c reductases. In: The Enzymes (Ed.: P. D. BOYER, H. LARDY and K. MYRBÄCK.) Vol. 7, 495—515. New York: Academic Press 1963.

HAWLEY, E. S., WAGNER, R. P.: Synchronous mitochondrial division in *Neurospora crassa*. J. Cell Biol. **35**, 489—499 (1967).

IWASHIMA, A., RABINOWITZ, M.: Partial purification of mitochondrial and supernatant DNA polymerase from *Saccharomyces cerevisiae*. Biochem. biophys. Acta (Amst.) **283**—293 (1969).

JAYARAMAN, J., COTMAN, C., MAHLER, H. R., SHARP, C. W.: Biochemical correlates of respiratory deficiency. 7. Glucose repression. Arch. Biochem. **116**, 224—251 (1966).

KADENBACH, B.: Synthesis of mitochondrial proteins. Demonstration of a transfer of proteins from microsomes into mitochondria. Biochem. biophys. Acta (Amst.) **134**, 430—442 (1967a).

— Synthesis of mitochondrial proteins. The synthesis of cytochrome c *in vitro*. Biochem. biophys. Acta (Amst.) **138**, 651—654 (1967b).

KAISER, W.: Incorporation of phospholipid precursors into isolated rat liver mitochondria. Europ. J. Biochem. **8**, 120—127 (1969).

— BYGRAVE, F. L.: Incorporation of choline into the outer and inner membranes of isolated rat liver mitochondria. Europ. J. Biochem. **4**, 582—585 (1968).

KALF, G. F.: Deoxyribonucleic acid in mitochondria and its role in protein synthesis. Biochemistry **3**, 1702—1706 (1964).

— CH'IH, J. J.: Purification and properties of deoxyribonucleic acid polymerase from rat liver mitochondria. J. biol. Chem. **243**, 4904—4916 (1968).

KAROL, M. H., SIMPSON, M. V.: DNA biosynthesis by isolated mitochondria: A replicative rather than a repair process. Science **162**, 470—473 (1968).

KIRSCHNER, R. H., WOLSTENHOLME, D. R., GROSS, N. J.: Replicating molecules of circular mitochondrial DNA. Proc. nat. Acad. Sci. (Wash.) **60**, 1466—1472 (1968).

KLEITKE, B., WOLLENBERGER, A.: On the site of synthesis of enzymes tightly bound to mitochondrial structure in rat liver. FEBS Letters **1**, 187—190 (1968).

KROON, A. M.: Protein synthesis in mitochondria. 3. On the effect of inhibitors on the incorporation of amino acids into protein by intact mitochondria and digitonin fractions. Biochem. biophys. Acta (Amst.) **108**, 275—284 (1965).

KÜNTZEL, H.: Mitochondrial and cytoplasmic ribosomes from *Neurospora crassa*: characterization of their subunits. J. molec. Biol. **40**, 315—320 (1969a).

— Proteins of mitochondrial and cytoplasmic ribosomes from *Neurospora crassa*. Nature (Lond.) **222**, 142—146 (1969b).

— NOLL, H.: Mitochondrial and cytoplasmic polysomes from *Neurospora crassa*. Nature (Lond.) **215**, 1340—1345 (1967).

LAFONTAINE, J. G., ALLARD, C.: A light and electron microscope study of the morphological changes induced in rat liver cells by the azo dye 2-Me-DAB. J. Cell Biol. **22**, 143—172 (1964).

LANSING, A. I., HILLIER, J., ROSENTHAL, T. B.: Electron microscopy of some marine egg inclusions. Biol. Bull. **103**, 294 (1952).

LEHNINGER, A. L.: The mitochondrion. New York-Amsterdam: Benjamin 1964.

Lloyd, D.: The development of organelles concerned with energy production. In: Microbial growth. (Ed.: P. Meadow and S. J. Pirt). Cambridge: Cambridge University Press 1969.

Longo, G. P., Scandalios, J. G.: Nuclear gene control of mitochondrial malic dehydrogenase in maize. Proc. nat. Acad. Sci. (Wash.) **62**, 104—111 (1969).

Luck, D. J. L.: Formation of mitochondria in *Neurospora crassa*. A quantitative radioautographic study. J. Cell Biol. **16**, 483—499 (1963).

— Reich, E.: DNA in mitochondria of *Neurospora crassa*. Proc. nat. Acad. Sci. (Wash.) **52**, 931—938 (1964).

Mehrotra, B. D., Mahler, H. R.: Characterization of some unusual DNA's from the mitochondria from certain "petite" strains of *Saccharomyces cerevisiae*. Arch. Biochem. **128**, 685—703 (1968).

Meyer, R. R., Simpson, M. V.: DNA biosynthesis in mitochondria. Partial purification of a distinct DNA polymerase from isolated rat liver mitochondria. Proc. nat. Acad. Sci. (Wash.) **61**, 130—137 (1968).

— Simpson, M. V.: DNA biosynthesis in mitochondria. Differential inhibition of mitochondrial and nuclear DNA polymerases by the mutagenic dyes ethidium bromide and acriflavin. Biochem. biophys. Res. Commun. **34**, 238—244 (1969).

Müller, M., Hogg, J. F., de Duve, C.: Distribution of tricarboxylic acid cycle enzymes and glyoxylate cycle enzymes between mitochondria and peroxisomes in *Tetrahymena pyriformis*. J. biol. Chem. **243**, 5385—5395 (1968).

Nass, M. M. K.: The circularity of mitochondrial DNA. Proc. nat. Acad. Sci. (Wash.) **56**, 1215—1222 (1966).

— Buck, C. A.: Comparative hybridization of mitochondrial and cytoplasmic amino-acyl transfer RNA with mitochondrial DNA from rat liver. Proc. nat. Acad. Sci. (Wash.) **62**, 506—513 (1969).

— Nass, S., Afzelius, B. A.: The general occurrence of mitochondrial DNA. Exp. Cell Res. **37**, 516—539 (1965).

Neubert, D., Helge, H., Merker, H. J.: Zum Nachweis einer RNS-Polymerase-Activität in Mitochondrien tierischer Zellen. Biochem. Z. **343**, 44—69 (1965).

Neupert, W., Brdiczka, D., Bücher, T.: Incorporation of amino acids into the outer and inner membrane of isolated rat liver mitochondria. Biochem. biophys. Res. Commun. 488—489 (1967).

North, R. J., Pollak, J. K.: An electron microscope study on the variation of nuclear-mitochondrial proximity in developing chick liver. J. Ultrastruct. Res. **5**, 497—503 (1961).

Novikoff, A. B.: Mitochondria (chondriosomes). In: The cell; biochemistry, physiology, morphology. (Ed.: J. Brachet and A. E. Mirsky) Vol. 2. New York-London: Academic Press 1961.

O'Brien, T. W., Kalf, G. F.: Ribosomes from rat liver mitochondria. J. biol. Chem. **242**, 2172—2179 (1967a).

— — Ribosomes from rat liver mitochondria. 2. Partial characterization. J. biol. Chem. **242**, 2180—2185 (1967b).

Parsons, D. F., Williams, G. R., Thompson, W., Wilson, D., Chance, B.: Improvements in the procedure for purification of mitochondrial outer and inner membrane. Comparison of the outer membrane with smooth endoplasmic reticulum. In: Mitochondrial structure and compartmentation (Ed.: E. Quagliariello, S. Papa, E. C. Slater, and J. M. Tager.) Bari: Adriatica Editrice (1967).

Penniall, R., Davidian, N.: Origin of mitochondrial enzymes. Cytochrome c synthesis by endoplasmic reticulum. FEBS Letters **1**, 38—41 (1968).

Rifkin, M. R., Wood, D. D., Luck, D. J. L.: Ribosomal RNA and ribosomes from mitochondria of *Neurospora crassa*. Proc. nat. Acad. Sci. (Wash.) **58**, 1025—1032 (1967).

Ritchie, D., Hazeltine, P.: Mitochondria in *Allomyces* under experimental conditions. Exp. Cell Res. **5**, 261—274 (1953).

Rogers, P. J., Preston, B. N., Titchener, E. B., Linnane, A. W.: Differences between the sedimentation characteristics of the ribonucleic acids prepared from yeast cytoplasmic ribosomes and mitochondria. Biochem. biophys. Res. Commun. **27**, 405—411 (1967).

ROBERTSON, J. D.: The ultra-structure of cell membranes and their derivatives. Biochem. Soc. Symp. **16**, 3—43 (1959).

ROODYN, D. B.: Protein synthesis in mitochondria. 3. The controlled disruption and sub-fractionation of mitochondria labelled *in vitro* with radioactive valine. Biochem. J. **85**, 177—189 (1962).

— Mitochondrial biogenesis in germ free mitochondria. FEBS Letters **1**, 203—205 (1968).

— REIS, P. J., WORK, T. S.: Protein synthesis in mitochondria. Requirements for the in-corporation of radioactive amino acids into mitochondrial protein. Biochem. J. **80**, 9—21 (1961).

— WILKIE, D.: The biogenesis of mitochondria. London: Methuen 1968.

SAGAN, L.: The origin of mitosing cells. J. theor. Biol. **14**, 225—274 (1967).

SCHATZ, G., SALTZGABER, J.: Identification of denatured mitochondrial ATPase in "struc-tural protein" from beef heart mitochondria. Biochem. biophys. Acta (Amst.) **180**, 186—189 (1969).

SCHNAITMAN, C., GREENAWALT, J. W.: Enzymatic properties of the inner and outer mem-branes of rat liver mitochondria. J. Cell Biol. **38**, 158—175 (1968).

SCHNEIDER, W. C.: Intracellular distribution of enzymes. 13. Enzymatic synthesis of deoxycytidine diphosphatecholine and lecithin in rat liver. J. biol. Chem. **238**, 3572—3578 (1963).

SEBALD, W., HOFSTOTTER, T., HACKER, D., BÜCHER, T.: Incorporation of amino acids into mitochondrial protein of the flight muscle of *Locusta migratoria in vitro* and *in vivo* in the presence of cycloheximide. FEBS Letters **2**, 177—180 (1969).

SINCLAIR, J. H., STEVENS, B. J.: Circular DNA filaments from mouse mitochondria. Proc. nat. Acad. Sci. (Wash.) **56**, 508—514 (1966).

SLATER, E. C., TAGER, J. M., PAPA, S., QUAGLIARIELLO, E. (editors): Biochemical aspects of the biogenesis of mitochondria. Bari: Adriatica Editrice 1968.

SMITH, A. E., MARCKER, K. A.: N-formylmethionyl transfer RNA in mitochondria from yeast and rat liver. J. molec. Biol. **38**, 241—243 (1968).

SMITH, D., TAURO, P., SCHWEIGER, E., HALVORSON, H. O.: The replication of mitochon-drial DNA during the cell cycle in *Saccharomyces lactis*. Proc. nat. Acad. Sci. (Wash.) **60**, 936—942 (1968).

SO, A. G., DAVIE, E. W.: The incorporation of amino acids into protein in a cell-free system from yeast. Biochemistry **2**, 132—136 (1963).

SOTTOCASA, G. L., ERNSTER, L., KUYLENSTIERNA, B., BERGSTRAND, A.: Occurrence of an NADH-cytochrome c reductase system in the outer membrane of rat liver mitochon-dria. In: Mitochondrial structure and compartmentation. (Ed.: E. QUAGLIARIELLO, S. PAPA, E. C. SLATER, and J. M. TAGER). Bari: Adriatica Editrice 1967.

STEMPAK, J.: Serial section analysis of mitochondrial form and membrane relationships in the neonatal rat liver cell. J. Ultrastruct. Res. **18**, 619—636 (1967).

SUYAMA, Y.: The origins of mitochondrial ribonucleic acids in *Tetrahymena pyriformis*. Biochemistry **6**, 2829—2839 (1967).

— MIURA, K.: Size and structural variations of mitochondrial DNA. Proc. nat. Acad. Sci. (Wash.) **60**, 235—242 (1968).

SWIFT, H.: Nucleic acids of mitochondria and chloroplasts. Amer. Naturalist **99**, 201—227 (1965).

TANDLER, B., ERLANDSON, R. A., SMITH, A. L., WYNDER, E. L.: Riboflavin and mouse hepatic cell structure and function. 2. Division of mitochondria during recovery from simple deficiency. J. Cell Biol. **41**, 477—493 (1969).

TUPPY, H., BIRKMAYER, G. D.: Cytochrome oxidase apoprotein in "petite" mutant yeast mitochondria. Europ. J. Biochem. **8**, 237—243 (1969).

VAZQUEZ, D.: Antibiotics which affect protein synthesis: the uptake of ^{14}C-chloramphenicol by bacteria. Biochem. biophys. Res. Commun. **12**, 409—413 (1963).

VESCO, C., PENMAN, S.: The cytoplasmic RNA of HeLa cells. New discrete species associated with mitochondria. Proc. nat. Acad. Sci. (Wash.) **62**, 218—225 (1969).

VIGNAIS, P. V., HUET, J. H., ANDRÉ, J.: Isolation and characterization of ribosomes from yeast mitochondria. FEBS Letters **3**, 177—181 (1969).

Wagner, R. P.: Genetics and phenogenetics of mitochondria. Science **163**, 1026—1031 (1969).

Wallace, P. G., Huang, M., Linnane, A. W.: The biogenesis of mitochondria. II. The influence of medium composition on the cytology of anaerobically grown *Saccharomyces cerevisiae*. J. Cell Biol. **37**, 207—220 (1968).

Wheeldon, L. W., Lehninger, A. L.: Energy-linked synthesis and decay of membrane proteins in isolated rat river mitochondria. Biochemistry **5**, 3533—3545 (1966).

Wilgram, G. F., Kennedy, E. P.: Intracellular distribution of some enzymes catalyzing reactions in the biosynthesis of complex lipids. J. biol. Chem. **238**, 2615—2619 (1963).

Wilson, R. H., Hanson, J. B., Mollenhauer, H. H.: Ribosome particles in corn mitochondria. Plant Physiol. **43**, 1874—1877 (1968).

Wintersberger, E.: DNA-abhängige RNA-Synthese in Rattenleber-Mitochondrien. Hoppe Seylers Z. physiol. Chem. **336**, 285—288 (1964).

— Protein-Synthese in isolierten Hefe-Mitochondrien. Biochem. Z. **341**, 409—419 (1965).

— Synthesis and function of mitochondrial ribonucleic acid. In: Regulation of metabolic processes in mitochondria. (Ed.: J. M. Tager, S. Papa, E. Quagliariello, and E. C. Slater). Amsterdam-London-New York: Elsevier 1966a.

— Occurrence of a DNA polymerase in isolated yeast mitochondria. Biochem. biophys. Res. Commun. **25**, 1—7 (1966b).

— A distinct class of ribosomal RNA components in yeast mitochondria as revealed by gradient centrifugation and by DNA-RNA hybridization. Hoppe-Seylers Z. physiol. Chem. **348**, 1701—1704 (1967).

Wolstenholme, D. R., Gross, N. J.: The form and size of mitochondrial DNA of the red bean, *Phaseolus vulgaris*. Proc. nat. Acad. Sci. (Wash.) **61**, 245—252 (1968).

Work, T. S., Coote, J. L., Ashwell, M.: Biogenesis of mitochondria. Fed. Proc. **27**, 1174—1179 (1968).

Wood, D. D., Luck, D. J. L.: Hybridization of mitochondrial ribosomal RNA. J. molec. Biol. **41**, 211—224 (1969).

Origin and Continuity of Plastids

Wilfried Stubbe*

Botanisches Institut der Universität, Düsseldorf

I. Introduction

The term plastid, first used by Schimper (1883), is applied to a category of organelles of the eucaryotic plant cell. Plastids exist in different forms known as chloroplasts, chromoplasts, and leucoplasts according to the type or absence of pigmentation. The cells of animals, fungi, and procaryotes (bacteria and blue-green algae) lack plastids. The electron microscope, however, reveals distinct lamellar structures (thylakoids) in the cell of blue-green algae and of photosynthetic bacteria. These lamellae are fairly similar to those found within chloroplasts (Figs. 1–5), and are believed to be functionally equivalent. This must be taken into account when considering the origin of plastids.

Obviously the *chloroplast* represents the fundamental and functionally most versatile type of plastid. In algae, particularly in unicellular species, one or two large chloroplasts per cell are frequently found. They generally exhibit a distinctive shape like a cup, plate, band, or a more complicated three dimensional form (e.g. in most *Desmidiaceae*). Even at the unicellular level of organization (e.g. in *Euglena*), however, many small lentil-shaped chloroplasts per cell are found. This type is common in mosses, ferns, and flowering plants.

The most conspicuous function of the chloroplast is photosynthesis; in eucaryotic cells this depends upon the presence of chlorophyll a. Additional pigments are usually present (e.g. different kinds of chlorophyll, carotenoids, xanthophylls or biliproteins) and sometimes mask the green pigmentation as in the brown and red algae. The plastids of these algae are thus called phaeoplasts and rhodoplasts, respectively. The many biochemical and ultrastructural researches of the past two decades have elucidated in large part the molecular basis of the photosynthetic apparatus (cf. e.g. Menke, 1962, 1964; Goodwin, 1966; Kirk and Tilney-Bassett, 1967). In addition, these investigations have shown that the plastids of certain classes of the plant kingdom exhibit distinctive structural characteristics (Menke, 1962; Manton, 1966).

The *chromoplast* generally appears yellow or orange and occurs in certain organs of higher plants, for example in petals, fruits, and roots. Chromoplasts are photosynthetically more or less inactive because of their reduced chlorophyll content and loss of thylakoid structure. They are able to accumulate starch and lipids. Chloroplasts and chromoplasts are collectively termed chromatophores.

* The author is indebted to Dr. H. Kutzelnigg for his valuable assistance in compiling the literature and preparing this paper. He is also grateful to Drs. U. Hallier, G. Mosig, and Erich E. Steiner for their constructive criticism of the English manuscript.

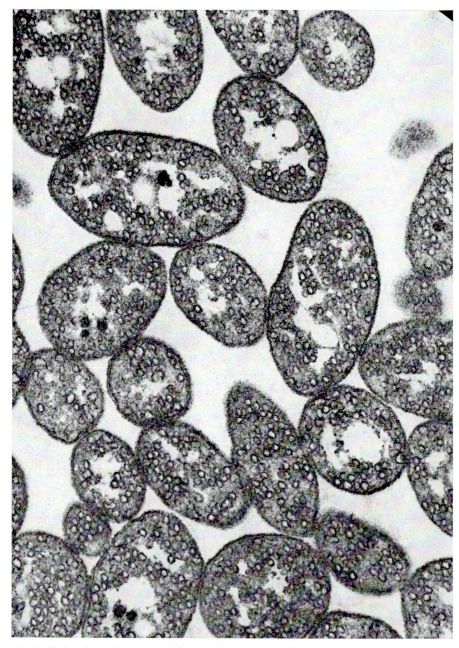

Fig. 1. Cells of the bacterium *Rhodopseudomonas spheroides* with vesicular thylakoids. By courtesy of Prof. Dr. W. Menke

The *leucoplast* is colorless. Various kinds of leucoplasts occur depending upon their function and the tissues in which they are found. Meristems often contain very small and undifferentiated "proplastids" which undergo multiplication. In storage tissues leucoplasts accumulate starch and are called amyloplasts. Less frequent are those types of leucoplasts which accumulate protein (proteinoplasts) or fat (elaio-plasts) and more specialized types such as those producing essential oils (AMELUNXEN and GRONAU, 1969). This variety of specialized types illustrates the manifold syn-thesizing capacities of the plastids.

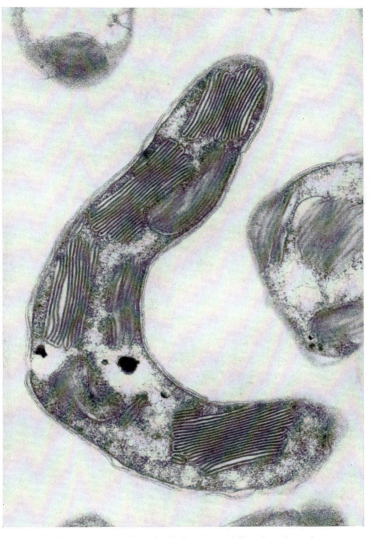

Fig. 2. Longitudinal section of *Ectothiorhodospira mobilis*, showing the arrangement of thylakoids into stacks. From REMSEN et al. (1968) by permission

The classification of plastid types based on pigmentation and function is superior to a morphological classification, since there is an enormous diversity in size and shape of plastids among the different taxa of the plant kingdom. Even in different tissues of an individual plant variation of plastid structure may be considerable (e.g. LAETSCH and PRICE, 1969).

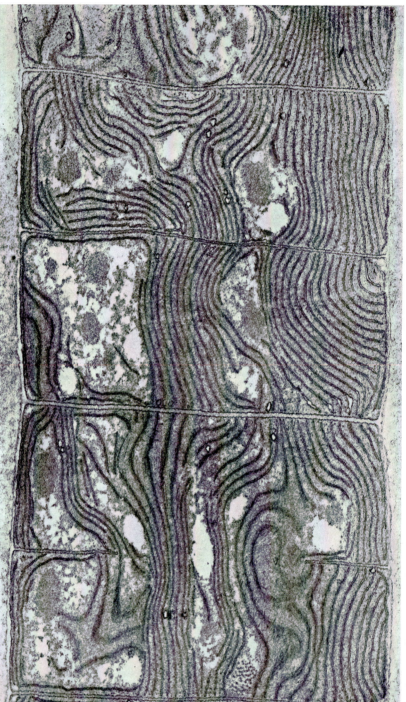

Fig. 3. Part of a trichome of the blue-green alga *Oscillatoria chalybea* with thylakoids. Longitudinal section. By courtesy of Prof. Dr. W. Menke

For more detailed information see SCHÜRHOFF (1924), KÜSTER (1956), KIRK and TILNEY-BASSETT (1967).

The idea that all these cellular components showing such a wide variety of phenotypes belong to the same category of organelle arises from the fact that transitional types exist between them, for example between chloroplasts and leucoplasts or chloroplasts and chromoplasts. The cytologists of the 19th century called attention to the fact that each type may be transformed into any other type of plastid ("metamorphosis" of the plastids according to SCHIMPER, 1885).

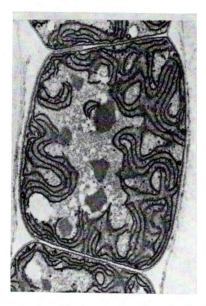

Fig. 4. Cell of the blue-green alga *Nostoc muscorum*. Longitudinal section. By courtesy of Prof. Dr. W. MENKE

II. Arguments for the Continuity of Plastids

A. Multiplication by Binary Fission

That the chloroplasts of all plants are homologous has never been questioned seriously; it has not been clear, however, whether or not plastids possess continuity in the same sense that chromosomes do. In other words, do plastids carry genetic information? After extensive studies on the development of plastids SCHMITZ (1882), SCHIMPER (1883, 1885), and MEYER (1883) expressed the view that plastids propagate only by fission. These authors thus proposed the theory of plastid continuity.

The division of chloroplasts which normally precedes cell division, or at least occurs simultaneously with it, can be observed fairly well in algae and mosses. Particularly impressive are the cinematographic studies of GREEN (1964) on the division of chloroplasts in growing cells of *Nitella*. In meristems of higher plants, however, *in vivo* observation of dividing plastids is extremely difficult if not impossible, since "proplastids" with a diameter of approximately $0.3-0.5\,\mu$ can hardly be distinguished from mitochondria in such cells. This explains attempts to relate

Fig. 5. Chloroplast in a leaf cell of *Oenothera*, the thylakoids differentiated into grana stacks and stroma lamellae. By courtesy of Prof. Dr. F. Schötz

mitochondria and plastids genealogically by some workers. (For literature see SCHÜR-HOFF, 1924; WEIER, 1963; DIERS, 1969).

Even experienced investigators have difficulty in recognizing proplastids in meristematic cells. In electron micrographs the undifferentiated plastids may be distinguished from mitochondria because they contain starch grains or short pieces of lamellae. Indirect evidence of their replication can be obtained from a statistical analysis of electron micrographs (DIERS, 1966, 1969; ANTON-LAMPRECHT, 1967). However, in 1962 MÜHLETHALER and BELL concluded from their electron microscopic investigation of archegonia of the fern *Pteridium aquilinum* that plastids (and mitochondria) are eliminated during maturation of the egg cell and are replaced by newly developed organelles. They believed that these organelles were derived from the nuclear membrane. These observations could not be confirmed by other investigators (DIERS, 1965, 1966; MENKE and FRICKE, 1964) and were challenged on genetical grounds (SCHÖTZ, 1962; STUBBE, 1962).

Apart from multiplication in meristematic tissues the plastids of higher plants also divide in nearly mature tissues, when the plastids have achieved a more advanced stage of development. This is shown by the difference in numbers of plastids in juvenile and mature cells and also by direct observation of fission configurations in the latter.

B. Differentiation and Dedifferentiation

Another problem exists with respect to the developmental aspect of plastid continuity. The dogma states that different modifications of plastids in various tissues of an individual plant can be transformed from one type into another. This has been questioned by FREY-WYSSLING, RUCH, and BERGER (1955). They postulated the following orthotropic development:

Proplastid → leucoplast → chloroplast → chromoplast

However, various investigators have reported contradictory evidence showing that development may also occur in the opposite direction (cf. THOMSON, LEWIS, and COGGINS, 1967; SCHÖTZ and SENSER, 1961). Especially important is the fact that under certain circumstances mature plastids can be transformed into proplastids, thus permitting mature cells to recover their meristematic (embryonic) status. BEN-SHAUL, EPSTEIN, and SCHIFF (1965) reported that in *Euglena* chloroplasts are transformed into proplastids stepwise during relatively few cell divisions in cultures kept in the dark. The possibility of reversible plastid differentiation is compatible with the theory of continuity, but sometimes plastid development may indeed be irreversible, especially if it is accompanied by a degeneration of the normal structure. Destruction of plastids interrupts cell development and prevents, as far as we know, further divisions.

In certain cases complete elimination of plastids by disintegration occurs regularly when zygotes form. The disintegration of the paternal chloroplast in the zygote of *Spirogyra* (CHMIELEWSKI, 1890) is a well known example*. In other cases the paternal cytoplasm is already eliminated prior to the fusion of the gametes (DIERS, 1967a).

---

* Observations on plastid fusion will be mentioned in chapter D.

The continuity of plastids through successive generations is maintained in these cases by maternal inheritance. Such uniparental transmission of plastids is supposed to occur in most flowering plants. It must be pointed out, however, that biparental plastid transmission has been proved for certain species of flowering plants and algae (Hagemann, 1964; Fritsch, 1965; Kirk and Tilney-Bassett, 1967; Diers, 1967 b). If plastids are endowed with such continuity we have to ask if they fulfill the criterion that they cannot be formed *de novo* if lost. Experiments to answer this question have been done with *Euglena gracilis* (cf. de Deken-Grenson and Messin, 1958; Grenson, 1964; McCalla, 1965; Schiff and Epstein, 1966; Granick and Gibor, 1967).

C. Elimination Experiments

In these investigations attempts were made to free flagellates of plastids using high temperatures, chemicals or UV irradiation. Special strains of *Euglena* are well suited for this purpose because they are able to live both heterotrophically and autotrophically. If cultured in the dark, the chloroplasts become reduced to proplastids. The question is whether the cells can dispence with the proplastids and whether they are able to replace the plastids when they are removed experimentally.

If we assume that in the course of evolution green phytoflagellates have turned into zooflagellates by accidental loss of their plastids, the experiments mentioned above should be promising. The finding of Schiff and Epstein (1966) that some of their experimentally bleached strains of *Euglena* no longer contained plastids was challenged by Granick and Gibor (1967): Electron micrographs presented by Schiff and Epstein reveal several plasmic structures which can be considered modified proplastids or mutated plastids. Further, all bleached *Euglena* mutants contain the polysaccharide paramylum which is supposed to be synthesized by plastids.

In view of these objections the elimination of plastids in *Euglena* does not appear to have been demonstrated unequivocally; nor has it been shown that plastids do not form *de novo*. In higher plants growing tissues which are free of plastids are unknown. As stated by Bauer (1943) the division of the nucleus and the cell seems to be strictly correlated with the presence of intact plastids.

D. Evidence for Heritable Differences

A more convincing demonstration that plastids show genetic continuity and are bearers of genetic information is shown by proof of specific and heritable differences between plastids contained in the same cell, the so called *mixed cell* („Mischzelle"). Van Wisselingh (1920) noted such mixed cells in *Spirogyra*. These cells contained two different kinds of chloroplasts, a normal one with, and one without pyrenoids. Since such differences are maintained for many generations, the assumption that these two plastid types (which live in the same cellular environment) are genetically different appears justified.

In higher plants the demonstration of mixed cells is similarly used as an argument for the existence of plastid inheritance (see cases listed by Hagemann, 1964). If genetically different plastids exhibit phenotypic markers the genetic analysis is facilitated. This method is especially important in tracing genetic effects of plastids which are expressed outside of the plastids.

Mixed cells can be established in different ways:

1. By mutation. *Euglena* is regarded as a favourable object for inducing plastid mutations (GIBOR and GRANICK, 1964; SCHIFF and EPSTEIN, 1966)*.

2. By grafting or injection. In *Acetabularia* the cells are large enough to produce mixed cells by grafting or injection of alien protoplasm.

3. By hybridization. The most convenient method for producing mixed cells is by hybridizing plants which permit biparental plastid transmission as in *Oenothera*.

Since in higher plants the sperm is relatively small in comparison with the egg cell, the majority of plastids in the heteroplastidic zygote descends from the seed parent; therefore, the progeny of reciprocal crosses is quantitatively different with respect to the percentages of the different kinds of plastids derived from each parent. In rare cases the egg cell is already a mixed cell. Thus three different plastid types can be combined within the zygote (STUBBE, 1957). Analysis of somatic segregation in the descendants of a mixed cell is necessary to ascertain whether or not phenotypically different plastids are indeed genetically different.

MICHAELIS extended the work of E. BAUR (1909) and CORRENS (1928) and refined the cytogenetic method of variegation pattern analysis (MICHAELIS, 1955a, b; 1962). This technique can be used for localizing extrachromosomal hereditary units within specific cell components. Unfortunately the application of this method requires extensive anatomical and cytological investigations (STUBBE, 1966). Because of the many demands of such a study most investigators depend upon circumstantial evidence. In *Oenothera*, for example, the dependence of certain chlorophyll deficiencies on genetic factors within the plastids is considered proven because:

1. biparental inheritance which leads to a mixed zygote is coupled with reciprocal differences;

2. the sorting out of the two plastid phenotypes takes place by somatic segregation as rapidly as one might expect (according to the number of cell constituents).

E. Plastid DNA**

Since genetic experiments have amply demonstrated that plastids are heritably different, one must ask what the chemical basis of the genetic continuum of the plastid is. The DNA found in the plastid is believed to be at least partially involved as the genetic material of this organelle***.

That plastids contain DNA remained uncertain until about 1963. Subsequently, convincing evidence of its presence in plastids was obtained from chemical and physical techniques. The DNA content of a single chloroplast is reported to vary among different species from 1 to 150×10^{-16} gm. The DNA is concentrated in certain regions of the plastid. Electron micrographs show fibrils $25-30$ Å thick in these regions which disappear following DNA-ase treatment.

---

* *Euglena* strains cannot be crossed since sexual reproduction has not yet been demonstrated.

** Reference is made to the extensive literature cited in the following reviews: GIBOR and GRANICK (1964), GRANICK and GIBOR (1967), KIRK and TILNEY-BASSETT (1967), SCHÖTZ (1967), HAGEMANN (1968).

*** Other possibilities of transmitting genetic properties should be kept in mind as for example the cortical structure in *Paramecium* (BEISSON and SONNEBORN, 1965).

The genetic problem is to demonstrate that the DNA within the plastids is responsible for specific inherited characters of the organism. Therefore it is necessary to compare different plastid types with respect to their DNA to prove the constancy of a recognized difference of plastid DNA and to demonstrate its genetic function. Valuable work has been done in this field during the last few years. By using a variety of techniques plastid DNA has been shown to be different from nuclear and mitochondrial DNA. Moreover, the plastid DNA replicates within the plastids as demonstrated by incorporation of ^{3}H-thymidine. Recent investigations have shown that replication is semi-conservative. SCHIFF and EPSTEIN (1966) reported that different strains of bleached *Euglena* contain different amounts of plastid DNA. Specific differences between the plastid DNA of related species are supposed to exist e.g. in the subgenus *Oenothera*. At present techniques are not sufficiently sensitive to detect such fine differences in DNA molecules. Up to now the differences have been recognized only by their genetic activity. It has been demonstrated that plastids possess an independent protein synthesizing system. In this connection it should be noted that the ribosomes within the plastids are different from those in the cytoplasm. They appear to be smaller and have a different $(A + U) : (G + C)$ ratio. Further, they seem to be more sensitive to certain antibiotics than cytoplasmic ribosomes. In this respect they resemble those of bacteria and mitochondria.

III. The Genetic Information of the Plastids (Plastome)

A. The Correlation between Genome and Plastome

In summary, one is led to the conclusion of SCHIFF and EPSTEIN (1966): "The chloroplast resembles a cell within a cell." It must be pointed out, however, that plastids are far from being autonomous. This has been effectively demonstrated by hybridization experiments in *Oenothera*. Plastids of different species were combined with various alien genotypes. Certain combinations led to various disturbances of normal development in the plants. RENNER (1922) explained these as disharmony (incompatibility) between nuclear and plastid genes. For the latter RENNER (1934) introduced the term "plastome" by which is meant the total genetic information within the plastids. The experiments also showed that the plastome, cooperating with the genome, controls a number of characters of the plant.

In the experiments of STUBBE (1959, 1960, 1964) the five wild type plastomes of *Euoenothera* were combined with 6 genotypes representing different phylogenetic groups. Only some of the 30 possible combinations (about 12) turned out to be compatible. The other combinations resulted in various abnormalities, many of which affected plastid development and led to different degrees of chlorophyll deficiency (see Fig. 6). In addition the following features were shown to depend upon genome-plastome interaction: plastid multiplication rate (first investigated by SCHÖTZ, 1954), cell division correlated with plastid development, lamina growth of the leaf, germination of pollen grains and growth of pollen tubes, and shape of starch grains within the pollen plastids (see also ARNOLD, 1963).

It is likely that the variety of features controlled by the plastome depends upon the number of hereditary sub-units (plastid genes). The nucleus presumably contains the bulk of the genetic information compared with that of the plastids. Nevertheless, it can be stated that the two components of the cell, nucleus and plastids, depend upon the activities of both the genome and plastome as well as any remaining hereditary components of the cell. This is a challenging field for cooperative research between geneticists and biochemists.

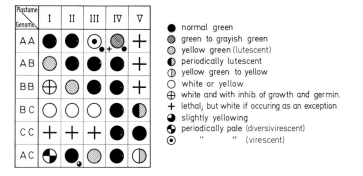

Plastome \ Genome	I	II	III	IV	V
A A	●	●	⊙+⊙		+
A B	◑	●	●	●	+
B B	⊕	◑	●	●	+
B C	○	○	○	●	◐
C C	+	+	+	●	●
A C	◕	●.	◑	●	◑

● normal green
◍ green to grayish green
◒ yellow green (lutescent)
◖ periodically lutescent
◑ yellow green to yellow
○ white or yellow
⊕ white and with inhib. of growth and germin.
+ lethal; but white if occuring as an exception
◐ slightly yellowing
◕ periodically pale (diversivirescent)
⊙   "   "   (virescent)

Fig. 6. Compatibility relations between different genotypes and plastid types in *Oenothera* (after STUBBE, 1964). (The use of more than one sign in some squares depends on slight differences between the A-complexes)

B. Mutation and Recombination

Plastid mutations usually can be recognized only if they lead to a chlorophyll deficiency; it has to be proven that the determinant is localized within the plastome (see Chapter II D). Few *quantitative* data on the spontaneous rate of the plastome are available (see MALY, 1958), although spontaneous plastome mutations have been observed frequently and in many plant species (see HAGEMANN, 1964). In *Oenothera* cultures plants with pale sectors or spots caused by mutations of the plastome occur at a rate of about 0.3 percent. We would expect that other flowering plants behave similarly.

It is surprising that many efforts to induce plastid mutations in higher plants have more or less failed (MALY, 1958; MICHAELIS, 1958, 1965; RÖBBELEN, 1962; ARNOLD and KRESSEL, 1965; DULIEU, 1967; KUTZELNIGG, 1968). In this respect lower plants generally seem likewise to be resistant to mutagenic agents (BAUER, 1943; HERBST, 1952; VAN BAALEN, 1965). Exceptions may occur, however; for example PRINGSHEIM (1964) mentioned that some strains of *Euglena* tend to mutate more readily than others.

The difficulties of inducing and detecting plastid mutations may possibly be due to a specific organization of the genetic material within the plastids. This might differ considerably from that of the chromosomes. In this connection HERRMANN (1968, 1969) raises the question whether chloroplasts are polyploid. Granted that each of the various DNA-regions of the chloroplast contains several copies of the plastome the manner of their distribution will determine whether a single mutation will manifest itself phenotypically. It is unknown to what extent repair mechanisms are involved in plastome mutation.

The problem of polyploidy of the plastome could possibly better be tackled if recombination between different plastid types could be observed. SAGER and RAMAnis (1968a, b; 1969) assume that in *Chlamydomonas reinhardi* the observed recombination of nonchromosomal genes (NC genes) depends on exchange of genetic substance between the two parental chloroplasts within the zygote. Gillham (1965), Gillham and Fifer (1968) and Schimmer and Arnold (1969) working also on recombinaiton of NC genes in *Chlamydomonas reinhardi* did not come to the same conclusion. Unfortunately the localization of NC genes in *Chlamydomonas* is not yet critically established. Since the recombination experiments cannot solve the problem of localization of the NC genes in *Chlamydomonas* another approach must be sought. Sager and Ramanis (1969) mention that they "are actively searching for an altered chloroplast protein resulting from mutation of one of these linked (NC) genes in order to localize this linkage group definitely".

In higher plants, especially in the genus *Oenothera* where localization of plastid genes is no longer a problem, recombination of plastome differences is not known. Theoretically it is possible that this phenomenon has escaped the geneticists' notice because the hybridization experiments were not sufficiently focussed on it.

Only a few cytological observations have been reported, which can be linked with plastid recombination (i.e. fusion of plastids). Menke (1960) discussed electron microscopic observations in meristematic cells of *Elodea* showing two plastids in close contact ("Kußszenen").

As Schürhoff (1924) mentioned, fusion of plastids is known to occur in the zygote of certain algae. This has been described by Schiller (1907) in *Ulva lactuca*. Fusion of plastids in the zygote of *Chlamydomonas reinhardi* has been observed by Cavalier-Smith and in *C. moewusii (syn. C. eugametos)* by Brown, Johnson, and Bold (1968). On the other hand, Ettl (1969) found the male chloroplast eliminated in *C. geitleri*. Fusion of chloroplasts of the gametes of *Acetabularia* has been reported by Crawley (1966). However, Kellner and Werz (1969) could not confirm this observation*.

Such cytological work is paralelled by tracer experiments on the physical conservation and distribution of plastid DNA during meiosis in *Chlamydomonas reinhardi*. Chiang (1968) found biparental transmission of plastid and mitochondrial DNA and pointed out that this contradicts the maternal inheritance of NC-genes. Sager and Lane (1969) report, however, preliminary results showing elimination of the male plastid DNA during maturation of the zygotes. This fits into the picture of maternal inheritance.

In view of the current situation important progress in this field can be expected in the near future.

IV. Plastid Phylogeny

Certain spontaneous plastome mutants which are detected by their chlorophyll deficiency prove the mutability of the plastome. The differentiation of distinct plastid wild-types on *Oenothera* must be based on mutation as well. Stubbe (1959, 1963, 1964) arranged the different wild-types of the plastome in *Euoenothera* into a phylogenetic scheme. This was based partly on the adaptation of the plastotypes to different genotypes, and independently thereof, on differences of the multiplication rates of the different sorts of plastids. It is evident that in outbreeding populations a newly

---

* We are indebted to Dr. C. G. Arnold who informed us regarding recent publications.

arising plastid type with a higher multiplication rate will gradually replace the pre-existing type, provided that mixing of these plastids occurs in the zygote.

With our present knowledge of the continuity and the genetic information of the plastids one can assume a plastid phylogeny for the whole plant kingdom including those primitive forms in which the structures for photosynthesis first arose.

Considerations about the phylogeny of plastid types in the larger taxa of the plant kingdom have been published very rarely (PASCHER, 1927; GEITLER, 1934; CHADE-FAUD, 1935; VISCHER, 1945; TEILING, 1952; MEYER, 1962; WEIER, 1963) though the manifold variation in shape suggests such speculations. We assume that the division of the plastidome into a number of lentil-shaped chloroplasts and the consequent increase in surface may be the reason for the prevalence of this type among higher plants. The larger manifold shaped plastids of most algae would represent the more primitive form according to this evolutionary concept. Apparently in *Euglena* and in some other algae a more advanced evolutionary state has been achieved. It would be worthwhile testing this hypothesis through biochemical investigation.

Tracing the phylogenetic pathway of plastids raises the problem of the origin of the typical chloroplast which is separated by a double membrane from the remaining protoplasm (see also SCHNEPF and BROWN in this volume). Obviously the organization of blue-green algae and some photosynthetic bacteria can be regarded as precursor, since in these organisms the thylakoid membranes are distributed within the cell and the double membrane envelope is not present. (A parallel question is the evolution of the nucleus of the eucaryotic cell, since it, like the plastid is wrapped in a double membrane.) Because there is no fossil record pertinent to this problem, the matter is entirely speculative. Two concepts should be mentioned:

1. MERESCHKOWSKY (1905) and FAMINTZIN (1907) following a suggestion of SCHIMPER (1885), developed the hypothesis that the chloroplasts correspond to symbiotic blue-green algae. This idea was revived by RIS and PLAUT (1962). Since then many investigators, especially electron microscopists, have accepted this view (e.g. ECHLIN, 1966; SCHNEPF, 1966).

2. The alternative possibility is that plastids, especially chloroplasts, originated stepwise within these primitive organisms. Theoretically a separation of the thylakoids together with parts of the genetic material could have occurred by their becoming wrapped in a double membrane, derived from the cell membrane. Subsequently this newly separated unit had to establish its independent replication by fission.

Most of the arguments in favour of each of the above speculations are based on similarity of structure. It is evident that such evidence can be used for spupporting both concepts. Therefore the problem reduces to the question: Which evolutionary steps are easier to achieve and therefore more probable?

The first hypothesis can be supported by examples of symbiosis with varying degrees of mutual dependence.*

The fact that especially in algae the chloroplasts are of a rather complicated shape and structure, whereas in higher plants the plastids are externally simpler argues against this concept. Moreover, in flowering plants, as in the root cells of *Gunnera*, an endosymbiosis with *Nostoc* can be observed (MIEHE, 1924). The work of REMSEN et al. (1968) on the fine structure of the bacterial cell of *Ectothiorhodospira mobilis*

---

* For details see SCHNEPF and BROWN (this volume p. 299 ff.).

(Fig. 3) is pertinent to the second hyopthesis on the origin of plastids. In this bacterium the photosynthetic lamellae are stacked in a manner resembling the grana stacks in the chloroplast of higher plants. These stacks are surrounded by a double membrane which belongs to the lamella system of the stacks. The lamellae are derived originally from the outer plasma membrane by invagination. Compared with simpler bacteria, for instance, *Rhodopseudomonas* (Fig. 1) or *Rhodospirillum rubrum* (cf. Drews and Giesbrecht, 1963; Menke, 1965) the lamellar system of *Ectothiorhodospira* shows a far more advanced compartmentation. One can imagine such a development leading to a typical chloroplast of the eucaryotic cell. In our opinion the problem of the origin of chloroplasts remains unsolved.

References

Amelunxen, F., Gronau, G.: Elektronenmikroskopische Untersuchungen an den Ölzellen von *Acorus calamus* L. Z. Pflanzenphysiol. **60**, 156—168 (1969).

Anton-Lamprecht, I.: Anzahl und Vermehrung der Zellorganellen im Scheitelmeristem von *Epilobium*. Ber. dtsch. bot. Ges. **80**, 747—754 (1967).

Arnold, C. G.: Die Plastiden als Erbträger extraplastidaler Merkmale. Ber. dtsch. bot. Ges. **76**, 3—12 (1963).

—, Kressel, M.: Versuche zur Auslösung von Plasmamutationen bei *Oenothera*. Z. Vererbungsl. **96**, 213—216 (1965).

Bauer, L.: Untersuchungen zur Entwicklungsgeschichte und Physiologie der Plastiden von Laubmoosen. Flora **136**, 30—84 (1943).

Baur, E.: Das Wesen und die Erblichkeitsverhältnisse der „*varietates albimarginatae hort.*" von *Pelargonium zonale*. Z. indukt. Abstamm.- u. Vererbungsl. **1**, 330—351 (1909).

Beisson, J., Sonneborn, T. M.: Cytoplasmic inheritance of the cell cortex in *Paramecium aurelia*. Proc. nat. Acad. Sci. (Wash.) **53**, 275—282 (1965).

Ben-Shaul, Y., Epstein, H. T., Schiff, J. A.: Studies on chloroplast development in *Euglena*. 10. The return of the chloroplast to the proplastid condition during dark adaptation. Canad. J. Botany. **43**, 129—136 (1965).

Brown, R. M., Johnson, S. C., Bold, H. C.: Electron and phasecontrast microscopy of sexual reproduction in *Chlamydomonas moewusii*. J. Phycol. **4**, 100—120 (1968).

Cavalier-Smith, T.: Organelle development in *Chlamydomonas reinhardi*. Ph. D. Thesis University of London (1967).

Chadefaud, M.: Le cytoplasme des algues vertes et brunes, ses éléments figurés et ses inclusions (Thèse Paris). Rev. algologique **8**, 5 (1936).

Chiang, K. S.: Physical conservation of parental cytoplasmic DNA through meiosis in *Chlamydomonas reinhardi*. Proc. nat. Acad. Sci. (Wash.) **60**, 194—200 (1968).

Chmielewski, V.: Eine Notiz über das Verhalten der Chlorophyllbänder in den Zygoten der *Spirogyra*-Arten. Botan. Ztg. **48**, 773 (1890).

Correns, C.: Über nichtmendelnde Vererbung. Z. indukt. Abstamm.- u. Vererbungsl. Suppl. I, 131—168 (1928).

Crawley, J. C. W.: Some observations on the fine structure of the gametes and zygotes of *Acetabularia*. Planta (Berl.) **69**, 365—376 (1966).

de Deken-Grenson, M., Messin, S.: La continuité génétique des chloroplastes chez les Euglènes. I. Mécanisme de l'apparition des lignées blanches dans les cultures traitées par la streptomycin. Biochim. biophys. Acta (Amst.) **27**, 145—155 (1958).

Diers, L.: Bilden sich während der Oogenese bei Moosen und Farnen die Mitochondrien und Plastiden aus dem Kern? Ber. dtsch. bot. Ges. **77**, 369—371 (1965).

— On the plastids, mitochondria, and other cell constituents during oogenesis of a plant. J. Cell Biol. **28**, 527—543 (1966).

— Der Feinbau des Spermatozoids von *Sphaerocarpos donnellii* Aust. (*Hepaticae*). Planta (Berl.) **72**, 119—145 (1967a).

— Übertragung von Plastiden durch den Pollen bei *Antirrhinum majus*. Molec. Gen. Genetics **100**, 56—62 (1967b).

DIERS, L.: Origin of plastids. 24. Symp. Soc. Exp. Biol. London (1969) (in preparation).

DULIEU, H.: Sur les différents types de mutations extranucléaires induites par le méthan sulfonate d'éthyle chez *Nicotiana tabacum* L. Mutation Res. **4**, 177—189 (1967).

DREWS, G., GIESBRECHT, P.: Zur Morphogenese der Bakterien-„Chromatophoren" (Thylakoide) und zur Synthese des Bacteriochlorophylls bei *Rhodopseudomonas spheroides* und *Rhodospirillum rubrum*. Zbl. Bakt. **190**, 508—536 (1963).

ECHLIN, P.: The cyanophytic origin of higher plant chloroplasts. Br. phycol. Bull. **3**, 150—151 (1966).

ETTL, H.: Über einen gelappten Chromatophor bei *Chlamydomonas geitleri nova spec.*, seine Entwicklung und Vereinfachung während der Fortpflanzung. Öst. bot. Z. **116**, 127—144 (1969).

FAMINTZIN, A.: Die Symbiose als Mittel der Synthese von Organismen. Biol. Zbl. **27**, 353—364 (1907).

FREY-WYSSLING, A., RUCH, F., BERGER, X.: Monotrope Plastidenmetamorphose. Protoplasma **45**, 97—114 (1955).

FRITSCH, F. E.: The structure and reproduction of the algae. Cambridge: Cambridge University Press 1965.

GEITLER, L.: Grundriß der Cytologie. Berlin: Gebr. Bornträger 1934.

GIBOR, A., GRANICK, S.: Plastids and mitochondria: inheritable systems. Science **145**, 890—897 (1964).

GILLHAM, N. W.: Linkage and recombination between nonchromosomal mutations in *Chlamydomonas reinhardi*. Proc. nat. Acad. Sci. (Wash.) **54**, 1560—1566 (1965).

— FIFER, W.: Recombination of non-chromosomal mutations: A three point cross in the green alga *Chlamydomonas reinhardi*. Science **162**, 683—684 (1968).

GOODWIN, T. W. (Editor): Biochemistry of chloroplasts, Vol. 1 and 2. London-New York: Academic Press 1966.

GRANICK, S., GIBOR, A.: The DNA of chloroplasts, mitochondria and centrioles. Progr. Nucl. Acid Res. **6**, 143—186 (1967).

GRENSON, M.: Physiology and cytology of chloroplast formation and "loss" in *Euglena*. Int. Rev. Cyt. **16**, 37—59 (1964).

GREEN, P. B.: Cinematic observations on the growth and division of chloroplasts in *Nitella*. Amer. J. Botan. **51**, 334—342 (1964).

HAGEMANN, R.: Plasmatische Vererbung. In: Genetik, Grundlagen, Ergebnisse und Probleme in Einzeldarstellungen (Ed.: H. STUBBE), Vol. 4. Jena: VEB Gustav Fischer Verlag 1964.

— Extrachromosomale Vererbung. Fortschr. Botanik **30**, 225—241 (1968).

HERBST, F.: Strahlenbiologische und cytologische Untersuchungen an Blaualgen. Dissertation, Universität Köln (1952).

HERRMANN, R. G.: Chloroplastengröße und inkorporierte ³H-Thymidinmenge. Autoradiographische Studien zur Frage: Gibt es genetisch mehrwertige Chloroplasten? Ber. dtsch. bot. Ges. **81**, 332 (1968).

— Are plastids polyploid? Exp. Cell Res. **55**, 414—416 (1969).

KELLNER, G., WERZ, G.: Die Feinstruktur des Augenflecks bei *Acetabularia*-Gameten und sein Verhalten nach der Gametenfusion. Protoplasma **67**, 117—120 (1969).

KIRK, J. T. O., TILNEY-BASSETT, R. A. E.: The plastids. Their chemistry structure and inheritance. London-San Francisco: Freeman 1967.

KÜSTER, E.: Die Pflanzenzelle. Jena: VEB Gustav Fischer 1956.

KUTZELNIGG, H.: Versuche zur Auslösung von Plastommutationen bei *Oenothera*. Dissertation Universität Düsseldorf (1968).

LAETSCH, W. M., PRICE, I.: Development of the dimorphic chloroplasts of sugar cane. Amer. J. Botan. **56**, 77—87 (1969).

MALY, R.: Die Mutabilität der Plastiden von *Antirrhinum majus* L. Sippe 50. Z. Vererbungsl. **89**, 692—696 (1958).

MANTON, I.: Some possibly significant structural relations between chloroplasts and other cell components. In: Biochemistry of chloroplasts (Ed.: T. W. GOODWIN), I, 23—47. London-New York: Academic Press 1966.

McCALLA, D. R.: Chloroplast mutagenesis: Effect of N-methyl-N′-nitro-N-nitrosoguanidine and some other agents on *Euglena*. Science **148**, 497—499 (1965).

Menke, W.: Einige Beobachtungen zur Entwicklungsgeschichte der Plastiden von *Elodea canadensis*. Z. Naturforsch. **15**b, 800—804 (1960).
— Structure and chemistry of plastids. Ann. Rev. Plant Physiol. **13**, 27—44 (1962).
— Feinbau und Entwicklung der Plastiden. Ber. Dtsch. Bot. Ges. **77**, 340—354 (1965).
— Fricke, B.: Beobachtungen über die Entwicklung der Archegonien von *Dryopteris filix mas*. Z. Naturforsch. **14**b, 520—524 (1964).
Mereschkowsky, C.: Über Natur und Ursprung der Chromatophoren im Pflanzenreiche. Biol. Zbl. **25**, 593—604 (1905).
Meyer, A.: Das Chlorophyllkorn in chemischer, morphologischer und biologischer Beziehung. Leipzig: A. Felix 1883.
Meyer, K. I.: On the evolution of the chromatophor in algae. Bull. Moskovsk. Obsch. Ispit. Prirodi, Otdel biol. **67**, 53—68 (1962) (zit. nach Hagemann, 1964).
Michaelis, P.: Über Gesetzmäßigkeiten der Plasmonumkombination und über eine Methode zur Trennung einer Plastiden-, Chondriosomen- resp. Sphaerosomen-(Mikrosomen-) und einer Zytoplasmavererbung. Cytologia **20**, 315—318 (1955a).
— Modellversuche zur Plastiden- und Plasmavererbung. Züchter **25**, 209—221 (1955b).
— Untersuchungen zur Mutation plasmatischer Erbträger, besonders der Plastiden. 1. Planta **51**, 600—634 (1958).
— Über Zahlengesetzmäßigkeiten plasmatischer Erbträger, insbesondere der Plastiden. Protoplasma **55**, 177—231 (1962).
— Beiträge zum Problem der Plastidenabänderung. 1. Z. Botan. **52**, 382—426 (1965).
Miehe, H.: Entwicklungsgeschichtliche Untersuchungen der Algensymbiose bei *Gunnera macrophylla*. Flora **17** (1924).
Mühlethaler, K., Bell, R. P.: Untersuchungen über die Kontinuität von Plastiden und Mitochondrien in der Eizelle von *Pteridium aquilinum* (L.) Kuhn. Naturwissenschaften **49**, 63—64 (1962).
Pascher, A.: Süßwasserflora Deutschlands. H. 4 *Volvocales*. Jena: Fischer 1927.
Pringsheim, E. G.: Die verwandtschaftlichen Beziehungen zwischen Lebewesen mit und ohne Blattgrün. Naturwissenschaften **51**, 154—157 (1964).
Remsen, C. C., Watson, S. W., Waterbury, J. B., Trüper, H. G.: Fine structure of *Ectothiorhodospira mobilis* Pelsh. J. Bact. **95**, 2374—2392 (1968).
Renner, O.: Eiplasma und Pollenschlauchplasma als Vererbungsträger bei den Oenotheren. Z. indukt. Abstamm.- u. Vererbungsl. **27**, 235—237 (1922).
— Die pflanzlichen Plastiden als selbständige Elemente der genetischen Konstitution. Ber. Sächs. Akad. Wiss., Math.-phys. Kl. **86**, 241—266 (1934).
Ris, H., Plaut, W.: Ultrastructure of DNA containing areas in the chloroplast of *Chlamydomonas*. J. Cell Biol. **13**, 383—391 (1962).
Röbbelen, G.: Plastommutationen nach Röntgenbestrahlung von *Arabidopsis thaliana* (L.) Heynh. Z. Vererbungsl. **93**, 25—34 (1962).
Sager, R., Lane, D.: Replication of chloroplast DNA in zygotes of *Chlamydomonas*. Fed. Proc. **28**, 347 (1969).
— Ramanis, Z.: The particulate nature of nonchromosomal genes in *Chlamydomonas*. Proc. nat. Acad. Sci. (Wash.) **50**, 260—268 (1963).
— — Recombination of nonchromosomal genes in *Chlamydomonas*. Proc. nat. Acad. Sci. (Wash.) **53**, 1053—1061 (1965).
— — The pattern of segregation of cytoplasmic genes in *Chlamydomonas*. Proc. nat. Acad. Sci. (Wash.) **61**, 324—331 (1968a).
— — Recombination analysis of cytoplasmic genes in *Chlamydomonas*. Genetics **60**, 219 (1968b).
— — A genetic map of non-Mendelian genes in *Chlamydomonas* (1969). In press.
Schiff, J. A., Epstein, H. T.: The replicative aspect of chloroplast continuity in *Euglena*. In: Biochemistry of chloroplasts (Ed.: T. W. Goodwin), I, 341—353. London-New York: Academic Press 1966.
Schiller, J.: Beiträge zur Kenntnis der Entwicklung der Gattung *Ulva*. Sitz.-Ber. d. mat. nat. Kl. Kgl. Ak. Wien **116** (1907).
Schimmer, O., Arnold, C. G.: Untersuchungen zur Lokalisation eines außerkaryotischen Gens bei *Chlamydomonas reinhardi*. Arch. Mikrobiol. **66**, 199—200 (1969).

SCHIMPER, A. T. W.: Über die Entwicklung der Chlorophyllkörner und der Farbkörper. Bot. Zeit. **41**, 105, 809 (1883)
— Untersuchungen über die Chlorophyllkörner und die ihnen homologen Gebilde. Jb. wiss. Botan. **16**, 1—247 (1885).
SCHMITZ, F.: Die Chromatophoren der Algen. Verh. Naturhist. Ver. preuß. Rheinlande **40** (1882).
SCHNEPF, E.: Organellen-Reduplikation und Zell-Kompartimentierung. In: Probleme der biologischen Reduplikation (Ed.: P. SITTE). Berlin-Heidelberg-New York: Springer 1966.
SCHÖTZ, F.: Über Plastidenkonkurrenz bei *Oenothera*. Planta **43**, 182—240 (1954).
— Zur Kontinuität der Plastiden. Planta **58**, 333—336 (1962).
— Extrachromosomale Vererbung. Ber. dtsch. bot. Ges. **80**, 523—538 (1967).
— SENSER, F.: Reversible Plastidenumwandlung bei der Mutante Weißherz von *Oenothera suaveolens* Desf. Planta **57**, 235—238 (1961).
SCHÜRHOFF, P. N.: Die Plastiden. In: Handbuch der Pflanzen-Anatomie, 1. Berlin: Gebr. Borntraeger 1924.
STUBBE, W.: Dreifarbenpanaschierung bei *Oenothera*. 1. Entmischung von 3 in der Zygote vereinigten Plastidomen. Ber. dtsch. bot. Ges. **70**, 221—226 (1957).
— Genetische Analyse des Zusammenwirkens von Genom und Plastom bei *Oenothera*. Z. Vererbungsl. **90**, 228—298 (1959).
— Untersuchungen zur genetischen Analyse des Plastoms von *Oenothera*. Z. Botan. **48**, 191—218 (1960).
— Sind Zweifel an der genetischen Kontinuität der Plastiden berechtigt? Eine Stellungnahme zu den Ansichten von MÜHLETHALER und BELL. Z. Vererbungsl. **93**, 175—176 (1962).
— Die Rolle des Plastoms in der Evolution der Oenotheren. Ber. dtsch. bot. Ges. **76**, 154—167 (1963).
— The role of the plastome in evolution of the genus *Oenothera*. Genetica **35**, 28—33 (1964).
— Die Plastiden als Erbträger. In: Probleme der biologischen Reduplikation (Ed.: P. Sitte), 273—286. Berlin-Heidelberg-New York: Springer 1966.
TEILING, E.: Evolutionary studies on the shape of the cell and of the chloroplast in desmids. Botaniska Notiser 1952, 264—306 (1952) (zit. nach HAGEMANN 1964).
THOMSON, W. W., LEWIS, L. N., COGGINS, C. W.: The reversion of chromoplasts to chloroplasts in Valencia oranges. Cytologia **32**, 117—124 (1967).
VAN BAALEN, C.: Mutation of the blue-green alga *Anacystis nidulans*. Science **149**, 70 (1965).
VAN WISSELINGH, C.: Über Variabilität und Erblichkeit. Z. indukt. Abstamm.- u. Vererbungsl. **22**, 65 (1920).
VISCHER, W.: Über einen pilzähnlichen autotrophen Organismus, *Chlorochytridion* und die systematische Bedeutung der Chloroplasten. Verh. Naturf. Ges. Basel **56**, 41 (1945).
WEIER, T. E.: Changes in the fine structure of chloroplasts and mitochondria during phylogenetic and ontogenetic development. Amer. J. Botan. **50**, 604—611 (1963).

Origin and Continuity of Golgi Apparatus

D. James Morré, H. H. Mollenhauer and C. E. Bracker

Departments of Botany and Plant Pathology and of Biology
Purdue University, Lafayette, Indiana
and
Charles F. Kettering Research Laboratory
Yellow Springs, Ohio

I. Introduction

The Golgi apparatus in its most familiar form is that part of the cell's endo-membrane system[1] consisting of regions of stacked cisternae (dictyosomes) which lack ribosomes. It is a complex structure with unique functions in compartmentalizing products of synthesis, serving as a site of cytomembrane differentiation and producing exocytotic vesicles whose membranes are capable of fusing with plasma membrane. Unlike semiautonomous organelles such as chloroplasts and mitochondria, the function of the Golgi apparatus in secretion depends on functional continuity with other components of the endomembrane system.

Much is known about the form and function of the Golgi apparatus, but under-standing of its origin and continuity is in its infancy. The few studies that have dealt directly with this problem have relied heavily on information from morphology and functional patterns. Composite schemes for Golgi apparatus formation are speculative and without extensive documentation, in part due to limitations set by the investigative methods.

In this chapter we summarize information on form and function to provide a basis for analyzing specific problems of origin and continuity. Sites of synthesis of membrane proteins and other constituents are discussed as well as transfer of the materials from sites of synthesis to sites of utilization and transformation. Nuclear envelope and endoplasmic reticulum are considered as sources of new membrane

---

1 We use the term endomembrane system (Figs. 1 and 6) to denote the functional continuum of membranous cell components consisting of the nuclear envelope, endoplasmic reticulum and Golgi apparatus as well as vesicles and other structures such as annulate lamellae derived from the major components. Documented continuity between each of the separate components leads to the view that they are all local specializations of a single interassociated membrane system peculiar to eukaryotic cells (PALADE, 1956; GRIMSTONE, 1958; MANTON, 1960; PORTER, 1961; ESSNER and NOVIKOFF, 1962; AGRELL, 1966; CLAUDE, 1968; KESSEL, 1968a).

constituents subject to influences of genome transcription and translation. Finally, hypotheses of Golgi apparatus origin and continuity formulated from less complex systems are discussed in relation to complex Golgi apparatus.

II. Organization of the Golgi Apparatus

The smooth membrane-bounded compartments which comprise the Golgi apparatus can be considered at three levels of organization: 1. the cisterna, 2. the dictyosome, and 3. the Golgi apparatus (Figs. 1 and 2). These elements vary in form and extent with the cell type and the metabolic state of the cell. But their continuity through succeeding generations is maintained (the parts either persist or reappear) and their reproduction must occur.

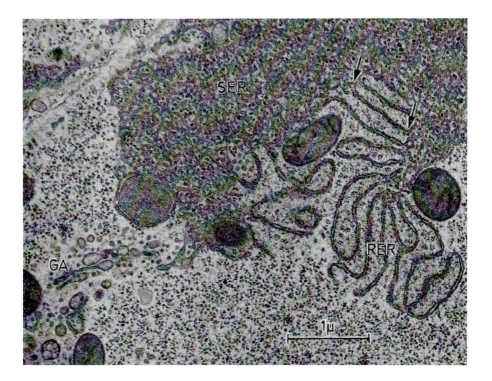

Fig. 1. Portion of the endomembrane system of a rat hepatocyte modified by cortisone treatment (MAHLEY et al., 1968). A massive region of smooth endoplasmic reticulum (SER) is positioned between rough-surfaced (with attached ribosomes) endoplasmic reticulum (RER) and the Golgi apparatus (GA). Smooth endoplasmic reticulum is most apparent in cells such as hepatocytes following the administration of drugs or chemicals that specifically result in its proliferation (see also JONES and FAWCETT, 1966). Continuity between rough-surfaced and smooth endoplasmic reticulum is shown at the single arrows. Both the Golgi apparatus and the region of association between Golgi apparatus and smooth endoplasmic reticulum are surrounded in the cytoplasm by a ribosome- and glycogen-free zone of exclusion. Glutaraldehyde-OsO$_4$ fixation. Electron micrograph courtesy of Dr. ROBERT W. MAHLEY, Department of Anatomy, Vanderbilt University Medical School, Nashville, Tennessee

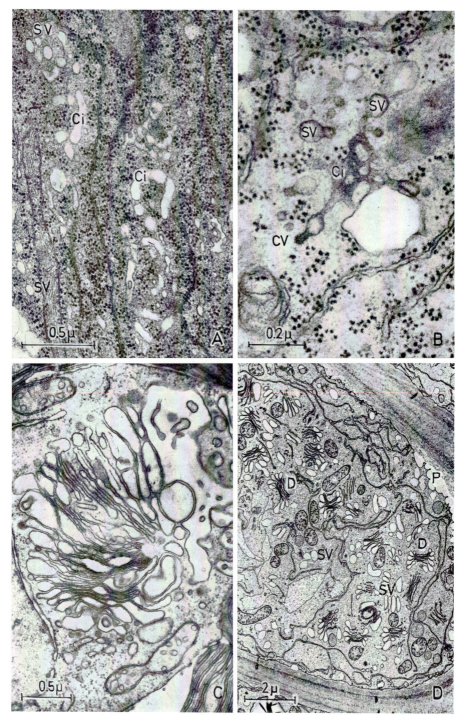

Fig. 2 A—D

Fig. 2 A—E. *Golgi apparatus of increasing complexity.* A. and B. Vesicle-producing cisternae (Ci) from subapical regions of young hyphae of fungi. A. *Aspergillus niger.* Formaldehyde-glutaraldehyde-OsO₄ fixation. B. *Armillaria mellea.* Glutaraldehyde-OsO₄ fixation. CV = coated vesicle; SV = secretory vesicle. These structures exist as single cisternae and show no tendency to form stacks. Yet, they appear to carry out functions usually associated with more

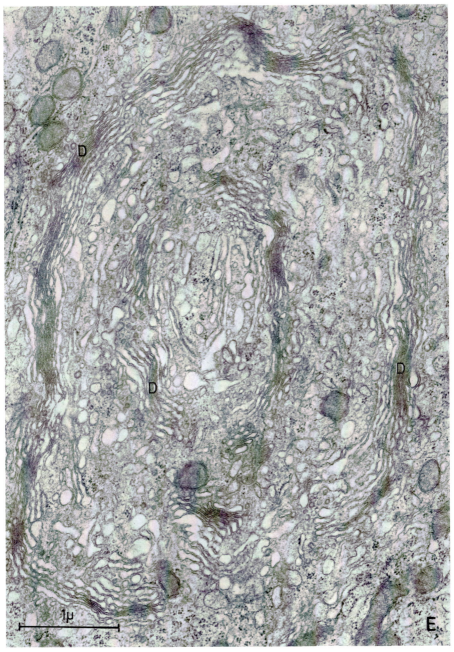

Fig. 2E

complex Golgi apparatus. Electron micrographs courtesy of S. N. GROVE, Purdue University. C. Golgi apparatus consisting of a single stack of cisternae in the marine alga *Prymnesium parvum*. Glutaraldehyde-OsO$_4$ fixation. From MANTON (1966a). Reprinted by permission of Cambridge University Press. D. Portion of the Golgi apparatus of an outer root cap cell of *Zea mays* consisting of groups of separated dictyosomes (D). SV = secretory vesicles. P = accumulations of secreted product adjacent to the cell wall. The secreted product is a polysaccharide. KMnO$_4$ fixation. From MORRÉ, JONES and MOLLENHAUER (1967). Reprinted by permission of Springer-Verlag, Berlin. E. Portion of the Golgi apparatus of rat epididymis showing numerous closely-spaced dictyosomes (D). Karnovsky's fixative. From FLICKINGER (1969c). Electron micrograph courtesy of Dr. C. J. FLICKINGER, University of Colorado, Boulder. Reprinted by permission of Academic Press, Inc., New York

A. Cisterna

A cisterna, by definition, is a sac or cavity filled with fluid contents within a cell or organism. Cisternae of the Golgi apparatus are bounded by smooth-surfaced membranes, are usually flattened, and most often consist of a central plate-like region ("saccule") continuous with a peripheral system of tubules and vesicles. In some cisternae, the plate-like regions predominate; others consist mainly of tubular elements. Both types of cisternae may exist within a single stack or dictyosome

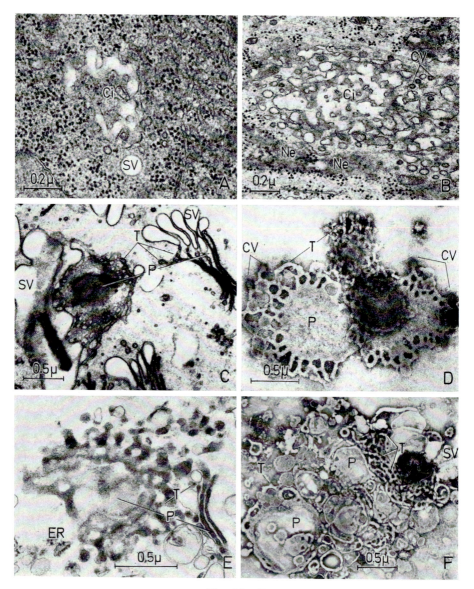

Fig. 3 A—F

(MOLLENHAUER and MORRÉ, 1966b; MOLLENHAUER, MORRÉ, and BERGMAN, 1967; MORRÉ, KEENAN, and MOLLENHAUER, in press).

Although they vary in form, Golgi apparatus cisternae have many common features (Figs. 3 and 4). The plate-like region is typically 0.5−1 μ in diameter and

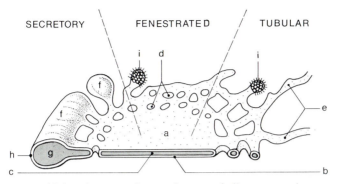

Fig. 4. Diagram combining features of several types of dictyosome cisternae: a. central plate like region. b. cisternal membrane. c. cisternal lumen. d. fenestrae (perforations). e. peripheral tubules. f. secretory vesicles. g. vesicle lumen. h. vesicle membrane. i. coated vesicles (600 to 750 Å diameter) have a nap-like electron-dense coating (see Figure 7B). They are distinct from secretory vesicles and are a consistent feature of Golgi apparatus cisternae (See Fig. 3D)

is often fenestrated (FLICKINGER, 1969a; Figs. 2E, 3B, and 4). Tubules (300 to 500 Å diameter) are continuous with the periphery of the plate-like region and may extend for several microns from the edge of the plate. The peripheral tubules

Fig. 3A—F. *Golgi apparatus cisternae in face view.* A. Single vesicle-producing cisterna of the fungus *Aspergillis niger.* Compare with Fig. 2A. Formaldehyde-glutaraldehyde-OsO₄ fixation. B. From a perinuclear dictyosome of the fungus *Pythium aphanidermatum.* Portions of the nuclear envelope (NE) are seen in tangential section. CV = coated vesicles. Glutaraldehyde-OsO₄ fixation. A and B courtesy of S. N. GROVE, Purdue University. C. From a dictyosome of a root cap cell of *Zea mays.* The central plate-like region (P), the system of peripheral tubules (T) and secretory vesicles (SV) are shown. The secretory vesicles are attached to the plate-like region via the system of peripheral tubules. The dictyosome on the right was sectioned transversely and shows these same features of cisternae in cross section. KMnO₄ fixation. From MOLLENHAUER and MORRÉ (1966b). Reprinted by permission of the Rockefeller Press, New York. D. Negatively stained from an isolated dictyosome of germinating lily *(Lilium longiflorum)* pollen. The dictyosome is partially unstacked to show the variation of cisternal features. The small cisterna near the forming face (top) is almost entirely tubular, while the cisternae near the maturing face (bottom) are more plate-like. CV = coated vesicles. Sodium phosphotungstate, pH 6.4. E. From an isolated dictyosome of rat liver showing the central plate-like region (P) and the system of peripheral tubules (T). The dictyosome on the right was sectioned transversely and shows these same features of cisternae in cross section. A fragment of rough-surfaced endoplasmic reticulum (ER) is shown at the cisternal periphery. Glutaraldehyde-OsO₄ fixation. From MORRÉ et al. (1970). F. Negative contrast of an isolated rat liver dictyosome partially unstacked as in D to reveal the form of successive cisternae. It varies from typically plate-like (left) with a few peripheral tubules to almost entirely tubular (right). Compare with Figure 3D. SV = secretory vesicles. From MORRÉ, KEENAN, and MOLLENHAUER (in press). Sodium phosphotungstate, pH 6.5

often form a complex anastomosing network (Manton, 1960; Essner and Novikoff, 1962; Cunningham, Morré, and Mollenhauer, 1966; Mollenhauer and Morré, 1966a, b; Schnepf and Koch, 1966a, b; Mollenhauer, Morré, and Bergman, 1967; Maul, 1969). Secretory vesicles are attached to tubules at the periphery of the cisternae (Mollenhauer and Morré, 1966b; Schnepf, 1968c; Morré, Keenan, and Mollenhauer, in press; Ovtracht, Morré, and Merlin, 1969; see also Fig. 3C).

B. Dictyosome

When Golgi apparatus cisternae are organized into stacks, the stacks are called dictyosomes (Figs. 2C—E; 3C—F). The number of cisternae per stack is variable, usually in the range of 5 to 8, but 30 or more are not unusual for dictyosomes of lower organisms.

Typically, dictyosomes are polarized so that cisternae at one pole or face of the stack differ from those at the opposite pole or face. Dictyosome polarity is evidenced by differences in the form and composition of successive cisternae and by their associations with other components of the endomembrane system. One pole of each dictyosome is associated with the nuclear envelope or with endoplasmic reticulum. This is the proximal pole or forming face. The membranes of cisternae near this pole are morphologically and histochemically similar to endoplasmic reticulum. Toward the opposite pole (distal pole or maturing face), there is a transition in the morphology and staining characteristics of the cisternae. The cisternal membranes become progressively more like plasma membrane, i. e. denser, thicker and showing the dark-light-dark pattern more clearly (Grove, Bracker, and Morré, 1968; see also Fig. 5). They become involved with the formation of secretory vesicles. Although dictyosomes of most cells conform to this basic plan, polarities of a more complex nature are sometimes encountered such as those where secretory vesicles are produced at both poles of the stack (Bainton and Farquhar, 1966; Manton, 1966a, b; Wise, 1969; see also Fig. 9).

Cisternae are held together in the stack but are separated from each other by an intercisternal region having a minimum thickness of about 100—150 Å (Mollenhauer, 1965a; Mollenhauer and Morré, 1966a). In some cells, a layer of parallel fibers, called intercisternal elements, occurs within the intercisternal region midway between the surfaces of adjacent cisternae (Mollenhauer, 1965a; Turner and Whaley, 1965; Cunningham, Morré, and Mollenhauer, 1966; Amos and Grimstone, 1968).

Fig. 5 A—G. In the fungus *Pythium ultimum*, dictyosome membranes are differentiated within the stack. They are nuclear envelope-like or endoplasmic reticulum-like at the pole adjacent to nuclear envelope or endoplasmic reticulum. They resemble plasma membrane at the opposite pole. A. Single paranuclear dictyosome (N = nucleus; NE = nuclear envelope; Dp = proximal or forming face; Dd = distal or maturing face; SV = secretory vesicle. B. Nuclear envelope. C. Endoplasmic reticulum. D. Membrane of a secretory vesicle adjacent to a dictyosome. E. Membrane of a secretory vesicle free in the cytoplasm. F. Plasma membrane (CW = cell wall). G. Dictyosome membranes showing progressive increase in membrane thickness and staining intensity across the stack. B—G are at the same magnification. Glutaraldehyde-OsO$_4$ fixation with a Ba(MnO$_4$)$_2$ post stain. From Grove, Bracker, and Morré (1968). Reprinted by permission of the American Association for the Advancement of Science

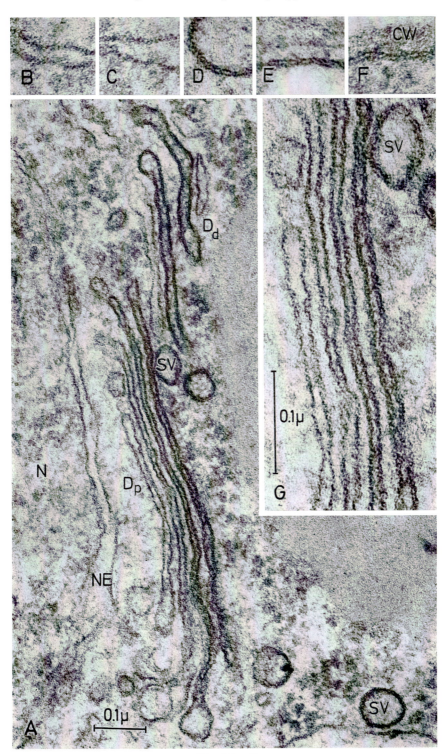

Fig. 5 A—G

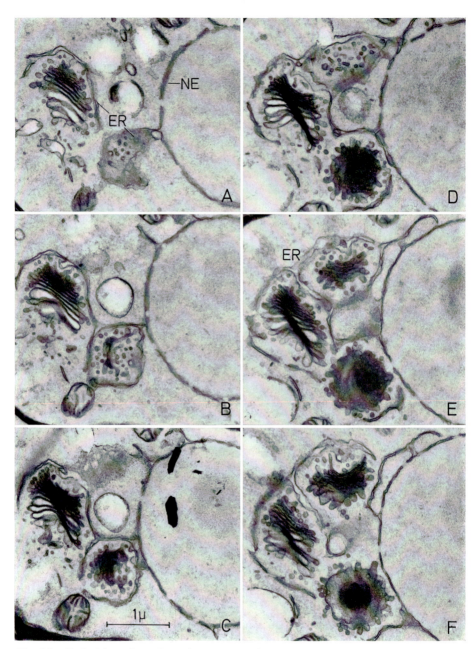

Fig. 6 A—F. Serial sections through a portion of the endomembrane system of the alga *Tetracystis excentrica* including three adjacent dictyosomes. Sheets of endoplasmic reticulum (ER) continuous with the nuclear envelope (NE) and known as the amplexus, surround each dictyosome, except for a region at the maturing face through which secretory vesicles are discharged. The dictyosome at the lower center is sectioned tangentially and shows successive cisternae in face view beginning with the endoplasmic reticulum at the forming face in A. To the left, a second dictyosome is sectioned transversely and shows the entire stack in cross section. A third dictyosome becomes evident in D—F at upper center and C provides a tangential view of the sheet of endoplasmic reticulum associated with the proximal face of the dictyosome. $KMnO_4$ fixation. Unpublished electron micrographs courtesy of Drs. R. MALCOLM BROWN, JR. and HOWARD J. ARNOTT, Cell Research Institute, The University of Texas, Austin

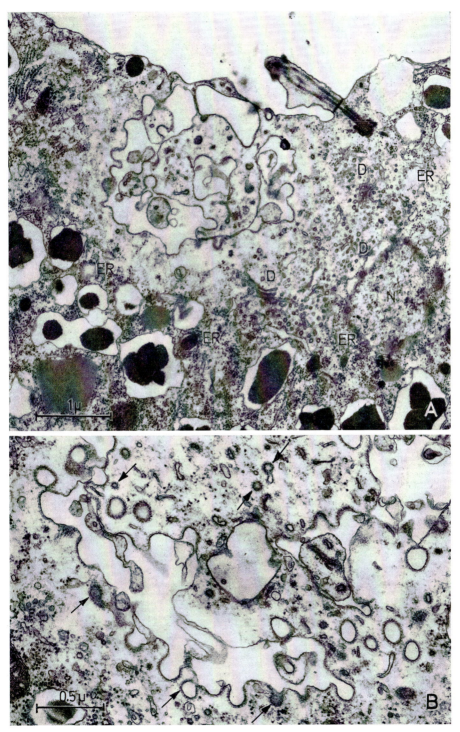

Fig. 7 A and B   Caption see p. 92

Dictyosomes are surrounded by a differentiated region of cytoplasm where ribosomes, glycogen, and organelles such as mitochondria and chloroplasts are scarce or absent (zone of exclusion: Fig. 1 and 7; see also Mollenhauer, 1965a; Golgi ground substance: Sjöstrand and Hanzon, 1954). Endoplasmic reticulum within the zone of exclusion has a smooth surface (lacking ribosomes), and coated vesicles of the Golgi apparatus are restricted to this region (Figs. 2B, 3B, unpublished observations). Similar zones of exclusion are associated with microtubules (Porter, 1966), centrioles (Bainton and Farquhar, 1966), and regions of centriole formation (Sorokin, 1968).

C. Golgi Apparatus (Singular or Plural)

The hierarchy of structural organization is completed by the association of dictyosomes to form Golgi apparatus (Grassé, 1957; Mollenhauer and Morré, 1966a; Schnepf and Koch, 1966b; see also Fig. 2). The individual dictyosomes of a Golgi apparatus may be compactly arranged into distinct Golgi apparatus zones as in mammalian cells (Fig. 2E), or arranged in clusters that are dispersed throughout the cytoplasm, as in plants and some invertebrates (Fig. 2D). If all the dictyosomes of a particular cell are interassociated, there is only one Golgi apparatus, but cells may have several Golgi apparatus each consisting of 1 to many dictyosomes (Mollenhauer and Morré, 1966a). When cells contain a single dictyosome (Manton, 1966a; Brown, 1969; see also Fig. 2C), this is the Golgi apparatus. The number of dictyosomes per cell ranges from none (certain fungi, cf. Bracker, 1967; Girbardt, 1969) to more than 25,000 (algal rhizoids; Sievers, 1965). Instead of stacked cisternae, some cells may contain single cisternae (Bracker, 1968; Girbardt, 1969; Matile and Moor, 1969; Figs. 2A, 3A) or systems of vesicle-producing tubules which function as the Golgi apparatus.

III. The Golgi Apparatus as Part of the Endomembrane System

Nuclear envelope, endoplasmic reticulum, Golgi apparatus, secretory vesicles, and plasma membrane are interassociated and appear to form a functional continuum. The infrequency of observation of direct, clear line membrane continuity between Golgi apparatus and endoplasmic reticulum or plasma membrane emphasizes that Golgi apparatus are discrete cell components (Grassé, Carasso, and Favard, 1955; Zeigel and Dalton, 1962). But adjacent components need not have permanent structural continuity to have functional continuity, since intermittent tubular or vesicular bridges would insure transfer of materials.

Fig. 7 A and B. Zoospores of the fungus *Pythium aphanidermatum*. A. Portion of an undifferentiated Golgi apparatus adjacent to a nucleus (N). Dictyosome (D) cisternae are small and poorly developed. Conspicuous secretory vesicles are absent. At this stage of development, the entire Golgi apparatus as well as the system of peripheral tubules is surrounded by an extensive zone of exclusion with profiles of rough-surfaced endoplasmic reticulum (ER) at the periphery of the zone of exclusion. B. Coated vesicles of the water expulsion apparatus (arrows) within the zone of exclusion. Glutaraldehyde-OsO$_4$ fixation. Electron micrographs courtesy of S. N. Grove, Purdue University

A. Associations with Endoplasmic Reticulum or Nuclear Envelope

Functional continuity between endoplasmic reticulum and Golgi apparatus has been shown by autoradiographic (CARO and PALADE, 1964; JAMIESON and PALADE, 1966; 1967a, b; 1968a, b; LEBLOND, 1965; WARSHAWSKY, LEBLOND, and DROZ, 1963), histochemical (cf. NOVIKOFF et al., 1962; FRIEND and MURRAY, 1965) and biochemical analyses (cf. REDMAN, SIEKEVITZ, and PALADE, 1966; REDMAN and SABATINI, 1966; SIEKEVITZ et al., 1967; JAMIESON and PALADE, 1967a, b, 1968a, b). Here, vectorial transfer of secretory proteins from endoplasmic reticulum to Golgi apparatus to secretory vesicles is accomplished via luminal continuity between conjoining cell components. At no time are the secretory proteins transported in soluble form through the cell sap.

Continuity is most often deduced, however, from direct observations of the morphological relationships between the two structures. In electron micrographs, the close relationship between endoplasmic reticulum and Golgi apparatus (or individual dictyosomes) is most conspicuous in the form of a cisterna of endoplasmic reticulum aligned along the forming face (see for example, FRIEND, 1965; Figs. 6, 8, 10, 13). This cisterna of endoplasmic reticulum provides continuity among dictyosomes (DRUM, 1966; MOLLENHAUER, 1965b; MANTON, 1967b) as well as a potential source of membrane material for Golgi apparatus function and replication. Regions of the endoplasmic reticulum immediately adjacent to the Golgi apparatus lack ribosomes (see for example FRIEND, 1965 and Fig. 13).

Small evaginations or blebs of the cisternal walls of the endoplasmic reticulum extend into the intercisternal space between the endoplasmic reticulum and the Golgi apparatus (Figs. 6, 8, 13, see diagrams of Fig. 10). The blebs resemble those of small vesicular profiles between the endoplasmic reticulum and the nearest cisterna of the dictyosome. For perinuclear dictyosomes, similar small vesicular projections occur from the outer membrane of the nuclear envelope adjacent to dictyosomes (NOVIKOFF et al., 1962; ZEIGEL and DALTON, 1962; MOORE and McALEAR, 1963; BOUCK, 1965; SCHNEPF and KOCH, 1966b; WHALEY, 1966; GROVE, BRACKER, and MORRÉ, 1968; FAWCETT and McNUTT, 1969)[2]. Observations of FRIEND (1965), MOLLENHAUER (1965b), FAURÉ-REMIET, FAVARD, and CARASSO (1962), HIRSCH (1963), NOVIKOFF and SHIN (1964), SCHNEPF (1966), JAMIESON and PALADE (1967, 1968a, b, c), KESSEL (1968b), FLICKINGER (1969c), GROVE, BRACKER, and MORRÉ (1970) and others support the proposals of DALTON (1961), ZEIGEL

---

2 The nuclear envelope is the most universal representative of the endomembrane system (PORTER, 1961). The frequency of perinuclear positioning of Golgi apparatus in lower organisms, in germ cells, and during early stages of embryogenesis (KESSEL, 1968b) may reflect an evolutionary trend in the development of complex endomembrane systems (see also MOORE and McALEAR, 1963). Perinuclear positioning of Golgi apparatus is rarely encountered in higher plants (MOLLENHAUER and MORRÉ, 1966a) and is less frequently observed in somatic cells of vertebrates. In these latter cell types, associations of Golgi apparatus with endoplasmic reticulum are most prevalent. An interesting example of dictyogenesis is described by BUCIARELLI (1966) for neoplastic mammalian cells. In these abnormal cells, structures resembling normal dictyosomes appear inside nuclei. Extensions of the nuclear envelope appear to protrude into the nucleus and may provide the material for intranuclear dictyogenesis in a manner similar to that suggested for normal dictyogenesis outside the nucleus. Both intra- and extranuclear annulate lamellae have been shown to arise by fusion of small vesicles derived from the nuclear envelope (KESSEL, 1968a), the manner resembling that described here for origin of Golgi apparatus cisternae.

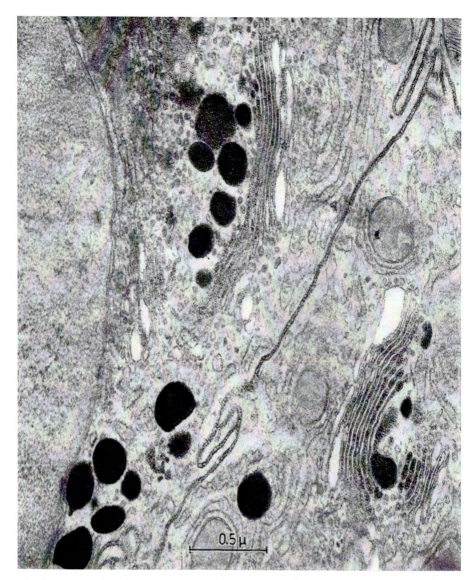

Fig. 8. Functional polarity of the Golgi apparatus of a mucopolysaccharide secreting mammalian cell (Brunner's gland of the mouse duodenum). Osmium fixation and stained for polysaccharides using the thiocarbohydrazide-silver proteinate technique of THIÉRY (1967). Large secretory granules at the maturing face of the apparatus stain heavily, indicating the presence of the mucopolysaccharide secretory products. Cisternal lumina also stain with a progressive increase in staining from the forming to the maturing face across the stacks of cisternae. The opposite or forming face shows the usual associations with profiles of endoplasmic reticulum and small vesicular profiles. The lumen of the intracellular space also stains, verifying the presence of sugars associated with the cell surface (THIÉRY, 1969). Unpublished electron micrograph courtesy of Dr. J.-P. THIÉRY, Laboratoire de Biologie Cellulaire, Faculté des Sciences, Ivry, France

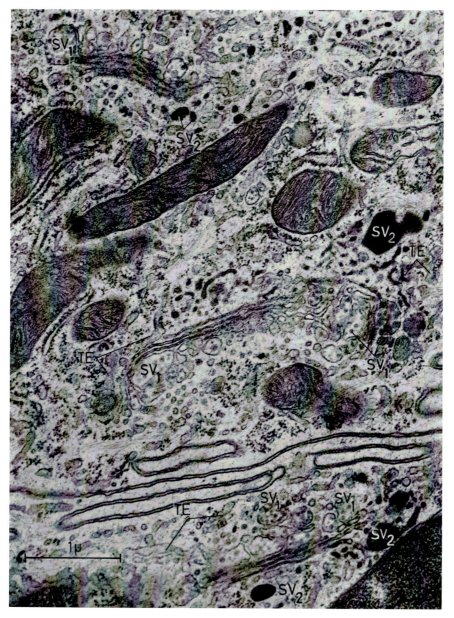

Fig. 9. Complex functional polarity illustrated by the Golgi apparatus of an intestina absorptive cell (epithelium of rat jejunum). Vesicles at one face contain lipid particles (SV$_1$). At the opposite face, a dark-staining type of vesicle is found (SV$_2$). Transitional elements (TE) of the endoplasmic reticulum are located at the periphery of the apparatus. OsO$_4$ fixation; uranyl acetate plus REYNOLD's lead ctirate post stains. Unpublished electron micrograph courtesy of Dr. RALPH A. JERSILD, JR., Department of Anatomy, Indiana University Medical School, Indianapolis

and DALTON (1962), and NOVIKOFF et al. (1962) that these small endoplasmic reticulum- or nuclear envelope-derived vesicles fuse to form new Golgi apparatus cisternae. For the opposite point of view see MANTON (1960) and SAKAI and SHIGENAKA (1967).

Evidence for involvement of endoplasmic reticulum in the formation of new Golgi apparatus cisternae was first provided by GRIMSTONE (1959). In starved *Triconympha*, a near elimination of rough endoplasmic reticulum was accompanied by failure of new dictyosome cisternae to form and by a diminution in the number

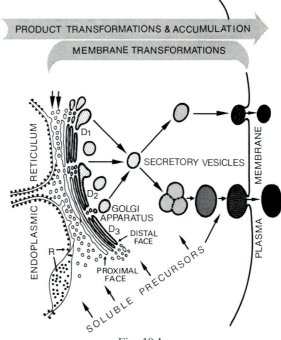

Fig. 10 A

Fig. 10 A and B. Diagrams summarizing the structural and functional relationships among Golgi apparatus, endoplasmic reticulum, secretory vesicles and the plasma membrane. The diagrammed concept of Golgi apparatus function involves the input of new membrane material from endoplasmic reticulum at a proximal or forming face and its progressive transformation and utilization in the elaboration of membranes of secretory vesicles which are plasma membrane-like and capable of fusing with plasma membrane. These transformations appear to be a general Golgi apparatus function and involve metabolism of membrane lipids and additions of sugars. Proteins of both the secretory products and for the formation of new membranes are synthesized on polyribosomes and associated messenger RNA of the rough-surfaced endoplasmic reticulum. Small vesicles which arise by blebbing of the granular surfaces of rough endoplasmic reticulum may function in the transfer of materials between endoplasmic reticulum and the Golgi apparatus.

Fig. 10 A. In polysaccharide- or mucopolysaccharide-secreting cells, product formation is associated with progressive changes in both cisternal and vesicle contents (cf. Fig. 8). Here, synthetic activities are detected within the plate-like portions of cisternae as well as in the forming vesicles. Studies with fungi and certain plant cells suggest the appearance of product to be associated with the transformation of Golgi apparatus membranes from endoplasmic reticulum-like to plasma membrane-like. In some cells, the secretory vesicles themselves may act as organelles of synthesis (VANDERWOUDE, MORRÉ, and BRACKER, 1969).

of cisternae per stack through continued production of secretory vesicles (secretory vesicle production *per se* can occur at least for a limited time in the absence of protein synthesis; cf. JAMIESON and PALADE, 1968a). Upon refeeding starved *Triconympha*, rough endoplasmic reticulum reappeared along with the capacity to form new dictyosome cisternae (GRIMSTONE, 1959).

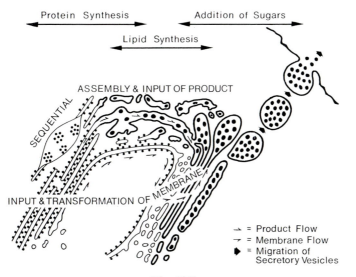

Fig. 10 B

Fig. 10 B. Based on studies with hepatocytes (MORRÉ, KEENAN, and MOLLENHAUER, in press), the synthesis and transport of lipoprotein particles (depicted as solid black circles) is visualized as a continuous multistep process with the addition of lipids, sterols, and poly-saccharides to the protein component. At the same time, the particles migrate from their sites of origin beginning at or near transition elements between rough and smooth endo-plasmic reticulum to the secretory vesicles of the Golgi apparatus. Migration is through channels of smooth membrane-bounded tubules, often corresponding to elements of smooth endoplasmic reticulum. These tubules provide direct luminar continuity between cisternae of rough endoplasmic reticulum and the secretory vesicles. In this scheme, the processes of vesicle formation and of entry of secretory product are spatially separated allowing a multiplicity of Golgi apparatus function in the segregation of secretory products.

Although the pattern of secretory vesicle formation and product input of polysaccharide- or mucopolysaccharide-secreting cells may vary from that of protein- or lipoprotein-secreting cells, a common feature of Golgi apparatus functioning remains. We suggest this to be the transformation of membranes from endoplasmic reticulum-like to plasma membrane-like in the formation of secretory vesicle membranes.

In addition to the general notions relating to concentration and transport (c.f. ZEIGEL and DALTON, 1962), two specific functions may be ascribed to complex Golgi apparatus to explain why secretory proteins go through the apparatus (c.f. LEBLOND, 1965). The first of these is the well established role of the Golgi apparatus in the synthesis of poly-saccharides or the polysaccharide component of mucopolysaccharides. The second is a less well established and completely undefined role of the Golgi apparatus in effecting changes in the tertiary or quaternary structure (maturation) of proteins destined for export. Both of these functions, as well as a role in lipid metabolism, are likely to be instrumental to product changes as well as to the process of endomembrane differentiation. Of the two main Golgi apparatus functions, i.e. product transformations vs. membrane transforma-tions (cf. MOLLENHAUER and MORRÉ, 1966a), the role in membrane transformations appears to be the most general

Potential sites of membrane input for formation of Golgi apparatus cisternae are not restricted to the forming face (Mollenhauer, 1965b; Manton, 1966b, 1967; Novikoff, 1967; Maul, 1969; Morré, Keenan, and Mollenhauer, in press). Associations between endoplasmic reticulum and Golgi apparatus occur at the cisternal peripheries (Buvat, 1958; Flickinger, 1969b) and may involve tubular connections about the same diameter (600 Å or less) as the thickness of sections employed for electron microscopy (Maul, 1969; Morré, Keenan, and Mollenhauer, in press). The existence of such connections is critical to Whaley's (1966) suggestion that membrane components might be added not only at the immature face of the dictyosome but also to successive cisternae across the stack (see also Flickinger, 1969c). As emphasized by Flickinger (1969c), most of these configurations are best described as regions of exceptionally close contact or apposition between Golgi apparatus and endoplasmic reticulum rather than examples of direct membrane continuity.

The association of Golgi apparatus and endoplasmic reticulum with lysosomes (Gerl) at the distal or maturing face of the Golgi apparatus is well established (Essner and Novikoff, 1962; Novikoff, 1967). Evidence for direct continuity between elements of smooth endoplasmic reticulum and forming secretory vesicles, as well as other regions of Golgi apparatus cisternae, has been provided for an alga (Manton, 1967), melanocytes (Maul, 1969) and hepatocytes (Morré, Keenan, and Mollenhauer, in press; Claude, in press). The significance of these relationships to compartmentation of secretory products is discussed with Fig. 10 (page 96—97).

Thus, there are numerous examples from a variety of cell types which indicate a close relationship between Golgi apparatus and endoplasmic reticulum. The evidence favors the interpretation that the Golgi apparatus depends on continuity with endoplasmic reticulum for its maintenance and continuity, and any consideration of the formation of Golgi apparatus, then, must take into account the endoplasmic reticulum as source for input of new membrane.

B. Continuity with Plasma Membrane

Continuity between Golgi apparatus and plasma membrane is provided by secretory vesicles which serve to compartmentalize materials, usually for transport to the cell surface (Mollenhauer and Morré, 1966a; Beams and Kessel, 1968; Figs. 1—5, 8—10, 13). Membranes of the secretory vesicles are morphologically similar to those of plasma membranes and are capable of fusing with plasma membranes (Sjöstrand, 1963; Schnepf and Koch, 1966b; Porter et al., 1967; Grove, Bracker, and Morré, 1968; Helminen and Ericsson, 1968; Matile and Moor, 1968[3]; Falk, 1969, Table 1, Fig. 5). As the membranes fuse, the vesicle contents are discharged, and the vesicle membranes are incorporated into the plasma membrane.

IV. Origin of Cisternae within Preexisting Dictyosomes

The generalized morphological and functional polarity of dictyosomes in both plant and animal cells (Grassé, 1957; Policard et al., 1958; Grimstone, 1959;

---

3 These authors report that in frozen-etched root tip cells of maize, particles on plasma membrane could not be distinguished from those of dictyosome-derived membranes but were easily distinguished from those of endoplasmic reticulum and homologous membranes.

SCHNEPF, 1961, 1968a, b; ZEIGEL and DALTON, 1962; MOLLENHAUER and WHALEY, 1963; CARO and PALADE, 1964; DANIELS, 1964; NOVIKOFF and SHIN, 1964; BOUCK, 1965; BRUNI and PORTER, 1965; FRIEND and MURRAY, 1965; BAINTON and FARQUHAR, 1966; HICKS, 1966; MANTON, 1966a,b, 1967; WHALEY, 1966; BERLIN, 1967; GROVE, BRACKER, and MORRÉ, 1968; THIÉRY, 1968, 1969; BROWN, 1969; RAMBOURG, HERNANDEZ, and LEBLOND, 1969) is in accord with the concept that stacked cisternae are in a state of turnover. According to this hypothesis, cisternae are formed at one pole adjacent to endoplasmic reticulum or nuclear envelope. They mature as they are displaced toward the opposite pole or maturing face (GRIMSTONE, 1959; SCHNEPF, 1961; MOLLENHAUER and WHALEY, 1963; SCHNEPF and KOCH, 1966a,b; WHALEY, 1966; BROWN, 1969). Here, cisternae are lost as secretory vesicles are produced. Thus, the formation of cisternae at the proximal pole is compensated for by loss of cisternae at the distal pole, so that a constant number of cisternae is maintained in each stack. Stages in cisternal formation are shown in serial sections through single algal dictyosomes in Fig. 6; see also MANTON (1960) for serial sections through a single dictyosome.

This concept of dictyosome function implies that the direction of membrane flow (BENNETT, 1956; NOVIKOFF et al., 1962) is from the Golgi apparatus to the cell surface.[4] The best evidence for this hypothesis comes from the algae studied by MANTON (1966a, b, 1967), SCHNEPF and KOCH (1966a, b), and BROWN (1969). Here, as proximal cisternae are formed, large vesicles are discharged at the distal pole, for incorporation into the plasma membrane with the loss of the entire distal cisterna. For additional examples or suggestions of loss or breakdown of distal dictyosome cisternae see GRIMSTONE (1959), KARRER and COX (1960), MOLLENHAUER and WHALEY (1963), and MOLLENHAUER (1965b).

Under steady state conditions, loss of cisternae at the distal pole compensates for generation of cisternae at the proximal pole. However, during a period of rapid stimulation of secretory vesicle production, cisternal generation may lag appreciably resulting in smaller dictyosomes with fewer cisternae (GRIMSTONE, 1959; SCHNEPF, 1961, 1968a; WEISBLUM, HERMAN, and FITZGERALD, 1962; WERZ, 1964; FLICKINGER, 1968a, b, 1969b). SCHNEPF (1968a) favors the interpretation that renewal of Golgi apparatus membranes may limit secretory vesicle production during periods of hypersecretion. Additional evidence for the dynamics of cisternal formation is provided by studies in which secretory vesicle production is stopped experimentally. Here, the dictyosomes enlarge, and the number of cisternae increase

---

4 This is opposite to the route suggested by DANIELS (1964) in *Amoeba* where the Golgi apparatus cisternae are suggested to be formed from invaginated cell membranes. Studies with ferritin markers suggest transfer from the cell surface to Golgi apparatus in absorptive cells of the rat intestine (BENEDETTI, 1958) but discount the generality of this route in *Amoeba* (WISE, 1969; FLICKINGER, 1969b). Plasma membrane resorption in the form of small, smooth-surfaced vesicles and its eventual reutilization by the Golgi apparatus in secretory vesicle elaboration has been suggested for acinar cells of the rat pituitary (AMSTERDAM, OHAD, and SCHRAMM, 1969; see also SCHNEPF, 1968a for a similar suggestion for plant cells). No morphological evidence for recycling of membrane is evident in nongrowing secretory cells in the maize root cap where we estimate that enough new membrane is contributed to the plasma membrane by secretory vesicles to replace the entire plasma membrane every 4 to 8 hours. We assume that the excess membrane breaks down to maintain the smooth contour of the cell surface and that the membrane materials are reutilized in the form of small molecules or subunits.

7*

(Hall and Witkus, 1964; Whaley, Kephart, and Mollenhauer, 1964; Mollenhauer and Morré, 1966a; Whaley, 1966; Coombs et al., 1968; Schnepf, 1968a). It is estimated that during periods of active secretory vesicle production each dictyosome turns over within 20—40 minutes and that cisternae are released at the rate of one every 1—4 minutes (Neutra and Leblond, 1966; Schnepf and Koch, 1966a, b; Brown, 1969; see also Schnepf, 1961).

A. Golgi Apparatus as a Site of Endomembrane Differentiation

If endoplasmic reticulum is the source of membrane that will ultimately become plasma membrane through the production of Golgi apparatus vesicles, a conversion from endoplasmic reticulum-like to plasma membrane-like membranes must occur somewhere enroute. Several lines of evidence indicate that this transformation occurs at the Golgi apparatus (Hicks, 1966; Porter, Kenyon, and Badenhausen, 1967; Grove, Bracker, and Morré, 1968; Helminen and Ericsson, 1968; Falk, 1969).

The existence of different membrane types is emphasized in the reports of Sjöstrand (1956, 1963, 1968), Yamamoto (1963) and others (see Table 1). Membranes of endoplasmic reticulum are thinner, with a less pronounced dark-light-dark pattern than plasma membrane, whereas membranes of the Golgi apparatus are intermediate. In epithelial cells of the lactating mammary gland of the rat, Helminen and Ericsson (1968, see also Table 1) found that the membrane bounding the milk proteins underwent an increase in thickness of 30—40 Å during transport from endoplasmic reticulum to alveolar lumen. The change in membrane dimensions occurred in the Golgi apparatus.

In the fungus *Pythium ultimum*, dictyosome membranes are differentiated across the stack of cisternae so that those at the proximal pole appear similar to endoplasmic reticulum and nuclear envelope, whereas those at the distal pole (including vesicle membranes) are similar to plasma membrane (Grove, Bracker, and Morré, 1968; see also Fig. 5). The intercalary cisternae are morphologically intermediate; each successive cisterna progressing toward the distal pole is more like plasma membrane (that is, denser, thicker and showing the dark-light-dark pattern typical of plasma membrane more clearly). The occurrence of dissimilar membranes in dictyosomes has been observed in other organisms (Hicks, 1966; Whaley, 1966; Sakai and Shigenaka, 1967; Ovtracht, 1967; Falk, 1969; Vanderwoude, Morré, and Bracker, in press; Morré, Keenan, and Mollenhauer, in press) and may be of general occurrence.

The transitional nature of the Golgi apparatus of rat liver is evident both in its morphology and in its lipid and protein composition (Morré, Keenan, and Mollenhauer, in press). Phospholipids (Table 2) and sterols (Yunghans, Keenan, and Morré, 1970; Keenan and Morré, 1970) of Golgi apparatus are present in amounts intermediate between those of endoplasmic reticulum (or nuclear envelope) and plasma membrane. Gel electrophoresis patterns of Golgi apparatus proteins are intermediate between endoplasmic reticulum and plasma membrane proteins with respect to numbers, intensities and positions of bands, but more nearly resemble endoplasmic reticulum than plasma membrane (Yunghans, Keenan, and Morré, 1970). Additionally, a number of enzymes found in Golgi apparatus fractions have specific activities intermediate between those of endoplasmic

Table 1. *Endomembrane differentiation based on membrane dimensions*

Tissue	Reference	Fixative[a]	Average membrane thickness (Å)		
			Endoplasmic reticulum	Golgi apparatus	Secretory vesicles-plasma membrane
Mouse kidney	SJÖSTRAND (1963)	OsO$_4$ KMnO$_4$	50 60	62[b] 77[b]	93 96
Secretory epithelial cells of rat mammary gland	HELMINEN and ERICSSON (1968)	OsO$_4$	57	68	97
Vegetative hyphae of *Pythium ultimum*	GROVE, BRACKER and MORRÉ (1968)	Glutaraldehyde OsO$_4$[Ba(MnO$_4$)$_2$][a]	30—48[c, d]	37—94[d, e]	94[d]

[a] Section staining given in brackets.
[b] Dimensions also apply to smooth endoplasmic reticulum.
[c] Dimensions also apply to nuclear envelope
[d] These values were corrected by 20% from these in the original publication to account for section shrinkage under the electron beam.
[e] Membranes of the Golgi apparatus were differentiated accross the stack of cisternae so that those at the proximal pole appeared similar to endoplasmic reticulum and nuclear envelope, whereas those at the distal pole (including vesicle membranes) were similar to plasma membrane.

Table 2. *Phospholipid composition of subcellular fractions isolated from rat liver (% of total lipid phosphorus recovered)*

Phospholipid	Total liver			Nuclear membrane		Endoplasmic reticulum (rough microsomes)			Golgi apparatus	Plasma membrane		
	I	II	III	I	IV	IV	V	VI	VI	II	VI	VII
Phosphatidyl choline	58	52	52	52	54	57	53	61	45±3	38	40	37
Phosphatidyl ethanolamine	24	25	25	25	22	23	21	19	17±2	16	18	22
Phosphatidyl serine	4	3	4	6	7	6	8	3	4±1	8	4	—
Phosphatidyl inositol	7	9	9	4	8	8	11	9	9±3	7	7	13[a]
Sphingomyelin	6	4	5	6	5	5	7	4	12±3	19	19	17
Phosphatidic acid	T	—	T	—	2	—	—	—	—	—	—	—
Lysophosphatidyl choline	—	2	—	—	—	—	—	5	6	4	7	4
Lysophosphatidyl ethanolamine	—	—	—	—	—	—	—	n.d.	6	—	6	—
Cardiolipin	3	5	5	—	—	—	—	—	—	8	—	—
Other	T	—	5	—	—	—	—	—	—	8	—	3

I.   Gurr, Finean, and Hawthorne (1963).
II.   Skipski, Barclay, Archibald, Terebus-Kekish, Reichman, and Good (1965).
III. Rouser, Nelson, Fleischer, and Simon (1968).
IV.   H. Kleinig, University of Freiburg (unpublished).
V.   Glaumann and Dallner (1968).
VI.   Keenan and Morré (1970).
VII. Pfleger, Anderson, and Snyder (1968).
[a] Plus phosphatidylserine: T = Trace or less than 0.5%; n.d. = not detected; — = values not reported

reticulum and plasma membrane. However, certain enzymatic activities characteristic of endoplasmic reticulum appear to be absent from Golgi apparatus (Morré, Keenan, and Mollenhauer, in press). The plasma membrane is enriched in 5'-nucleotidase, and the Golgi apparatus is enriched in thiamine pyrophosphatase and certain sugar transferases. Thus, if proteins are transferred from endoplasmic reticulum to Golgi apparatus in the formation of secretory vesicles, the process must involve either a selective transfer so that some proteins become concentrated in the secretory vesicle membranes (Cheetham et al., 1969), or some enzymes derived from endoplasmic reticulum might be progressively activated or inhibited as changes in lipid and carbohydrate composition of the membranes occur.

B. Sites of Synthesis and Vectorial Transport of Membrane Constituents

Golgi apparatus are not known to be sites of protein synthesis (Warshawsky, Leblond, and Droz, 1963; Caro and Palade, 1964; Leblond, 1965; Jamieson and Palade, 1966, 1967a, b). They do not absorb ultraviolet light strongly *in vivo* (Hibbard and Lavin, 1945), and *in vitro* analyses (Kuff and Dalton, 1959; Yunghans, Keenan, and Morré, 1970) indicate the presence of little or no nucleic acids. In autoradiographic analyses and cell fractionation studies, early incorporation of amino acids is associated with ribosome-rich regions and to some extent with mitochondria and nuclei (cf. Leblond, 1965; Siekevitz et al., 1967; Ashley and Peters, 1969). Therefore, we assume localization of membrane protein synthesis within these regions[5]. Polyribosomes and associated messenger RNA of rough endoplasmic reticulum are potential sites of synthesis of membrane proteins (Siekevitz et al., 1967) as well as of proteins for export (Redman, Siekevitz, and Palade, 1966; Warshawsky et al., 1963; Caro and Palade, 1964; Jamieson and Palade, 1966; 1967a, b).

Unlike proteins, membrane phospholipids may be synthesized by a variety of cell components, including rough and smooth endoplasmic reticulum (Siekevitz et al., 1967; Stein and Stein, 1969). Alternatives for synthesis of the phospholipids of Golgi apparatus cisternae include: 1. synthesis in the endoplasmic reticulum, followed by transfer along with membrane proteins to the Golgi apparatus; 2. synthesis *in situ* at the Golgi apparatus; and 3. synthesis elsewhere in the cytoplasm followed by transfer to Golgi apparatus. Comparing endoplasmic reticulum-, Golgi apparatus-, and plasma membrane-rich fractions of intact cells, the time sequence of incorporation and short-term turnover of ^{3}H-glycerol (Chlapowski, 1969) and ^{14}C-labeled membrane precursors (Morré, 1970) is compatible with transfer of membrane lipids from endoplasmic reticulum to Golgi apparatus to plasma membrane.

Even if phospholipids are transferred from endoplasmic reticulum to the Golgi apparatus, conversion of membranes from endoplasmic reticulum-like to plasma membrane-like during the production of secretory vesicle membranes must involve changes affecting the lipid composition (Table 2, see also Hokin, 1968). Chief among these would be an increase in the proportion of sphingomyelin and sterols, with a

---

5 In view of the existence of specific membrane-associated RNA fractions (Moulé, 1968), a limited protein synthetic capacity for smooth membranes (including Golgi apparatus) cannot be completely ruled out.

corresponding decrease in the proportion of phosphatidyl choline (lecithin). In rat liver, smooth microsomes are active in steroid synthesis and metabolism (Conney et al., 1965; Jones and Armstrong, 1965; Jones and Fawcett, 1966), and Golgi apparatus fractions contain choline kinase and phosphorylcholine-cytidyl transferase (Morré, Keenan, and Mollenauer, in press) which are important in lecithin and sphingomyelin metabolism (Sribney and Kennedy, 1958).

The problem of vectorial transport of constituents from endoplasmic reticulum to Golgi apparatus has received some attention with secretory proteins (Caro and Palade, 1964; Redman, Siekevitz, and Palade, 1966; Redman and Sabatini, 1966; Jamieson and Palade, 1966, 1967a, b, 1968b). Continuous or intermittently continuous channels formed by lumina of endoplasmic reticulum and Golgi apparatus cisternae compartmentalize and direct the flow of these products (Fig. 10). Flow is apparently regulated at the transition elements between conjoining components. Transition elements[6] are characterized by active metabolism and marked sensitivity to inhibitors (Jamieson and Palade, 1968b; Imai and Coulston, 1968). Direct or intermittent continuities of the various components of the endomembrane system provide a similar mechanism for vectorial transport of membrane constituents.

C. Transition Elements of Fungi: Golgi Apparatus Equivalents

Before proceeding to the question of the origin of cisternae in the absence of pre-existing dictyosomes, it is well to consider the possibilities for a minimum functional unit. Insight has come from studies with fungi, many of which do not possess the stacks of cisternae that usually characterize the Golgi apparatus (Bracker, 1967; Girbardt, 1969), yet they contain secretory vesicles and exhibit surface growth by vesicular additions. Secretory vesicles are formed directly from short tubular or inflated cisternae which appear to serve as transition elements[6]. These transition elements function as equivalents of the Golgi apparatus (Bracker, 1968; see also Fig. 2A, 2B, 3A). Like dictyosomes, the fungal transition elements are surrounded in the cytoplasm by a zone of exclusion and show functional continuity with endoplasmic reticulum. The principal difference between these Golgi apparatus and those of other organisms is the absence of stacking, i.e., the transformations required to produce a secretory vesicle are accomplished by a single cisterna or sometimes even a single tubule. These structures may represent fundamental units of Golgi apparatus. Consideration should be given also to zones of exclusion as part of the unit, since they are consistently associated with Golgi apparatus and Golgi apparatus-like structures.

V. Origin of Cisternae in the Absence of Preexisting Dictyosomes

Evidence from electron microscopy suggests that dictyosomes originate through initiation of individual cisternae by fusion of vesicles derived from endoplasmic

---

6 Where different membrane components conjoin, membrane forms of morphology intermediate between the two conjoining components are observed. We refer to these as transition elements (Zeigel and Dalton, 1962), although Jamieson and Palade (1967a, b) have applied this term to tubular elements of endoplasmic reticulum adjacent to the periphery of the Golgi apparatus and which consist of part rough and part smooth-surfaced membranes.

reticulum or nuclear envelope. This is in keeping with concepts of membrane flow (BENNETT, 1956; NOVIKOFF et al., 1962) and of the Golgi apparatus as part of a system of discontinuous transition elements. We now consider questions of how sites of dictyogenesis are determined and how numbers of dictyosomes are increased and regulated.

A. Precisternal Stages of Dictyogenesis

In the few examples where dictyosome prestages have been described (WARD and WARD, 1968; WERZ, 1964; MARUYAMA, 1965; see also Figs. 11 and 12), new dictyosomes arise in a ribosome-free region of the cytoplasm or zone of exclusion. A similar region is associated with the origin of centrioles (SOROKIN, 1968). In early stages of oogenesis (WARD and WARD, 1968), this region is surrounded by endoplasmic reticulum and smooth-surfaced tubular membranous elements (Fig. 11). In dormant seeds of higher plants, the embryos contain zones of exclusion surrounding clusters of vesicular profiles prior to germination when normal dictyosomes are not seen (Fig. 12). Dictyosomes appear rapidly with the onset of germination, and their appearance coincides with the disappearance of the clusters of vesicles. Also, in young oocytes of *Oryzias* (YAMAMOTO, 1964) and the guinea pig (ADAMS and HERTIG, 1964), the Golgi apparatus develops from clusters of small vesicles within a zone of exclusion. WARD (1965) previously reported a pattern of dictyogenesis for the oocytes of *Rana pipiens* where vesicles apparently derived from annulate lamellae were aligned along the surface of a network of fine fibers. The nature of the zone of exclusion is not known. With frog oocytes, WARD and WARD (1968) later favored the interpretation that vesicles arise "de novo" within the mass of fibers, possibly by the combination of the fine fibers with lipid, and that the fibrous zones in which Golgi apparatus prestages arise are rich in ribonucleoproteins although free of ribosomes. This interpretation is in keeping with the observations of WERZ (1964) that actinomycin D prevents the formation of the dictyosome prestages in *Acetabularia* while puromycin prevents their transformation into dictyosomes.

Where precisternal stages have been described, they are encountered in generative or resting cells and are associated with a regional differentiation of the cytoplasm. If differentiation of the cytoplasm precedes dictyogenesis, what determines the sites of this differentiation? Observations with *Acetabularia* support the notion that these prestages are sites of previous dictyosomes. In this organism, a decrease in the numbers of dictyosomes during encystment is accompanied by an increase in the number of cytoplasmic regions resembling prestages (WERZ, personal communication). In oocytes, masses of filamentous and/or particulate material accumulate close to the nuclear membrane. The zones in which Golgi apparatus prestages arise have been suggested to originate from these regions (WARD and WARD, 1968).

B. Cisternal Stages of Dictyogenesis

Evidence from electron microscopy shows that single cisternae are an early intermediate in dictyogenesis (MARUYAMA et al., 1962; MARUYAMA, 1965). In studies with developing pollen grains of *Tradescantia*, three generations of synchronous development and degeneration of dictyosomes were described (MARUYAMA, GAY, and KAUFMANN, 1962; MARUYAMA, 1965). The first generation occurred in the

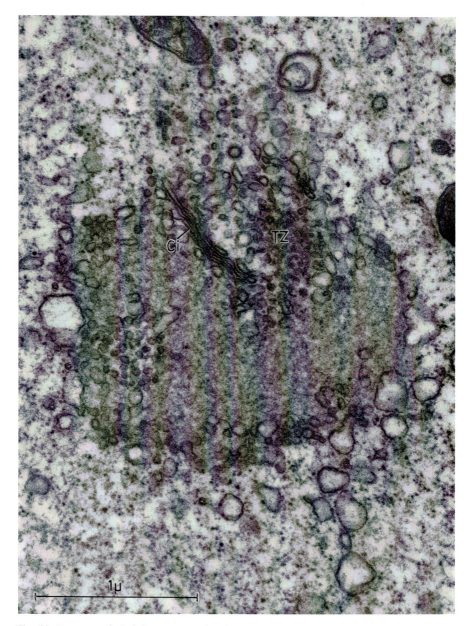

Fig. 11. Prestage of Golgi apparatus development in a frog oocyte where the first definitive adjacent Golgi apparatus cisternae (Ci) have appeared. Transition zones (TZ) containing numerous small vesicular profiles arise within the fine particles or filaments of the zone of exclusion in advance of definitive Golgi apparatus cisternae. OsO_4 fixation. From Ward and Ward (1968). Electron micrograph courtesy of Drs. R. T. Ward and E. Ward, Department of Anatomy, State University of New York, Downstate Medical Center, Brooklyn. Reprinted with the permission of the Société Française de Microscopie Electronique

pollen mother cells, the second in the microspore and the third in post meiotic mitosis. In each, the change in dictyosome generations took place during a period of cell divisions and appeared to involve single cisternae as well as ring structures composed of several concentric cisternae.

Curled or ring-like dictyosome cisternae have been observed in a number of quiescent or meristematic plant cells (SCHNEPF, 1961, 1964; HALL and WITKUS, 1964; CLOWES and JUNIPER, 1964; WHALEY et al., 1964; MARUYAMA, 1965; MOLLENHAUER, 1965b). In quiescent or meristematic cells of higher plants, dictyosome cisternae are few in number, curled or loosely associated (MOLLENHAUER and MORRÉ, 1966a), and secretory vesicles are not conspicuous. This morphological pattern is typical of resting, replicating, or regenerating dictyosomes in the maize root tip and is characteristic of the functional state designated as quiescent or replicating (MOLLENHAUER and MORRÉ, 1966a).

MARUYAMA (1965) observed that during pollen development, even single cisternae were absent between the first metaphase division and the end of meiosis. During this period, groups of small vesicular profiles similar to those reported here for embryos of seeds (Fig. 12) were present. When the microspores separated from the tetrad, ring-like cisternae were again encountered singly or in ring structures mostly among clusters of vesicular profiles. The size and number of cisternae within each ring structure increased as the microspore developed. Typical dictyosomes were eventually observed with intermediate stages involving fission of the ring structures and an opening of the stacked cisternae into typical dictyosome profiles. The origin of the precisternal vesicular or tubular structures appeared to involve components of the endoplasmic reticulum or nuclear envelope (MARUYAMA, 1965), but cisternae or precisternal stages were never absent.

The appearance and disappearance of stacked cisternae or dictyosomes in cytoplasmic regions previously occupied by single cisternae has also been described by ELLIOTT and ZIEG (1968) for *Tetrahymena pyriformis*. In logarithmic growth, normal vegetative cells of this organism did not contain dictyosomes, but in the oral region numerous, separate and smooth-surfaced cisternae were present with tubules between them. During starvation, before mixing the cells for mating, stacks of lamellae were formed in the region previously occupied by the smooth-surfaced cisternae and tubules. During late stages of conjugation, these structures were modified to resemble classical Golgi apparatus, but when the cells underwent fission following a period of feeding, the stacks disappeared and the situation characteristic of cells in logarithmic growth was restored. Similarly, in the yeast *Schizosaccharomyces pombe*, normal cells do not contain organized Golgi apparatus, but they appear in protoplasts during wall regeneration and again appear to be absent in the fully regenerated cells (HAVELKOVÁ and MENŠÍK, 1966).

C. Associations of Dictyosomes to Form Complex Golgi Apparatus

The formation of large, complex Golgi apparatus consisting of many closely aligned dictyosomes could result from: 1. multiplication of dictyosomes without extensive separation or 2. aggregation of widely spaced dictyosomes. The latter interpretation is favored in spermatids where rapid fusion of dictyosomes forms the large, aggregate Golgi apparatus of the acroblast (POLLISTER, 1930; GATENBY et al., 1958)

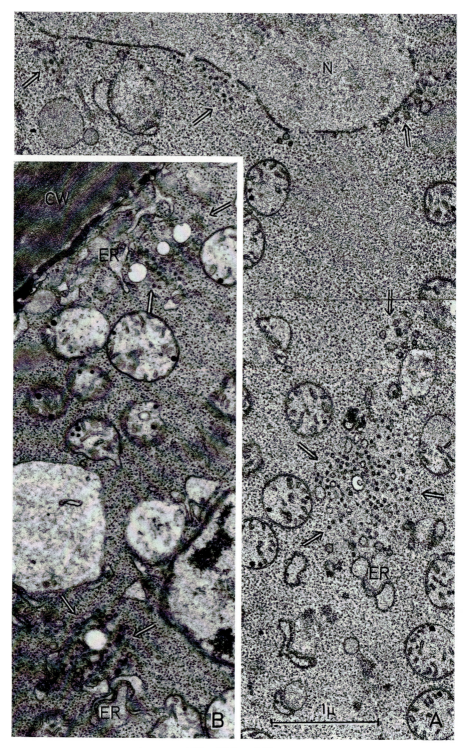

Fig. 12 A and B

and in the study of CORREIA (1964/1965). From an examination of 4 to 18 day old chick embryos, CORREIA concluded that the Golgi apparatus originates with the formation of small dictyosomes and proceeds through a progressive positioning of structures.

In general, younger embryo or generative cells are characterized by discrete dictyosomes often in a perinuclear arrangement (AFZELIUS, 1956; ADAMS and HERTIG, 1964; YAMAMOTO and ONOZATO, 1965; BEAMS and KESSEL, 1968; KESSEL, 1968b; FLICKINGER, 1969c), whereas in later stages, dictyosomes are distributed throughout the cytoplasm and associated with endoplasmic reticulum. Observations of living mammalian cells show that migration of Golgi apparatus components occurs (THIÉRY, personal communication), so that a perinuclear origin followed by migration to other parts of the cell is not an unreasonable expectation. Reversible fusion and separation of dictyosomes to yield alternating compact and dispersed Golgi apparatus was reported by POLLISTER (1930) during early stages of spermatogenesis in *Gerris*. In sporangia of the fungus *Pythium middletonii*, shortly before cell cleavage to form zoospores, the Golgi apparatus changes from individual dictyosomes scattered throughout the protoplast to clusters of closely associated dictyosomes (HEINTZ and BRACKER, unpublished results). Each cluster arises adjacent to a nucleus, where the dictyosomes function in concert to produce vesicles and cisternae that coalesce in a zone of exclusion to provide new plasma membrane during cell cleavage. Subsequently, and at other stages of the life cycle, these specialized formations of dictyosomes are lacking, and the Golgi apparatus consists of separated dictyosomes.

D. Growth and Differentiation of Golgi Apparatus

In an undifferentiated state, Golgi apparatus most often consist of small dictyosomes (FERREIRA, 1959; MOLLENHAUER and MORRÉ, 1966b; FLICKINGER, 1969c; see also Fig. 7A). They are then capable of growth and differentiation to give rise to the familiar Golgi apparatus forms encountered in fully differentiated cells. During embryonic development of liver and other organs of the rat, both FERREIRA (1959) and FLICKINGER (1969c) observed Golgi apparatus composed of tubular dictyosome cisternae in early developmental stages. As the cells differentiated, plate-like cisternae were acquired, and this was followed closely by the appearance of secretory vesicles. In general, the younger embryo cells were characterized by perinuclear arrangements of dictyosomes, whereas in later stages, dictyosomes were more randomly distributed throughout the cytoplasm and were characterized by cisternae becoming more plate-like and with conspicuous secretory vesicles (cf. YAMAMOTO and ONOZATO, 1965). A Golgi apparatus with tubular cisternae which

---

Fig. 12A and B. Portions of endosperm cells of mature, dry maize seeds. These seeds do not contain normal appearing dictyosomes with stacked cisternae at this stage of development. Instead, zones of exclusion (double arrows) containing numerous small vesicular profiles are scattered throughout the cytoplasm or are located adjacent to the nucleus (N) and near the cell surface. We suggest that these structures are Golgi apparatus prestages. Associations with fragments of rough-surfaced endoplasmic reticulum (ER) are frequently observed at the periphery of the zones of exclusion. CW = cell wall. A. Glutaraldehyde fixation followed by brief exposure to aqueous $KMnO_4$. B. Glutaraldehyde-OsO_4 fixation.

Ferreira (1959) first interpreted as an undifferentiated form of this cell component (see Fig. 7), may also exist in germ cells (Gatenby et al., 1958) and in resting cells during periods of secretory inactivity (Ovtracht, personal communication).

VI. Multiplication of Dictyosomes

In spite of conflicting viewpoints as to the behavior of dictyosomes during cell division (Wilson, 1925; Dalton, 1951; Gatenby, 1960; Bourne and Tewari, 1964; Dougherty, 1964; Robbins and Gonatas, 1964a, b; Roth, Wilson, and Chakraborty, 1966; Chang and Gibley, 1968) it is clear that the Golgi apparatus or its component dictyosomes multiply in both plant and animal cells, since their numbers do not decline as a result of cell division (cf. Buvat, 1963; Clowes and Juniper, 1964; Mollenhauer and Morré, 1966a; Ward and Ward, 1968) except in special examples such as spermatocytes of *Limax* where the primary spermatocyte has eight dictyosomes and the number is halved at each mitosis so that the spermatid receives only two (Gatenby, 1919). Lamellar components of the Golgi apparatus are retained through each of the phases of the mitotic cycle in hepatoma cells (Chang and Gibley, 1968). Spermatocytes and giant amebae both show typical Golgi apparatus in metaphase and anaphase (Roth, Wilson, and Chakraborty, 1966). Although Golgi apparatus components persist during division of a variety of animal cells (Wilson, 1925; Dougherty, 1964; Ward and Ward, 1968), there is no clear indication from these studies as to how they multiply (Ward and Ward, 1968). Information suggesting modes of dictyosome multiplication has come largely from plants where dictyosomes seem to persist or even increase during periods of cell division (Buvat, 1963; Murakami, Morimura, and Takamiya, 1963; Clowes and Juniper, 1964; Healy and Jensen, 1965; Whaley, 1966; Løvlie and Bråten, 1968) as well as participate in cell plate formation during cytokinesis (see reviews of Mollenhauer and Morré, 1966a; Beams and Kessel, 1968; Northcote, 1968).

An example of dictyosome doubling described by Nagy and Fridvalsky (1968) for a multinuclear alga *(Botrydium granulatum)* appears to be highly integrated with other events in cell division. During metaphase, they observed two dictyosomes on opposite sides of a nucleus, one at each of the two poles of the future spindle. At a later stage in metaphase the dictyosomes doubled, and centrioles occurred between the paired dictyosomes. Cleavage of the nuclear envelope occurred opposite the centrioles and between the paired dictyosomes.

Doubling of dictyosomes by fission (simultaneous division of the cisternae) was proposed on the basis of correlative light and electron microscopy (Gatenby, 1919; Gatenby et al., 1958; Gatenby, 1960) of germ cells (molluscs and insects) and has been widely used to explain electron microscope images of dictyosomes paired end-to-end in lower organisms (Grassé, 1957; Buvat, 1958 a, b; Gatenby, 1960; Gatenby and Tahmisian, 1960; Drawert and Mix, 1963; Daniels, 1964 and others; see also Carasso and Favard, 1961; Figs. 17 and 18 of Buvat, 1963; Whaley, 1966; Mollenhauer and Morré, 1966a; Beams and Kessel, 1968; Clowes and Juniper, 1968), higher plants (Diers, 1966), and animals (Afzelius, 1956; Dalton, 1961; Wischnitzer, 1962; Yamamoto and Onozato, 1965). Multiplication by fission, where the entire stack of cisternae divides perpendicular to

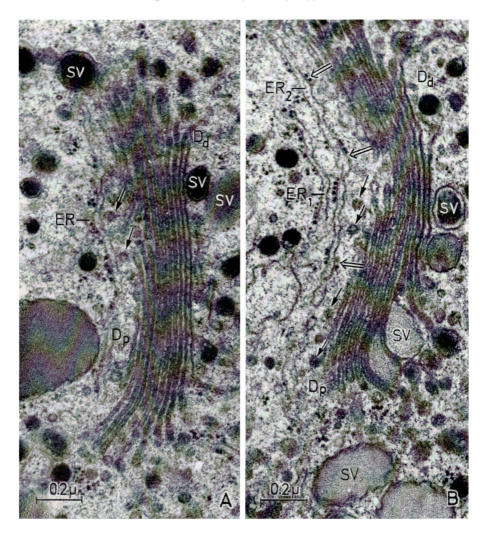

Fig. 13 A and B. Replication stages of dictyosomes of a growing cell of *Micrasterias denticulata* Bréb. The dictyosome on the left (A) shows a polar configuration with wide cisternae on the proximal pole or forming face (D_p) and somewhat narrower cisternae on the opposite or distal pole or maturing face (D_d). The cisternae at the proximal pole exhibit a median discontinuity indicating a stage of multiplication. Between the transitional endoplasmic reticulum lamellae (ER; ribosomes only on the membrane leaflet opposite the dictyosomes), small coated blebs or projections are found (arrows). The endoplasmic reticulum cisternae remain undivided at this stage. The larger vesicles with electron dense contents are secretory vesicles (SV). The dictyosome on the right (B) shows what may represent a more advanced stage of replication where wide cisternae are present in median positions upon which there are now two distinct stacks of shorter cisternae (daughter stacks). A few clusters of ribosomes are present on the endoplasmic reticulum membrane leaflet adjacent to the dictyosomes in positions where dictyosome lamellae no longer appear to be forming (double arrows). At the proximal pole, two endoplasmic reticulum cisternae (ER_1 and ER_2) are present. Glutaraldehyde-OsO$_4$ fixation. From KIERMAYER (1970).

the axis of functional polarity, might explain the appearance of two or more smaller dictyosomes where one large dictyosome was observed in earlier stages (Grassé, 1957; Gatenby et al., 1958; Yamamoto and Onozato, 1965). For the alga *Micrasterias*, Kiermayer (1967) suggested that dictyosomes were capable of growing in diameter while maintaining the usual number of cisternae. In some of the wider dictyosomes, he observed a discontinuity near the center of the cisterna adjacent to endoplasmic reticulum (Fig. 13) which suggested that, after a diameter increase, dictyosomes underwent progressive fission of each cisterna beginning with the cisterna adjacent to endoplasmic reticulum (see also Drawert and Mix, 1963).

Failure of portions of forming cisternae to fuse does not necessarily constitute evidence of fission (Grassé, 1957; Whaley, 1966). Separation of the forming region of endoplasmic reticulum into two segments as diagrammed in Fig. 14 would account equally well for the appearance of dictyosomes apparently in the process of fission (Fig. 13; see also Plate II of Grassé, 1957; Fig. 8 of Dalton, 1961; Fig. 14 of Drawert and Mix, 1963 and Figs. 16, 31, and 32 of Hemmes and Hohl, 1969). At the onset, two shorter cisternae would be formed where a longer cisterna appeared previously (Grassé, 1957). Continued separation of the two daughter cisternae would result in stacks having a skewed appearance as in Fig. 13 B with normal cisternal dimensions being restored by growth in extent of the forming regions.[7]

Increase in numbers of dictyosomes are not restricted to dividing cells (Clowes and Juniper, 1964; Healy and Jensen, 1965; Mollenhauer, 1965b; Whaley, 1966). For example, the growing oocyte or spermatocyte offers a favorable subject for dictyosome multiplication (Kater, 1928; Pollister, 1930; Johnson, 1931; Afzelius, 1956; Wartenburg, 1962; Wischnitzer, 1962; Adams and Hertig, 1964; Yamamoto and Onozato, 1965; Ward and Ward, 1968) although egg maturation may be accompanied by a reduction in dictyosome number (Afzelius, 1956). The location and even the existence of the Golgi apparatus has been questioned in rat oocytes during the first maturation division before fertilization (Odor, 1960). In oocytes, some form of dictyosome multiplication, usually involving fragmentation, has generally been assumed.

Dictyosome propagation by fragmentation of existing dictyosomes into replicating units of one or more cisternae has been suggested for at least one other developmental stage where increases in the number of dictyosomes per cell take place without cell division (Mollenhauer and Morré, 1966a). In cells adjacent to the root cap initials of maize, the number of dictyosomes per cell increases aproximately tenfold (Mollenhauer, 1965b). Endoplasmic reticulum also increases in amount during this developmental stage, and cisternae are present in the cytoplasm singly or in stacks of 2 or 3 (Mollenhauer, 1965b).

---

7 More recently, we have obtained evidence from the fungus *Pythium aphanidermatum* consistent with the scheme of Fig. 14 but indicating that dictyosome doubling is preceeded by an extension of the forming face regions. As a result, cisternae having twice normal diameter are produced. Following formation of a complete dictyosome having twice normal diameter, normal dimensions seem to be restored by separation of the forming face region into two as diagrammed in Figs. 14 F—G. The only point of disagreement between these observations and the diagrams of Fig. 14 is that growth apparently occurs before, not after, multiplication (Bracker, Grove and Morré, in preparation).

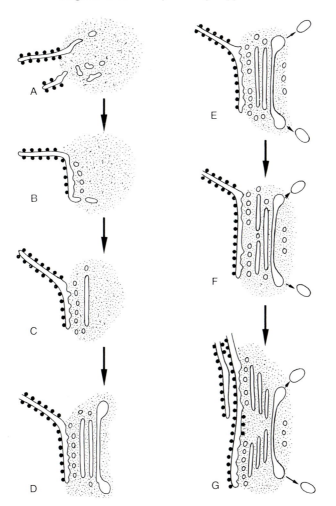

Fig. 14 A—G. Hypothetical scheme relating observations concerning the origin and continuity of Golgi apparatus. In the absence of preexisting dictyosomes, cisternae are presumed to arise in differentiated regions of the cytoplasm (or zones of exclusion) via prestages consisting of groups of small vesicules or tubular elements associated with endoplasmic reticulum or nuclear envelope (A). The elementary units which may be tubules or vesicles (B) further differentiate to give rise to flattened plate-like portions of cisternae (C). Additional cisternae are then formed by repetition of this basic generative event with endoplasmic reticulum or nuclear envelope providing one source of new membrane material (D). Multiplication of existing dictyosomes and formation of complex Golgi apparatus is explained on the basis of multiplication of the forming regions associated with endoplasmic reticulum and associated zones of exclusion. One possibility for multiplication during periods of active secretory vesicle production (E—G) is that two sets of somewhat shorter cisternae are formed where a longer cisterna appeared previously. Continued separation of the two daughter forming regions with simultaneous production of new cisternae (F) would result in somewhat skewed stacks (G). Dictyosome replication is completed when the last common cisterna is lost through formation of secretory vesicles. Ultimately, normal cisternal dimensions might be restored by growth in extent of the forming region.[7] Extension of this scheme for dictyosome replication in three dimensions would result in the eventual formation of complex Golgi apparatus consisting of numerous interassociated dictyosomes. Other possibilities for dictyosome multiplication, including dictyosomes not involved in secretory vesicle production, are provided in the text.

Some images of paired dictyosomes are inconsistent with fission or fragmentation of individual cisternae (Drum, 1966; Manton, 1967; Pickett-Heaps, 1968). In the diatoms studied by Drum (1966), pairs of dictyosomes are frequently oriented on opposite sides of paired strands of endoplasmic reticulum which pass between each dictyosome pair in a manner suggesting that multiplication of endoplasmic reticulum may accompany multiplication of dictyosomes. Alternatively, in cold treated root cells, dictyosomes are often aligned along a common strand of endoplasmic reticulum which is either continuous with or replaces an intercalary cisterna of the midregion of each dictyosome (unpublished results). Increases in amount of endoplasmic reticulum and multiplication of dictyosomes may coincide in other organisms (Grimstone, 1958; Clowes, and Juniper, 1964; Fig. 18 of Hemmes and Hohl, 1969; M. J. Brown, private communication; H. H. Mollenhauer, unpublished results; see also Fig. 13), although this apparent relationship has not been studied quantitatively.

Thus, the mode of dictyosome multiplication is far from certain. In addition to binary fission or division of the forming regions discussed in detail above and in Fig. 14, possibilities include *de novo* origin (Dalton, 1951; Buvat, 1958, 1963; Gatenby, et al., 1958; Carasso and Favard, 1961; Mercer, 1963; Ward and Ward, 1968; Flickinger, 1969b), multiplication by fragmentation (cf. Wilson, 1925; dictyokinesis: Perroncito, 1910) or cisternal separation (Gatenby, 1960; Mollenhauer and Morré, 1966a; Masuda, 1967; Ward and Ward, 1968; see Figure 62 of Clowes and Juniper, 1968) or origin at the Golgi apparatus periphery (see Figure 4 of Buvat, 1958). The latter proposal of Buvat (1958) would explain increases in dictyosomes in meristematic or elongating plant cells where neither prestages nor "division" figures are commonly observed (Buvat, 1963; Whaley, 1966; unpublished observations). In this connection, Whaley (1966) discussed the possibility of an "organizational field" in the vicinity of Golgi apparatus which might participate in the formation of new dictyosomes. It seems likely to us that Golgi apparatus exert some control over their own formation particularly if zones of exclusion, prestages and organizational fields are considered as Golgi apparatus components.

Whaley (1966) cautions that the manner of Golgi apparatus formation might vary among species and according to the developmental stage. Wischnitzer (1962), for example, mentions that in young oocytes Golgi apparatus might originate from the nuclear envelope, whereas later increases in Golgi apparatus might result from fission of existing dictyosomes. We consider that all modes of multiplication enumerated above are possible as variations on a common theme of cisternal formation. A more or less general mechanism for the origin of Golgi apparatus membranes is provided by organization of endoplasmic reticulum or nuclear envelope derived vesicles within a zone of exclusion or organizational field of an existing dictyosome or prestage.

VII. Nuclear Control of Golgi Apparatus Structure and Function

Specific characteristics of Golgi apparatus are not permanently established but vary with the developmental state of the cell. They vary with the history and metabolic state of the cell, differ for embryonic and adult cells, and change with resting and proliferating stages. Moreover, these specific characteristics of Golgi apparatus do not appear to be autonomously derived. The diversity of structural and functional properties according to cell type and tissue, and with respect to regions within a cell, is documented in the reviews by MOLLENHAUER and MORRÉ (1966a), WHALEY (1966), and BEAMS and KESSEL (1968), the papers by CAVAZOS et al. (1967), WIENKE et al. (1968), and FLICKINGER (1969c), and in Figs. 1—9. These observations suggest that nuclear genes ultimately control Golgi apparatus form and function including formation of new cisternae and the organization of dictyosomes within a Golgi apparatus; for a similar interpretation of diversity among plasma membranes see the review by BENEDETTI and EMMELOT (1968).

Direct evidence for nuclear control over dictyogenesis comes from studies by FLICKINGER (1967, 1968a) with *Amoeba proteus* and WERZ (1964) with *Acetabularia*. Golgi apparatus were not observed after the second day after enucleation of *Amoeba*, although extensive arrays of granular endoplasmic reticulum were retained longer (FLICKINGER, 1967, 1968a). Golgi apparatus persisted in nucleated portions of cut cells and in controls. The disappearance of Golgi apparatus in *Amoeba* was preceded by a reduction in size of the cisternae and in the number of cisternae per dictyosome (FLICKINGER, 1968a). The report of nuclear (DNA) dependence in *Acetabularia* was based on experiments with inhibitors (actinomycin D and puromycin) and on experiments comparing darkened and enucleated cells (WERZ, 1964).

A similar pattern was observed for the disappearance of Golgi apparatus in actinomycin D-treated (FLICKINGER, 1968b) or starved (DANIELS, 1964) amebae. In *Trichonympha*, the disappearance of dictyosomes due to starvation was reversed by feeding (GRIMSTONE, 1958). In amebae, after 3 to 4 days of treatment with actinomycin D, the number and size of the Golgi apparatus were reduced (FLICKINGER, 1968b). The number of cisternae decreased, and single cisternae were sometimes seen. The Golgi apparatus in rat parotid also diminished in size following actinomycin D treatment (HAN, 1967).

Enucleated and actinomycin D-treated cells are capable of some protein synthesis and perhaps of membrane synthesis for some time after the interruption of translation, depending on the stability of the messenger RNAs (FLICKINGER, 1968a). If dictyogenesis (or continued production of cisternae) depends on products of nuclear genes, studies with actinomycin D suggest that these products may turn over slowly relative to turnover of Golgi apparatus cisternae (cf. NEUTRA and LEBLOND, 1966; SCHNEPF and KOCH, 1966a, b; BROWN, 1969; see also SCHNEPF, 1961). Nevertheless, the results with inhibitors and enucleated cells are consistent both with a nuclear dependence of Golgi apparatus continuity and with Golgi apparatus breakdown through loss of peripheral elements and of entire cisternae. The absence of some form of the Golgi apparatus or Golgi apparatus associated material in these treated cells has not been established, making the interpretation of regeneration studies difficult. In FLICKINGER's (1969b) experiments, enucleated ameba cells were held for periods up to 5 days, until most of the normal appearing dictyosomes disappeared. The cells regenerated normal-appearing Golgi apparatus when supplied with a transplanted nucleus from another ameba. Although these observations establish a high degree of nuclear control, a *de novo* origin of the Golgi apparatus remains unproven.

8*

Examples of genetic disorders which may involve altered Golgi apparatus function include altered formation of lysosomes (de Duve and Wattiaux, 1966), melanosomes (Windhorst, Zelickson, and Good, 1966; Brumbaugh, 1968), and mucopolysaccharides (Dogson and Lloyd, 1968). Whether these disorders result from altered patterns of synthesis, assembly, or secretion within the Golgi apparatus is unknown. With Hurler and Hunter syndromes, for example, abnormal muco-polysaccharide deposition was a consequence of a defective catabolic sequence (Fratantoni, Hall, and Neufeld, 1968). In the fly *Chironomus pallidivittatus* however, the synthesis of secretory granules was correlated with a specific Balbiani ring of the polytene chromosome (Beermann, 1961; Clever, 1968).

VIII. Concluding Comments

The overall pattern of morphological change associated with Golgi apparatus origin and continuity suggests that multiplication follows the same pattern of hierarchy as organization. Golgi apparatus seem to be formed from dictyosomes, while dictyosomes arise one cisterna at a time. As summarized in Fig. 14, the origin of cisternae in the absence of an existing Golgi apparatus involves pres-tages where single cisternae are formed in a manner suggestive of fusion of tubular or vesicular elements within a cytoplasmic zone of exclusion. Additional cisternae may arise, not from replication of the newly formed cisternae as suggested by Hall and Witkus (1954), but by repetitive cycles of cisternal formation through fusion of vesicles (Maruyama, 1965).

Division of the Golgi apparatus or its parts has been assumed for many years (cf. Wilson, 1925; Pollister, 1930) but usually without adequate documentation. Dictyosomes must multiply, since the original Golgi apparatus compliment is restored as cells divide. Although the mechanisms remain obscure, fusion of endoplasmic reticulum- or nuclear envelope-derived vesicles within a zone of exclusion or organizational field of a preexisting dictyosome or prestage appears as a general feature of Golgi apparatus ontogeny.

Related to the problem of dictyosome multiplication is the question of the extent to which Golgi apparatus are self-perpetuating or control their own for-mation (Wilson, 1925 p. 714—717). At one extreme, extrachromosomal inheritance occurs when self-duplicating parts of the cytoplasm are passed on to daughter cells (Wilson, 1925; Clowes and Juniper, 1968). At the other extreme is *de novo* construction in each generation from information supplied by nuclear genes (Schiff and Epstein, 1965). This would include the view of Bell et al. (Bell and Mühlet-haler, 1964; Bell, Frey-Wyssling, and Mühlethaler, 1966) that early oocyte development is a time when various cell components are completely renewed in a nucleus-controlled process. Nuclear dependence is evidenced by the responsiveness of Golgi apparatus replication to overall signals for cell division and the many developmental changes reflected in altered Golgi apparatus structure and function. Nevertheless, examples exist for both somatic (Clowes and Juniper, 1964; Dou-gherty, 1964; Roth et al., 1966; Løvlie and Bråten, 1968) and generative cells (Gatenby, 1919; Wilson, 1925; Pollister, 1930; Afzelius, 1956; Wartenburg, 1962; Wischinitzer, 1962; Roth, Wilson and Chakraborty, 1966) where Golgi apparatus (as well as endoplasmic reticulum fragments; see Porter, 1961) appear to be passed on from one cell generation to the next. Even with mature spermatids, where the Golgi apparatus becomes detached from the acrosome and is eventually cast off with the cytoplasmic remnant (BEAMS and KESSEL, 1968), maternal transfer of oocyte dictyosomes to the fertilized egg would provide for Golgi apparatus continuity. Plastids, mitochondria, centrioles, and centriolar derivatives such as blepharoplasts may not be the only cell components which are self-multiplying. They are certainly not the only parts of cells to be passed on from a cell to its daughters. Although we cannot conclude that Golgi apparatus control their own formation, we can conclude that their formation is subject to regulation by nuclear genes and procedes in a manner dependent upon functional continuity with endoplasmic reticulum or nuclear envelope[8].

IX. Summary

The smooth membrane-bounded compartments or cisternae which characterize the Golgi apparatus are often flattened but with a continuous system of peripheral tubules and vesicles. Golgi apparatus: 1. are associated with the nuclear envelope and its extensions (principally endoplasmic reticulum); 2. produce secretory vesicles having membranes which are plasma membrane-like and capable of fusing with plasma membrane; 3. are potential sites of endomembrane differentiation since their membranes have morphological and biochemical properties intermediate between those of endoplasmic reticulum and plasma membrane; and 4. are surrounded by a region of differentiated cytoplasm nearly devoid of other cell components, including ribosomes (zone of exclusion).

Golgi apparatus cisternae occur in polarized stacks called dictyosomes, or they exist singly without stacking. Evidence from electron microscopy supports a dynamic concept of functioning for each dictyosome in which cisternae are lost, usually at one pole, as secretory vesicles are produced. A constant number of cisternae per dictyosome is maintained as new cisternae are produced at the opposite pole, from membrane materials derived from endoplasmic reticulum.

Formation of two or more dictyosomes at a region of endoplasmic reticulum-dictyosome association is a mechanism for increasing dictyosome numbers presented as an alternative to dictyosome division or fragmentation. Origin of cisternae in the absence of dictyosomes involves prestages in which single cisternae arise within a zone of exclusion where only tubular or vesicular elements existed previously. Evidence from plants suggests that dictyosomes then arise through successive formation of additional cisternae adjacent to the first. Associations among individual dictyosomes give rise to complex Golgi apparatus.

8 Portions of the studies reported here were made possible through grants to the authors from the NSF (GB 7078 and GB 03044) and from the USPH (GN 15492). Purdue University AES Journal Paper No. 3897. Charles F. Kettering Research Contribution No. C-359. We thank Drs. R. M. BROWN, Jr, H. FALK, and W. W. FRANKE, University of Freiburg; C. J. FLICKINGER, University of Colorado; S. N. GROVE, Purdue University; R. A. JERSILD, Indiana University Medical Center, Indianapolis; O. KIERMAYER, University of Köln; J.-P. THIÉRY, Faculté des Sciences, Ivry, France; and G. WERZ, Max-Planck-Institut für Meeresbiologie, Wilhelmshaven, for providing unpublished information and Drs. W. P. CUNNINGHAM; T. W. KEENAN, and C. A. LEMBI for helpful discussions.

References

Adams, E. C., Hertig, A. T.: Studies on guinea pig oocytes. I. Electron microscopic observations on the development of cytoplasmic organelles in oocytes of primordial and primary follicles. J. Cell Biol. **21**, 397—427 (1964).

Afzelius, B. A.: Electron microscopy of Golgi elements in sea urchin eggs. Exp. Cell Res. **11**, 67—85 (1956).

Agrell, I. P. S.: Continuity of membrane systems in the cells of imaginal discs. Z. Zellforsch. **72**, 22—29 (1966).

Amsterdam, A., Ohad, I., Schramm, M.: Dynamic changes in the ultrastructure of the acinar cell of the rat parotid gland during the secretory cycle. J. Cell Biol. **41**, 753—773 (1969).

Amos, W. B., Grimstone, A. V.: Intercisternal material in the Golgi body of *Trichomonas*. J. Cell Biol. **38**, 466—471 (1968).

Ashley, C. A., Peters, T.: Electron microscopic radioautographic detection of sites of protein synthesis and migration in liver. J. Cell Biol. **43**, 237—249 (1969).

Bainton, D. F., Farquhar, M. G.: Origin of granules in polymorphonuclear leukocytes. Two types derived from opposite faces of the Golgi complex in developing granulocytes. J. Cell Biol. **28**, 277—301 (1966).

Beams, H. W., Kessel, R. G.: The Golgi apparatus: structure and function. Int. Rev. Cytol. **23**, 209—276 (1968).

Beermann, W.: Ein Balbiani-Ring als Locus einer Speicheldrüsenmutation. Chromosoma (Berl.) **12**, 1—25 (1961).

Bell, P. R., Frey-Wyssling, A., Mühlethaler, K.: Evidence for the discontinuity of plastids in the sexual reproduction of a plant. J. Ultrastruct. Res. **15**, 108—121 (1966).

— Mühlethaler, K.: The degeneration and reappearance of mitochondria in the egg cells of a plant. J. Cell Biol. **20**, 235—248 (1964).

Benedetti, E. L.: Sulla presenza di granuli ferruginosi nell'Apparato di Golgi delle cellule erithroblastiche. Rend. Accad. Naz. Linciei **24**, 757—759 (1958).

— Emmelot, P.: Structure and function of plasma membranes isolated from liver. In: Ultrastructure in Biological Systems. The Membranes (Ed.: A. Dalton and F. Haguenau), pp. 33—120. New York-London: Academic Press 1968.

Bennett, H. S.: The concepts of membrane flow and membrane vesiculation as mechanisms for active transport and ion pumping. J. biophys. biochem. Cytol. **2** (Suppl.), 99—103 (1956).

Berlin, J. D.: The localization of acid mucopolysaccharides in the Golgi complex of intestinal goblet cells. J. Cell Biol. **32**, 760—766 (1967).

Bouck, G. B.: Fine structure and organelle associations in brown algae. J. Cell Biol. **26**, 523—537 (1965).

Bourne, G. H., and Tewari, H. B.: Mitochondria and the Golgi complex. In: Cytology and Cell Physiology (Ed.: G. H. Bourne), pp. 377—421. New York-London: Academic Press 1964.

Bracker, C. E.: Ultrastructure of fungi. Ann. Rev. Phytopathol. **5**, 343—374 (1967).

— The ultrastructure and development of sporangia in *Gilbertella persicaria*. Mycologia **60**, 1016—1067 (1968).

Brown, R. M.: Observations on the relationship of the Golgi apparatus to wall formation in the marine Chrysophycean alga, *Pleurochrysis scherffelii* Pringsheim. J. Cell Biol. **41**, 109—123 (1969).

Brumbaugh, J. A.: Ultrastructural differences between forming eumelanin and pheomelanin as revealed by the pink-eye mutation in the fowl. Develop. Biol. **18**, 375—390 (1968).

Bruni, C., Porter, K. R.: The fine structure of the parenchymal cell of the normal rat liver. I. General observations. Amer. J. Pathol. **46**, 691—755 (1965).

Bucciarelli, E.: Intranuclear cisternae resembling structures of the Golgi complex. J. Cell Biol. **30**, 664—665 (1966).

Buvat, R.: Nouvelles observations sur l'appareil de Golgi dans les cellules de végétaux vasculaires. C. R. Acad. Sci. (Paris) **246**, 2157—2160 (1958a).

Buvat, R.: Recherches sur les infrastructures du cytoplasme, dans les cellules du méristème apical, des ébauches foliares et des feuilles développées d'*Elodea canadensis*. Ann. Sci. nat. Bot. **19**, 121—161 (1958b).
— Electron microscopy of plant protoplasm. Int. Rev. Cytol. **14**, 41—155 (1963).
Carasso, N., Favard, P.: Les ultrastructures cytoplasmiques. II. Appareil de Golgi. In: Traité de Microscopie Electronique. (Ed.: C. Magnam) II, 963—997. Paris: Hermann 1961.
Caro, L., Palade, G. E.: Protein synthesis, storage, and discharge in the pancreatic exocrine cell. An autoradiographic study. J. Cell Biol. **20**, 473—495 (1964).
Cavazos, F., Green, J. A., Hall, D. G., Lucas, F. V.: Ultrastructure of the human endometrial glandular cell during the menstrual cycle. Amer. J. Obstet. Gynec. **99**, 833—854 (1967).
Chang, J. P., Gibley, C. W.: Ultrastructure of tumor cells during mitosis. Cancer Res. **28**, 521—534 (1968).
Cheetham, R. D., Keenan, T. W., Nyquist, S., Morré D. J.: Biochemical comparisons of endoplasmic reticulum-, Golgi apparatus-, and plasma membrane-rich cell fractions from rat liver in relation to cytomembrane differentiation. J. Cell Biol. **43**, 21a (1969).
Chlapowski, F. J.: Incorporation and short-term turnover of tritiated glycerol in various membrane systems and organelles of *Acanthamoeba palestinesis*. J. Cell Biol. **43**, 22a—23a (1969).
Claude, A.: Interrelation of cytoplasmic membranes in mammalian liver cells: endoplasmic reticulum and Golgi complex. J. Cell Biol. **39**, 25a—26a (1968).
— Growth and differentiation of endoplasmic and Golgi membranes in the course of synthesis and transport of lipo-protein granules. Proc. 1st Intern. Sym. Cell Biol. Cytopharmacol., Venice, Italy, 1969. New York: Raven Press (in press).
Clever, U.: Regulation of chromosome function. Ann. Rev. Genet. **2**, 11—30 (1968).
Clowes, F. A. L., Juniper, B. E.: The fine structure of the quiescent centre and neighbouring tissues in root meristems. J. exp. Botany **15**, 622—630 (1964).
— — Plant Cells. Oxford-Edinburg: Blackwell Sci. 1968.
Conney, A. H., Schneidman, K., Jacobson, M., Kuntzman, R.: Drug-induced changes in steriod metabolism. Ann. N. Y. Acad. Sci. **123**, 98—109 (1965).
Coombs, J., Lauritis, J. A., Darley, W. M., Volcani, B. E.: Studies on the biochemistry and fine structure of silica shell formation in diatoms. V.Effects of colchicine on wall formation in *Navicula pelliculosa* (Breb), Hilse. Z. Pflanzenphys. **58**, 124—152 (1968).
Correia, M. J. R.: La cytogénèse de l'appareil de Golgi étudiée dans les cellules ganglionnaires du poulet. Arch. Port. Sci. Biol. **15**, 77—82 (1964/1965).
Cunningham, W. P., Morré, D. J., Mollenhauer, H. H.: Structure of isolated plant Golgi apparatus revealed by negative staining. J. Cell Biol. **28**, 169—179 (1966).
Dalton, A. J.: Cytoplasmic changes during cell division with reference to mitochondria and the Golgi substance. Ann. N. Y. Acad. Sci. **51**, 1295—1302 (1951).
— Golgi apparatus and secretion granules. In: The Cell. (Ed.: Brachet, J., Mirsky, A. E.), Vol. 2, pp. 603—617. New York-London: Academic Press 1961.
Daniels, E. W.: Origin of the Golgi system in amoebae. Z. Zellforsch. **64**, 38—51 (1964).
De Duve, C., and Wattiaux, R.: Functions of lysosomes. Ann. Rev. Physiol. **28**, 435—492 (1966).
Diers, L.: On the plastids, mitochondria, and other cell constituents during oogenesis of a plant. J. Cell Biol. **28**, 527—543 (1966).
Dogson, K. S., and Lloyd, A. G.: Metabolism of acidic glycosaminoglycans (mucopolysaccharides). In: Carbohydrate Metabolism and Its Disorders (Ed.: Dickens, F., Randle, P. J., and Whelan, W. J.), I, 169—212. New York-London: Academic Press 1968.
Dougherty, W. J.: Fate of Golgi complex, lysosomes, and microbodies during mitosis of rat hepatic cells. J. Cell Biol. **23**, 25 A (1964).
Drawert, H., and Mix, M.: Licht- und elektronenmikroskopische Untersuchungen an Desmidiaceen. XI. Mitteilung: Die Struktur von Nucleolus und Golgi-Apparat bei *Micrasterias denticulata* Breb. Port. Acta Biol. **7**, 17—28 (1963).

Drum, R. W.: Electron microscopy of paired Golgi structures in the diatom *Pinnularia nobilis*. J. Ultrastruct. Res. **15**, 100—107 (1966).

Elliott, A. M., and Zieg, R. G.: A Golgi apparatus associated with mating in *Tetrahymena pyriformis*. J. Cell Biol. **36**, 391—398 (1968).

Emmelot, P., and Benedetti, E. L.: On the possible involvement of the plasma membrane in the carcinogenic process. In: Carcinogenesis: A Broad Critique, 471—533. Baltimore, Maryland: Wilkins and Wilkins Company 1967.

Essner, E., Novikoff, A.: Cytological studies on two functional hepatomas: interrelations of endoplasmic reticulum, Golgi apparatus and lysosomes. J. Cell Biol. **15**, 289—312 (1962).

Fawcett, D. W., and McNutt, N. S.: The ultrastructure of the cat myocardium. I. Ventricular papillary muscle. J. Cell Biol. **42**, 1—45 (1969).

Fauré-Fremiet, E., Favard, P., Carasso, N.: Étude au microscope électronique des ultrastructures d'*Epistylis anastatica* (Cilié Péritriche). J. Microsc. **1**, 287—312 (1962).

Ferreira, J. F. D.: A differencia cão do condrioma aparelho de Golgi e ergastoplasma. Thesis, 1—214, Faculty of Medicine, University of Lisbon, 1959.

Falk, H.: Fusiform vesicles in plant cells. J. Cell Biol. **43**, 167—174 (1969).

Flickinger, C. J.: Electron microscope study of enucleated *Amoeba proteus*. J. Cell Biol. **35**, 40 A—41 A (1967).

— The effects of enucleation on the cytoplasmic membranes of *Amoeba proteus*. J. Cell Biol. **37**, 300—315 (1968a).

— Cytoplasmic alterations in *Amebae* treated with actinomycin D. Comparison with the effects of surgical enucleation. Exp. Cell Res. **53**, 241—251 (1968b).

— Fenestrated cisternae in the Golgi apparatus of the epididymus. Anat. Rec. **163**, 39—54 (1969a).

— The development of Golgi complexes and their dependence upon the nucleus in *Amebae*. J. Cell Biol. **43**, 250—262 (1969b).

— The pattern of growth of the Golgi complex during the fetal and postnatal development of the rat epididymis. J. Ultrastruct. Res. **27**, 344—360 (1969c).

Fratantoni, J. C., Hall, C. W., Neufeld, E. F.: The defect in Hurler's and Hunter's syndromes: Faulty degradation of mucopolysaccharide. Proc. nat. Acad. Sci. (Wash.) **60**, 699—706 (1968).

Friend, D. S.: The fine structure of Brunner's gland in the mouse. J. Cell Biol. **25**, 563—576 (1965).

— Murray, M. J.: Osmium impregnation of the Golgi apparatus. Amer. J. Anat. **117**, 135—149 (1965).

Gatenby, J. B.: The cytoplasmic inclusions of germ cells: V. The gametogenesis and early development of *Limnacea stagnalis* L., with special reference to the Golgi apparatus and the mitochondria. Quart. J. Micr. Sci. **63**, 445—491 (1919).

— Notes on the gametogenesis of a pulmonate Mollusc. An electron microscope study. La Cellule **60**, 289—303 (1960).

— Tahmisian, T. N.: Centriole adjunct, centroles, mitochondria, and ergastroplasm in Orthopteran spermatogenesis. La Cellule **60**, 105—135 (1960).

— — Devine, R., Beams, H. W.: The Orthopteran dictyosome. An electron microscope study. Cellule Rec. Cytol. Histol. **59**, 27—56 (1958).

Girbardt, M.: Die Ultrastruktur der Apikalregion von Pilzhyphen. Protoplasma **67**, 413—441 (1969).

Glaumann, H., and Dallner, G.: Lipid composition and turnover of rough and smooth microsomal membranes in rat liver. J. Lipid Res. **9**, 720—729 (1968).

Grassé, P.-P.: Ultrastructure, polarité et reproduction de l'appareil de Golgi. C. R. Acad. Sci. (Paris) **245**, 1278—1281 (1957).

— Carasso, N.: Ultrastructure of the Golgi apparatus in protozoa and metazoa (somatic and germinal cells). Nature (Lond.) **179**, 31—33 (1957).

— — Favard, P.: Les dictyosomes (appareil de Golgi) et leur ultrastructure. C. R. Acad. Sci. (Paris) **241**, 1243—1245 (1955).

Grimstone, A. V.: Cytoplasmic membranes and the nuclear membrane in the flagellate *Trichonympha*. J. biophys. biochem. Cytol. **6**, 369—378 (1959).

GROVE, S. N., BRACKER, C. E., MORRÉ, D. J.: Cytomembrane differentiation in the endo-plasmic reticulum-Golgi apparatus-vesicle complex. Science **161**, 171—173 (1968).
— — — An ultrastructural basis for hyphal tip growth in *Pythium ultimum*. Amer. J. Botany **57**, 245—266 (1970).
GURR, M. I., FINEAN, J. B., HAWTHORNE, J. N.: The phospholipids of liver cell fractions. I. The phospholipid composition of the liver-cell nucleus. Biochim. biophys. Acta (Amst.) **70**, 406—416 (1963).
HALL, W. T., WITKUS, E. R.: Some effects on the ultrastructure of the root meristem of *Allium cepa* by 6 aza uracil. Exp. Cell Res. **36**, 494—501 (1964).
HAN, S. S.: An electron microscopic and autoradiographic study of the rat parotid gland after actinomycin D administration. Amer. J. Anat. **120**, 161—184 (1967).
HAVELKOVÁ, M., MENŠÍK, P.: The Golgi apparatus in the regenerating protoplasts of *Schizosaccharomyces*. Naturwissenschaften **53**, 562 (1966).
HEALY, P. L., JENSEN, W. A.: Changes in ultrastructure and histochemistry accompanying floral induction in the shoot apex of *Pharbitus*. Amer. J. Botany **52**, 622 (1965).
HELMINEN, H. J., ERICSSON, J. L. E.: Studies on mammary gland involution. I. On the ultrastructure of the lactating mammary gland. J. Ultrastruct. Res. **25**, 193—213 (1968).
HEMMES, D. E., HOHL, H. R.: Ultrastructural changes in directly germinating sporangia of *Phytophora parasitica*. Amer. J. Botany **56**, 300—313 (1969).
HIBBARD, H., LAVIN, G. I.: A study of the Golgi apparatus in chicken gizzard epithelium by means of the quartz microscope. Biol. Bull. **89**, 157—161 (1945).
HICKS, R. M.: The function of the Golgi complex in transitional epithelium. Synthesis of the thick cell membrane. J. Cell Biol. **30**, 623—643 (1966).
HIRSCH, G. C.: The "Golgi apparatus" or the lamellar-vacuolar field in the electron micro-scope. Sym. Soc. Cell. Chem. **14**, 197—206 (1963).
HOKIN, L. E.: Dynamic aspects of phospholipids during protein synthesis. Int. Rev. Cytol. **23**, 187—208 (1968).
IMAI, H., COULSTON, F.: Ultrastructural studies of absorption of methoxychlor in the jejunal mucosa of the rat. Exp. molec. Path. **8**, 135—158 (1968).
JAMIESON, J. D., PALADE, G. E.: Role of the Golgi complex in the intracellular transport of secretory proteins. Proc. nat. Acad. Sci. (Wash.) **55**, 424—431 (1966).
— — Intracellular transport of secretory proteins in the pancreatic exocrine cell. I. Role of the peripheral elements of the Golgi complex. J. Cell Biol. **34**, 577—596 (1967a).
— — Intracellular transport of secretory proteins in the pancreatic exocrine cell. II. Transport to condensing vacuoles and zymogen granules. J. Cell Biol. **34**, 597—615 (1967b).
— — Intracellular transport of secretory proteins in the pancreatic exocrine cell. III. Dissociation of intracellular transport from protein synthesis. J. Cell Biol. **39**, 580—588 (1968a).
— — Intracellular transport of secretory proteins in the pancreatic exocrine cell. IV. Metabolic requirements. J. Cell Biol. **39**, 589—603 (1968b).
JOHNSON, H. H.: Centrioles and other cytoplasmic components of the male germ cells of the *Gryllidae*. Z. Wiss. Zool. **140**, 115—116 (1931).
JONES, A. L., ARMSTRONG, D. T.: Increased cholesterol biosynthesis following phenobar-bital induced hypertrophy of agranular endoplasmic reticulum in liver. Proc. Soc. exp. Biol. Med. **119**, 1136—1139 (1965).
— FAWCETT, D. W.: Hypertrophy of the agranular endoplasmic reticulum in hamster liver induced by phenobarbital (with a review on the functions of this organelle in liver). J. Histochem. Cytochem. **14**, 215—232 (1966).
KARRER, H. E., COX, J.: Electron microscopic observations on developing chick embryo liver. The Golgi complex and its possible role in the formation of glycogen. J. Ultra-struct. Res. **4**, 149—165 (1960).
KATER, J. Mc A.: Morphological aspects of protoplasmic and deutoplasmic synthesis in oocytes of *Cambarus*. Z. Zellforsch. **8**, 186—221 (1928).
KEENAN, T. W., MORRÉ, D. J.: Phospholipid class and fatty acid composition of Golgi apparatus isolated from rat liver and comparison with other cell fractions. Biochemistry **9**, 19—25 (1970).

KESSEL, R. G.: Annulate lamellae. J. Ultrastruct. Res. **24** (Suppl. 10), 1—82 (1968a).
— Electron microscope studies on developing oocytes of a Coelenterate medusa with special reference to vitellogenesis. J. Morph. **126**, 211—248 (1968b).
KIERMAYER, O.: Dictyosomes in *Micrasterias* and their "division". J. Cell Biol. **35**, 68A (1967).
— Elektronenmikroskopische Untersuchungen zum Problem der Cytomorphogenese von *Micrasterias denticulata* Breb. I. Allgemeiner Überblick. Protoplasma **69**, 97—132 (1970).
KUFF, E. L., DALTON, A. J.: Biochemical studies of isolated Golgi membranes. In: Subcellular Particles (Ed.: HAYASHI, T.), pp. 114—127. New York: Ronald Press 1959.
LEBLOND, C. P.: General conclusions. In: The Use of Radioautography in Investigating Protein Synthesis (Ed.: LEBLOND, C. P., WARREN, K. B.) **4**, 321—339. New York-London: Academic Press 1965.
LØVLIE, A., BRÅTEN, T.: On the division of cytoplasm and chloroplast in the multicellular green alga *Ulva mutabilis* Føyn. Exp. Cell Res. **51**, 211—220 (1968).
MAHLEY, R. W., GRAY, M. E., HAMILTON, R. L., LEQUIRE, V. S.: Lipid transport in liver. II. Electron microscopic and biochemical studies of alterations in lipoprotein transport induced by cortisone in the rabbit. Lab. Invest. **19**, 358—369 (1968).
MANTON, I.: On a reticular derivative from Golgi bodies in the meristem of *Anthoceros*. J. biochem. biophys. Cytol. **8**, 221—231 (1960).
— Observations on scale production in *Prymnesium parvum*. J. Cell Sci. **1**, 375—380 (1966a).
— Observations on scale production in *Pyramimonas amylifera* Conrad. J. Cell Sci. **1**, 429—438 (1966b).
— Further observations on scale formation in *Chrysochromulina chiton*. J. Cell Sci. **2**, 411—418 (1967).
MARUYAMA, K.: Cyclic changes of the Golgi body during microsporogenesis in *Tradescantia paludosa*. Cytologia (Tokyo) **30**, 354—374 (1965).
— GAY, H., KAUFMANN, B. P.: Development of the Golgi body in the *Tradescantia* pollen grain. Amer. J. Botany **49**, 662 (1962).
MASUDA, H.: Structural localization of some phosphatases in the Golgi region of cultivated cells. Cytologia (Tokyo) **32**, 463—473 (1967).
MATILE, PH., MOOR, H.: Vacuolation: Origin and development of the lysosomal apparatus in root-tip cells. Planta (Berl.) **80**, 159—175 (1968).
— — ROBINOW, C. F.: Yeast cytology. In: The Yeasts (Ed.: ROSE, A. H., HARRISON, J. S., pp. 219—302. New York-London: Academic Press 1969.
MAUL, G. G.: Golgi-melanosome relationship in human melanoma *in vitro*. J. Ultrastruct. Res. **26**, 163—176 (1969).
MERCER, E. H.: The evolution of intracellular phospholipid membrane systems. In: The Interpretation of Ultrastructure (Ed.: HARRIS, R. J. C.), pp. 369—384. New York-London: Academic Press 1962.
MOLLENHAUER, H. H.: An intercisternal structure in the Golgi apparatus. J. Cell Biol. **24**, 504—511 (1965a).
— Transition forms of Golgi apparatus secretion vesicles. J. Ultrastruct. Res. **12**, 439—446 (1965b).
— MORRÉ, D. J.: Golgi apparatus and plant secretion. Ann. Rev. Plant Physiol. **17**, 27—46 (1966a).
— — Tubular connections between dictyosomes and forming secretory vesicles in plant Golgi apparatus. J. Cell Biol. **29**, 373—376 (1966b).
— — BERGMAN, L.: Homology of form in plant and animal Golgi apparatus. Anat. Rec. **158**, 313—318 (1967).
— WHALEY, W. G.: An observation on the functioning of the Golgi apparatus. J. Cell Biol. **17**, 222—225 (1963).
MOORE, R. T., McALEAR, J. H.: Fine structure of Mycota. 4. The occurrence of the Golgi dictyosome in the fungus *Neobulgaria pura* (Fr.) Petrak. J. Cell Biol. **16**, 131—141 (1963).
MORRÉ, D. J.: *In vivo* incorporation of radioactive metabolites by dictyosomes and other cell fractions of onion stem. Plant Physiol. **45**, 791—799 (1970).

Morré, D. J., Hamilton, R. L., Mollenhauer, H. H., Mahley, R. W., Cunningham, W. P., Cheetham, R. D., Lequire, V. S.: Isolation of a Golgi apparatus-rich cell fraction from rat liver. I. Method and morphology. J. Cell Biol. **44**, 484—490 (1970).

— Jones, D. D., Mollenhauer, H. H.: Golgi apparatus mediated polysaccharide secretion by outer rootcap cells of *Zea mays*. I. Kinetics and secretory pathway. Planta (Berl.) **74**, 286—301 (1967).

— Keenan, T. W., Mollenhauer, H. H.: Golgi apparatus function in membrane transformations and product compartmentalization: studies with cell fractions from rat liver. Proc. 1st Intern. Sym. Cell Biol. Cytopharm., Venice, Italy 1969 (in press).

— Merlin, L. M., Keenan, T. W.: Localization of glycosyl transferase activities in a Golgi apparatus-rich fraction isolated from rat liver. Biochem. biophys. Res. Commun. **37**, 813—819 (1969).

— Nyquist, S., Rivera, E.: Lecithin biosynthetic enzymes of onion stem and the distribution of phosphorylcholine-cytidyl transferase among cell fractions. Plant Physiol. **45**, 800—804 (1970).

Moulé, Y.: Biochemical characterization of the components of the endoplasmic reticulum in rat liver cell. In: Structure and Function of the Endoplasmic Reticulum in Animal Cells, Proc. Fed. European Biochem. Soc., 4th (Ed.: Gran, F. C.) 1—12. New York-London: Academic Press 1968.

Murakami, S., Morimura, Y., Takamiya, A.: Electron microscopic studies along cellular life cycle of *Chlorella ellipsoidea*. "Studies on Microalgae and Photosynthetic Bacteria". Plant Cell Physiol. 65—83 (1963).

Nagy, J., Fridvalazky, L.: Dictyosome-nuclei relationships in *Botrydium granulatum* (Crysophyta). In: Electron Microscopy 1968 (Ed.: Cocciarelli, D. J.) **2**, 423—424. European Regional Conference on Electron Microscopy, 4th Rome: Tipografia Poliglotta Vaticana 1968.

Neutra, M., Leblond, C. P.: Synthesis of the carbohydrate of mucus in the Golgi complex as shown by electron microscope radioautography of goblet cells from rats injected with glucose-H³. J. Cell Biol. **30**, 119—136 (1966).

Northcote, D. H.: Structure and function of plant cell membranes. Brit. med. Bull. **24**, 107—112 (1968).

Novikoff, A. B.: Enzyme localization, and ultrastructure of neurons. In: The Neuron (Ed.: Hyden, H.) 255—319. Amsterdam: Elsevier Publ. Co. 1967.

— Essner, E., Goldfischer, S., Heus, M.: Nucleosidediphosphatase activities of cytomembranes. In: The Interpretation of Ultrastructure (Ed.: Harris, R. J. C.) 149—192. New York-London: Academic Press 1962.

— Shin, W. Y.: The endoplasmic reticulum in the Golgi zone and its relations to microbodies, Golgi apparatus and autophagic vacuoles in rat liver cells. J. Microscop. **3**, 187—206 (1964).

Odor, D. L.: Electron microscopic studies on ovarian oocytes and unfertalized tubal ova in the rat. J. biophys. biochem. Cytol. **7**, 567—574 (1960).

Ovtracht, L.: Ultrastructure des cellules sécrétrices de la glande multifide de l'escargot. J. Microscop. **6**, 773—790 (1967).

— Morré, D. J., Merlin, L. M.: Isolement de l'appareil de Golgi d'une glande sécrétrice de mucopolysaccharides de l'escargot (*Helix pomatia*). J. Microscop. **8**, 989—1002 (1969).

Palade, G. E.: The endoplasmic reticulum. J. biophys. biochem. Cytol. **2** (Suppl.), 85—98 (1956).

Perroncito, A.: Contributions à chromidies et appareil réticulaire interne dans les cellules spermatiques: le phénomène de la dictyokinése. Arch. ital. Biol. **54**, 307—345 (1910).

Pickett-Heaps, J. D.: Ultrastructure and differentiation in *Chara* (Fibrosa). IV. Spermatogenesis. Aust. J. biol. Sci. **21**, 655—690 (1968).

Pfleger, R. C., Anderson, N. G., Snyder, F.: Lipid class and fatty acid composition of rat liver plasma membranes isolated by zonal centrifugation. Biochemistry **7**, 2826—2833 (1968).

Policard, A., Bessis, M., Breton, J., Thiéry, J. P.: Polarité de la centrosphere et des corps de Golgi dans les leucocytes des mammifères. Exp. Cell Res. **14**, 221—223 (1958).

Pollister, A. W.: Cytoplasmic phenomena in the spermatogenesis of *Gerris*. J. Morph. **49**, 455—506 (1930).

Porter, K. R.: Cytoplasmic microtubules and their functions. In: Principles of Biomolecular Organization, Ciba Foundation Symp. (Ed.: Wolstenholme, G. E. W., O'Conner, M.), pp. 308—345. London: J. and A. Churchill 1966.

— The ground substance: observations from electron microscopy. In: The Cell (Ed.: Brachet, J., Mirsky, A. E.) Vol. 2, pp. 621—675. New York: Academic Press 1961.

— Kenyon, K., Badenhausen, S.: Specialization of the unit membrane. Protoplasma **63**, 262—274 (1967).

Rambourg, A., Hernandez, W., Leblond, C. P.: Detection of complex carbohydrates in the Golgi apparatus of rat cells. J. Cell Biol. **40**, 395—414 (1969).

Redman, C. M., Sabatini, D. D.: Vectoral discharge of peptides released by puromycin from attached ribosomes. Proc. nat. Acad. Sci. (Wash.) **56**, 608—615 (1966).

— Siekevitz, P., Palade, G. E.: Synthesis and transfer of amylase in pigeon pancreatic microsomes. J. biol. Chem. **241**, 1150—1158 (1966).

Robbins, E., Gonatas, N. K.: Histochemical and ultrastructural studies on HeLa cell cultures exposed to spindle inhibitors with special reference to the interphase cell. J. Histochem. Cytochem. **12**, 704—711 (1964a).

— — The ultrastructure of a mammalian cell during the mitotic cycle. J. Cell Biol. **21**, 429—463 (1964b).

Roth, L. E., Wilson, H. J., Chakraborty, I.: Anaphase structure in mitotic cells typified by spindle elongation. J. Ultrastruct. Res. **14**, 460—483 (1966).

Rouser, G., Nelson, G. J., Fleischer, S., Simson, G.: Lipid composition of animal cell membranes, organelles and organs. In: Biological Membranes. Physical Fact and Function (Ed.: Chapman, D.) 5—69. New York-London: Academic Press 1968.

Sakai, A., Shigenaka, M.: Behavior of cytoplasmic membranous structures in spermatogenesis of the grasshopper, *Atractomorpha bedeli*. Bolivar. Cytologia (Tokyo) **32**, 72—86 (1967).

Schnepf, E.: Quantitative Zusammenhänge zwischen der Sekretion des Fangschleimes und den Golgi-Strukturen bei *Drosophyllum lusitanicum*. Z. Naturforsch. **16**b, 605—610 (1961).

— Transport by compartments. In: Transport and Distribution of Matter in Higher Plants (Ed.: Mothes, K., Muller, E., Nelles, A., Neumann, D.), pp. 39—49. Berlin: Akademie-Verlag 1968a.

— Zur Feinstruktur der schleimsezernierenden Drüsenhaare auf der Ochrea von *Rumex* und *Rheum*. Planta (Berl.) **79**, 22—34 (1968c).

— Sekretion und Exkretion bei Pflanzen. Protoplasmatologia, Handbuch der Protoplasmaforschung **8**, 1—181 (1969).

— Koch, W.: Golgi-Apparat und Wasserausscheidung bei *Glaucocystis*. Z. Pflanzenphys. **55**, 97—109 (1966a).

— — Über die Entstehung der pulsierenden Vacuolen von *Vacuolaria virescens* (Chloromonadophyceae) aus dem Golgi-Apparat. Arch. Mikrobiol. **54**, 229—236 (1966b).

Schiff, J. A., Epstein, H. T.: The continuity of the chloroplast in *Euglena*. In: Reproduction: Molecular, Subcellular and Cellular (Ed.: Locke, M.), pp. 131—189. New York-London: Academic Press 1965.

Siekevitz, P., Palade, G. E., Dallner, G., Ohad, I., Omura, T.: The biosynthesis of intracellular membranes. In: Organizational Biosynthesis (Ed.: Vogel, H. J., Lampen, J. O., Bryson, V.), pp. 331—362. New York-London: Academic Press 1967.

Sievers, A.: Elektronenmikroskopische Untersuchungen zur geotropischen Reaktion. I. Über Besonderheiten im Feinbau der Rhizoide von *Chara foetida*. Z. Pflanzenphys. **53**, 193—213 (1965).

Sjöstrand, F. S.: The ultrastructure of cells as revealed by the electron microscope. Int. Rev. Cytol. **5**, 455—533 (1956).

— A comparison of plasma membrane, cytomembranes, and mitochondrial membranes with respect to ultrastructural features. J. Ultrastruct. Res. **9**, 561—580 (1963).

— Ultrastructure and function of cellular membranes. In: Ultrastructure in Biological Systems. The Membranes. (Ed.: Dalton, A. J., Haguenau, F.), pp. 151—210. New York-London: Academic Press 1968.

SJÖSTRAND, F. S., HANZON, V.: Ultrastructure of Golgi apparatus of exocrine cells of mouse pancreas. Exp. Cell Res. **7**, 415—429 (1954).

SKIPSKI, V. P., BARCLAY, M., ARCHIBALD, F. M., TEREBUS-KEKISH, O., REICHMAN, E. S., GOOD, J. J.: Lipid composition of rat liver cell membranes. Life Sci. **4**, 1673—1680 (1965).

SOROKIN, S. P.: Reconstruction of centriole formation and ciliogenesis in mammalian lungs. J. Cell Sci. **3**, 207—230 (1968).

SRIBNEY, M., KENNEDY, E. P.: The enzymatic synthesis of sphingomyelin. J. biol. Chem. **233**, 1315—2322 (1958).

STEIN, O., STEIN, Y.: Lecithin synthesis, intracellular transport, and secretion in rat liver. IV. A radioautographic and biochemical study of choline-deficient rats injected with choline-^{3}H. J. Cell Biol. **40**, 461—483 (1969).

THIÉRY, J.-P.: Mise en évidence des polysaccharides sur coupes fines en microscopie électronique. J. Microscop. **6**, 987—1018 (1967).

— Mise en évidence de muco- et glycoproteines dans l'appareil de Golgi. In: Electron Microscopy 1968 (Ed.: BOCIARELLI, D. S.) **2**, 59—60. European Regional Conference of Electron Microscopy, 4th. Rome: Tipographica Poliglotta Vaticana 1968.

— Role de l'appareil de Golgi dans la synthèse des mucopolysaccharides étude cytochimique. I. Mise en évidence de mucopolysaccharides dans les vésicules de transition entre l'ergastoplasme et l'appareil de Golgi. J. Microscop. **8**, 689—708 (1969).

TURNER, F. R., WHALEY, W. G.: Intercisternal elements of the Golgi apparatus. Science **147**, 1303—1304 (1965).

VAN DER WOUDE, W. J., MORRÉ, D. J., BRACKER, C. E.: A role for secretory vesicles in polysaccharide biosynthesis. Proc. XI Int. bot. Congr. (Abstracts), Seattle, Washington, 226, 1969.

VAN DER WOUDE, W. J., MORRÉ, D. J., BRACKER, C. E.: Isolation and characterization of secretory vesicles in germinating pollen of *Lilium longiflorum*. J. Cell. Sci. (in press).

WARD, R. T.: Formation of Golgi bodies during maturation of oocytes in *Rana pipiens*. Anat. Rec. **151**, 430 (1965).

— WARD, E.: The multiplication of Golgi bodies in the oocytes of *Rana pipiens*. J. Microscop. **7**, 1007—1020 (1968).

WARSHAWSKY, H., LEBLOND, C. P., DROZ, B.: Synthesis and migration of proteins in the cells of the exocrine pancreas as revealed by the specific activity determination from radioautographs. J. Cell Biol. **16**, 1—23 (1963).

WARTENBURG, H.: Elektronenmikroskopische und histochemische Studien über die Oogenese der Amphibieneizelle. Z. Zellforsch. **58**, 427—486 (1962).

WEISBLUM, B., HERMAN, L., FITZGERALD, P. J.: Changes in pancreatic acinar cells during protein deprivation. J. Cell Biol. **12**, 313—327 (1962).

WERZ, G.: Elektronenmikroskopische Untersuchungen zur Genese des Golgi-Apparates (Dictyosomen) und ihrer Kernabhängigkeit bei *Acetabularia*. Planta (Berl.) **63**, 366—381 (1964).

WHALEY, W. G.: Proposals concerning replication of the Golgi apparatus. In: Probleme der Biologischen Reduplication (Ed.: P. SITTE), pp. 340—371. Berlin-Heidelberg-New York: Springer 1966.

— KEPHART, J. E., MOLLENHAUER, H. H.: The dynamics of cytoplasmic membranes during development. In: Cellular Membranes in Development (Ed.: LOCKE, M.) 135—173. New York-London: Academic Press 1964.

WIENKE, E. C., CAVAZOS, F., HALL, D. G., LUCAS, F. V.: Ultrastructure of the human endometrial stroma cell during the menstrual cycle. Amer. J. Obstet. Gynec. **102**, 65—77 (1968).

WILSON, E. B.: The cell in development and heredity, pp. 165—168. New York: Macmillan 1925.

WINDHORST, D. B., ZELICKSON, A. S., GOOD, R. A.: Shediak-Higashi syndrome: Hereditary gigantism of cytoplasmic organelles. Science **151**, 81—83 (1966).

WISCHINITZER, S.: An electron microscopic study of the Golgi apparatus of amphibian oocytes. Z. Zellforsch. **57**, 202—212 (1962).

Wise, G. E.: Cytochemistry of the Golgi apparatus in *Amoeba proteus*. J. Cell Biol. **43**, 159a (1969).

Yamamoto, K., Onozato, H.: Electron microscope study on the growing oocyte of the goldfish during the first growth phase. Mem. Fac. Fisheries, Hokkaido Univ. **13**, 79—106 (1965).

Yamamoto, M.: Electron microscopy of fish development. III. Changes in the ultrastructure of the nucleus and cytoplasm of the oocyte during its development in *Oryzias latipes*. J. Fac. Sci. Univ. (Tokyo) **10**, 335—346 (1964).

Yamamoto, T.: On the thickness of the unit membrane. J. Cell Biol. **17**, 413—422 (1963).

Yunghans, W., Keenan, T. W., Morré, D. J.: Isolation of a Golgi apparatus-rich cell fraction from rat liver. III. Lipid and protein composition. Exp. molecul. Path. **12**, 36—45 (1970).

Zaar, K., Schnepf, E.: Membranfluß und Nukleosiddiphosphatase-Reaktion in Wurzelhaaren von *Lepidium sativum*. Planta (Berl.) **88**, 224—239 (1969).

Zeigel, R. F., Dalton, A. J.: Speculations based on the morphology of the Golgi system in several types of protein secreting cells. J. Cell Biol. **15**, 45—54 (1962).

Origin and Continuity of Cell Vacuoles

Roger Buvat

*Laboratoire de Cytologie Végétale, Faculté des Sciences,
Université d'Aix — Marseille*

I. Introduction

The property of forming vacuoles has been recognized as early as 1835, by DUJARDIN, as one peculiar to the living cytoplasm (the "Sarcode" of DUJARDIN). Through NAEGELI's work (1844) vacuoles became definitively regarded as normal morphological components of the plant protoplasm. With the advent of modern microscopes, after ERNST ABBE's work, many cytologists have aimed at elucidating the structure of vacuoles, which formed the subject of a long and well-known essay by HUGO DE VRIES (1885).

DE VRIES demonstrated that the osmotic properties of the cells result from those of the vacuoles that they contain and that those vacuoles ("tonoplasts") are limited by a semipermeable membrane or "tonoplasma". He regarded the vacuoles as permanent and differentiated cytoplasmic components. This concept brings up the question of the *origin* of those organelles. DE VRIES and his student F. C. WENT (1888) found that small embryonic vacuoles are formed by strangulation of preexisting ones, and inferred that these organelles form continuous lineages. PFEFFER (1890) on the other hand observed *de novo* formation of vacuoles in plasmods of *Chondrioderma difforme*, and noticed that these newly formed vacuoles may fuse with each other or with other plasmodial vacuoles. He did not accept the idea of the continuity of vacuoles and regarded the tonoplasma as a perivacuolar modification of the cytoplasm. (Nowadays, we know that it is really a lipoprotein "plasma membrane").

At the beginning of the century, GUILLIERMOND (1903, 1906, 1908) and then P. A. DANGEARD (1916) showed that all plant cells have enclaves selectively stainable by vital stains such as bright cresil-blue and neutral red. This coloration is due to the presence in vacuoles of colloïdal substances, one of which, the *metachromatin*, is mainly found in algae and fungi.

The use of vital staining has then allowed to study *the vacuole changes* during cell differentiation in plants (P. A. DANGEARD, 1916; GUILLIERMOND, 1920, 1924, 1925, 1929, 1930; P. DANGEARD, 1923). These investigations give many examples of the continuity of vacuoles, but do not really deal with the problem of their origin in the meristematic cells.

In animal cells, beside the digestive vacuoles of the protozoa and beside the pulsatile vacuoles, the hydrophilous enclaves have often been regarded as transitory or non-existent. However, PARAT and PAINLEVÉ (1924, 1925), and PARAT (1928) have shown the constant existence of inclusions stained by neutral red in animal cells. These inclusions, however, never reach the size of plant vacuoles. Some of them were described as "secretion grains" (Arnold, 1914; von Mollendorf, 1918) without homology with vacuoles. Parat (1928) regards these animal cell vacuoles as really homologous with those in plant cells, and considers the Golgi apparatus or the reticular systems assimilated to it, as being nothing else but vacuoles. Later analyses, using the electron microscope, have shown that the Golgi apparatus does not confine itself to vacuoles, but that it probably shares the genesis of some of them.

About 1930, the morphological development of the vacuolar system was described satisfactorily enough and proved some possibilities of continuity in vacuoles, but the problem of the origin of vacuoles of the meristematic plant cells and of animal cells was left without any decisive answer.

II. The Normal Vacuolar Apparatus in Plant Cells

Differentiated plant cells generally contain a large central vacuole, the remaining protoplasm being reduced to a thin film, pushed aside against the cell wall.

This vacuole derives from the development of the very different vacuolar apparatus of the meristematic cell, as it is easily observed in a young root of wheat or barley, vitally stained with neutral red (Fig. 1). In the proliferating cells of the meristem, this apparatus is in the form of a thin reticular network, which swells in globular vacuoles. Later, these vacuoles fuse in a large single enclave (Fig. 1a—f).

A. Origin of the Vacuolar System in the Meristematic Cells; Possible Relations with the Endoplasmic Reticulum

Ultra-thin sections of highly meristematic plant cells, like those of foliar initia in the vegetative point, or those in the most rapidly proliferating parts of radicular meristems, are sometimes deprived of vacuole profiles, whatever fixation techniques are used. Therefore, some scientists have asserted that the most meristematic cells are deprived of any vacuolar system (Frey-Wyssling and Mühlethaler, 1965). This is one of the main differences between meristematic and embryonic cells. Indeed, many zygotes contain large vacuoles which become reduced in volume during segmentation (Schulz and Jensen, on *Capsella*, 1968; Jensen, on *cotton*, 1968). The mechanism of this reduction is as yet unknown (Jensen, 1968). Then, it would appear that the study of the derivatives of initiatory cells should give informations on the origin of the vacuoles. Indeed, many papers report the formation of vacuoles but from different origins.

The first suggestions are concerned with the behaviour of the endoplasmic reticulum (E. R.) in the squamules that develop at the base of young *Elodea* leaves (Buvat, 1958). The cells of these organs contain concentric or parallel layers of E. R., essentially of a *smooth* type. These saccules form series of dilatations, linked by narrow strands, that look like vacuoles (Fig. 2).

On the other hand, in various roots, osmium fixation allows to discern flat profiles, connected with ribosomes, displaying plainly the *rough* type of E. R.; these elements are closely related to other profiles of a *smooth* type which appear either

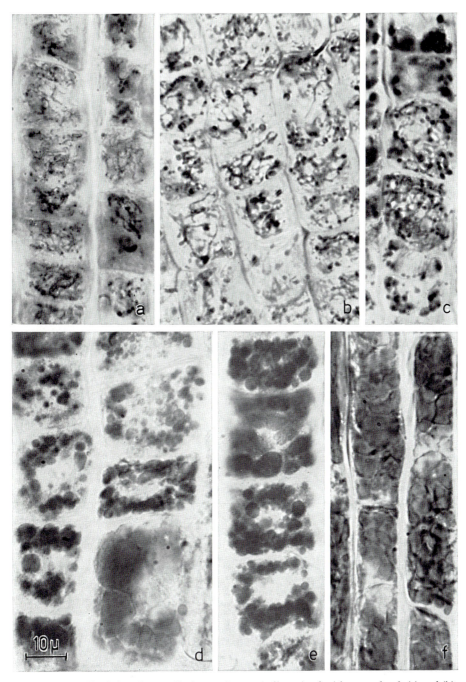

Fig. 1. a—f. Cells of the wheat radicular meristem vitally stained with neutral red. (a) and (b): zone of maximum proliferation; thinly reticulated vacuoles. (c): Beginning of partial swellings of the vacuole network. (d) and (e): Transformation of the network into globular vacuoles and beginning of fusions between these vacuoles. (f): Stage just preceding the fusion into one vacuole in the elongation zone of the root. (× 1,100).

narrow or more or less dilated (Buvat, 1965; Poux, 1962 a; Fig. 3, see also Fig. 4). Continuities between the two types of profiles, in agreement with the early descriptions of E. R. in animal cells by Palade (1956), have been noticed by Poux (1962 b) and more recently by Mesquita (1969) who concludes that vacuoles originate from the E. R. of the meristematic cells.

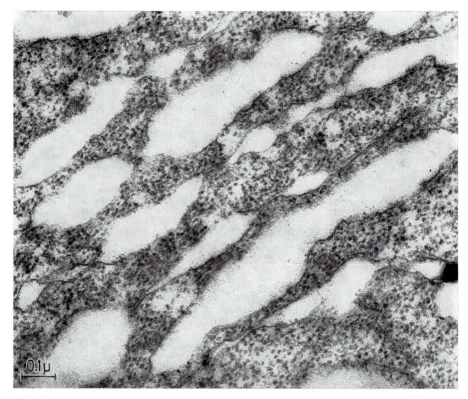

Fig. 2. Cytoplasmic area of cell of *Elodea canadensis* squamula. Roughly parallel profiles of smooth E. R., forming dilatations similar to meristematic vacuoles (from: Buvat, 1958). (× 90,000)

One may easily be convinced while examining the smooth profiles in cells of such roots in their somewhat restricted narrow meristematic region, that the enclaves limited by "smooth" membranes represent young vacuoles growing and assuming globular shapes at the very beginning of the differentiation. These statements lead to admit that cytoplasmic enclaves, saccule-shaped and flattened in the beginning and therefore similar to portions of *smooth* endoplasmic reticulum, represent a primordial inframicroscopic state of the vacuolar system.

In the same type of cells vital neutral red staining reveals *a thinly reticulated* vacuolar system. The profiles in question are therefore thin sections of a reticular system, likely to swell in many points, before it breaks up into independent, globular vacuoles.

It seems likely that these enclaves show, at the beginning, continuities with the typical rough E. R., but the rough E. R. is only in *exceptions* subjected to the swelling process and it seems that both systems become independent very early, even if we admit that they may have momentary, nonpermanent, relations here and there (see WHALEY,

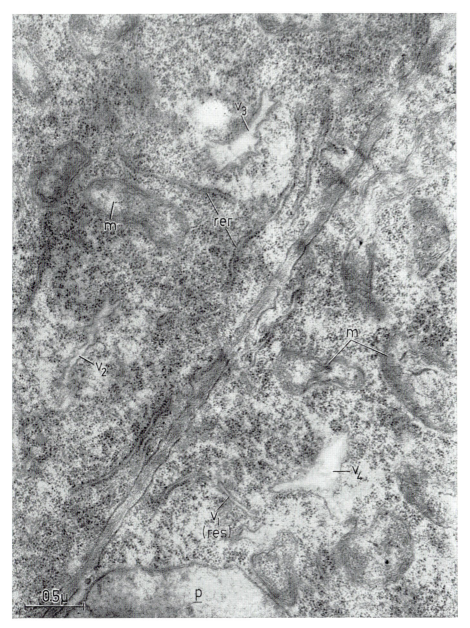

Fig. 3. *Cuburbita pepo* root: meristematic cells. Birth of vacuoles ($V_{1-2-3-4}$) by hypertrophy of cavities originally similar to the "smooth" E. R. (V_1, *res*); (*m*) mitochondria; (*p*) proplastid; (*rer*) rough E. R. $\times$ 32,000

9*

KEPHART, and MOLLENHAUER, 1964, p. 156). This idea is confirmed by GIFFORD and STEWART (1967) who think that the smooth E. R. is in fact the tonoplast, and not a kind of E. R. Such independence of the true (or rough) E. R. and the growing vacuolar system is also postulated by MANTON (1962) for the stellate vacuoles of the meristematic cells of *Anthoceros*.

A common origin of both systems on the other hand is suggested by preparations of roots fixed with potassium permanganate (BUVAT, 1961). It is known that this fixative tends to let the vacuoles exude part of their liquid and to give them amoeboid profiles that probably aggravate the stellate appearance that they may have in living cells (Fig. 4). The sharp or flat extensions of the observed profiles are not entirely artifacts and show the reticular origin of the young vacuoles. But it is known that permanganate fixation does not allow to distinguish between the rough and smooth

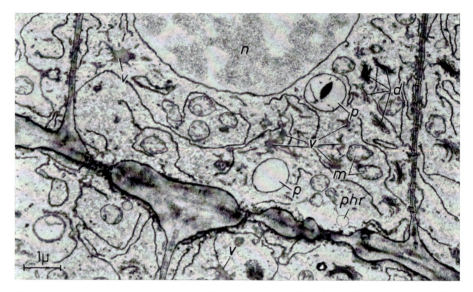

Fig. 4. Meristematic cell of barley root (*Hordeum sativum*) Fixation: KMnO₄. Many stellate profiles are sections of the thin network from which the vacuoles (*v*) will develop. These enclaves are easily distinguished from the very narrow profiles of E. R., from the proplastids (*p*) and from enclaves with a single membrane, the phragmosomes (*phr*). (*d*) dictyosomes; (*m*) mitochondria; (*n*) nucleus. × 10,000

types of endoplasmic reticulum. This reticular origin of meristematic vacuoles has been confirmed by BOWES (1965), in the shoot apex of *Glechoma hederacea* and by ENGLEMAN (1966) in cotton-embryo.

The permanganate method usually keeps the vacuolar content in the form of a homogeneous mass. Within a certain cell type, this material is the more electron-opaque the more the concentration of the vacuole content rises. These differences allow us to distinguish the profiles of young vacuoles from those of the typical endoplasmic reticulum. Thus ist is possible to follow, at the very beginning of the differentiation, the development of those young vacuoles through hypertrophy, dilution of their content, coalescence and, finally, fusion into a single vacuole.

B. Vacuoles Arising in the Phragmoplast

In the subterminal proliferating zone of the roots, after the *cell-plate* elaboration, at the end of the telophase the substance of the *phragmoplast* coming from the preceding achromatic spindle and that assumed a gel-like consistence progressively turns like the cytoplasm of the mother-cell that surrounded the mitotic system. During this transformation, in the mass of the phragmoplast, which was relatively homogeneous at the beginning, the different types of figurative components of the cytoplasm appear. Some of them, if not all, seem to form *in situ*, and especially tiny vacuoles become visible, beside typical profiles of "rough" endoplasmic reticulum. After permanganate fixation (Fig. 5) these enclaves reveal a more or less irregular shape with concave sides and flat extensions. *From the very beginning*, they appear surrounded with a plasma membrane which looks identical to those of the typical E. R. But their content is, from the start, electron-dense and more or less uniform, while the inner spaces of the E. R. are devoid of any manganophilous material.

These observations suggest that some vacuoles may indeed appear *de novo*, that is to say, independently from preexisting vacuoles (also compare GUILLIERMOND's observations on fungi (1925, 1929, 1930) and CASSAIGNE, 1931). But in the phragmoplast these vacuoles apparently come from preexisting closed plasma membranes, secreting, into the space that they surround, a manganophilous material that may be partly or totally identified with the electronegative or hydrophilous *vacuolar colloïd* which alone characterizes the vacuole system and distinguishes it from the E. R.

These new-born vacuoles of the phragmoplasts in the process of solation are like those which are found in the meristematic cells of the same organ. After osmium and glutaraldehyde-osmium fixations, they are found again either with the same appearances or as round-shaped profiles. These profiles are thin sections in a more or less continuous *reticulated* system, as it is proved by the vital staining with neutral red of young corn or barley roots. Evidently the young vacuoles dealt with in this section represent the vacuolar system classically studied by GUILLIERMOND, DANGEARD and their students. Later results (see also Section VI) indicate that these new-born vacuoles derive from the isolation of vesicles coming from the dilatation of smooth parts of the E. R. (provacuoles of MATILE and MOOR, 1968) and likely incorporating contributions from Golgi apparatus. This concept agrees with our own observations (BUVAT, 1957, 1958, 1961, 1965; BUVAT and MOUSSEAU, 1960).

C. Pinocytosis and Vacuoles

POUX (1962) showed that small globular vacuoles are formed by pinocytosis in plant cells. Later on, the formation of hydrophilous enclaves by pinocytosis has been observed frequently and more detailed in roots (POUX, 1962b; see also Fig. 6). These enclaves probably unload into preexisting typical vacuoles. Originally they arise unconnected with the preceding vacuolar system.

However, the problem of the heterogeneity of vacuoles has recently become complicated through some work with *animal cells*.

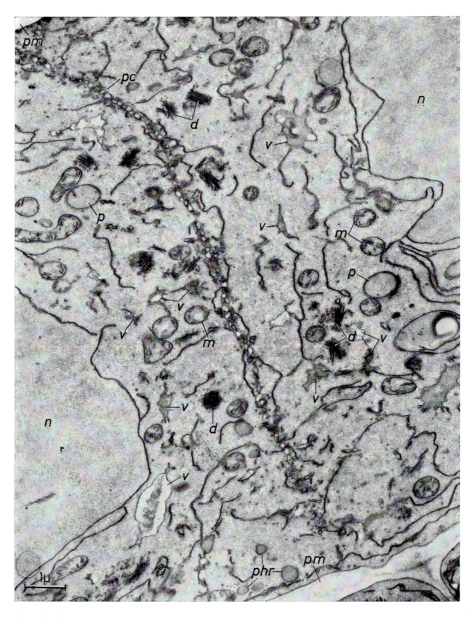

Fig. 5. Wheat root. Cell-plate (*pc*) of a cell in telophase, fixed before its joining with the walls of the mother-cell. (*v*) new-born vacuoles, within the phragmoplast. (*d*) dictyosomes forming a multitude of tiny vesicles that concentrate at the extremities of the cell-plate; (*m*) mito-chondria; (*n*) daughter nuclei; (*p*) proplastids; (*phr*) phragmosomes; (*pm*) walls of the mother-cell. Fixation with KMnO₄. × 11,000

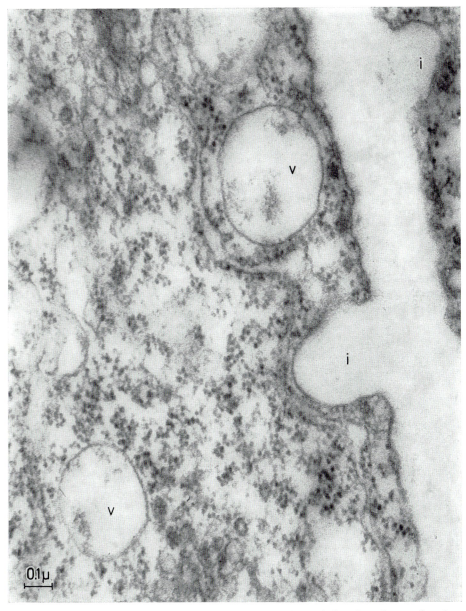

Fig. 6. Barley root cells (*Hordeum sativum*). (*i*) invaginations of the plasmalemma forming vacuoles of pinocytosis (*v*). (× 75,000)

III. Various Types of Vacuoles in Animal Cells

A. Earlier Observations

With a few exceptions, cytologists did not pay much attention to the vacuoles of animal cells. Some of them have even held that animal cells are deprived of vacuoles.

However, vital stains become localized selectively in definite enclaves in a great variety of animal cells (von Möllendorf, 1918). These enclaves have often been regarded as "secretion grains", sometimes solid or very concentrated (Arnold, 1914; von Möllendorf, 1918) or as structures coming from *Golgi zones* in animal cells (Parat, 1928). They remain very discrete and do not reach the huge sizes of plant vacuoles. Parat and Painlevé (1924, 1925) and Parat (1928) have shown that these neutral-red-stained enclaves are essentially located in "Golgi zones" and that they develop proportionally to the metabolic activity. Very precise comparisons between the results of vital staining and the techniques of metallic impregnations (silver and osmium impregnations) have led the authors to consider Golgi apparatus and the vacuole system as homologous of the other components then discernible in Golgi zones representing aggregations of substances (including lipids) resulting from the grouping of vacuoles.

From all the research with the electron microscope, we now know that the Golgi apparatus, made up with dictyosomes and the vesicles that they emit, cannot be reduced simply to a group of vacuoles. What remains of the earlier work, is that the animal vacuoles of Golgi zones, stainable with neutral red, are probably *products of dictyosomes*. Nowadays this result assumes a new importance, because several investigations in cytochemistry and electron microscopy demonstrate that animal cells contain hydrophilous inclusions, rich in hydrolases. This diversity and the relations of origin between the animal vacuoles and Golgi apparatus give a new start to the study of the origin and development of plant cell vacuoles, too.

B. Recent Results

1. The Golgi Vacuoles

Morphological and cytochemical research on dictyosomes, the constitutive units of the Golgi apparatus, demonstrate that these organelles have a well-defined *polarity*. In animals as well as in plants, the saccules of the dictyosome form more or less continually at the so-called proximal pole and scatter by vesiculation at the opposite (distal) pole (Grasse, 1957; Berkaloff, 1963; also compare Morré, Mollenhauer, and Bracker, this volume). This vesiculation produces fluid enclaves often called Golgian vacuoles in animal cells (e. g. Novikoff and Shin, 1964).

Cytoenzymological work confirms this polarity. Novikoff (1963), Goldfischer, Essner, and Novikoff (1964), among others, in animal cells, Poux (1963 a) and Coulomb (1969) in plant cells have shown that the saccules and the vesicles of the dictyosomes have an acid phosphatase activity, all the more marked as it is seen nearer the distal pole. In particular, the *Golgi* vacuoles resulting from the dictyosomes carry hydrolyzing enzymes such as acid phosphatase.

Biochemically at least, these "Golgian" vacuoles approximate the cytoplasmic components described by de Duve (1959) under the name of *lysosomes* that have been the object of intensive research in recent years (e. g. Friend and Farquhar, 1967).

2. The Lysosomes

This interesting period of investigations had its beginning with de Duve's work (1959). Through differential centrifugation of rat-liver homogenate, he obtained a fraction of particles resembling mitochondria in their sedimentation characteristics, but different in biochemical composition (DE DUVE, 1959) and fine structure (NOVIKOFF, BEAUFAY, and DE DUVE, 1956; BREWER and HEATH, 1963). These particles vary in size and are limited by a single-layered cytoplasmic membrane, from 40 to 70 Å thick; they contain a more or less dense material, especially rich in acid hydrolases.

Thus, they resemble a kind of enzyme *storage* (storage granules) probably synthesized by the endoplasmic reticulum. The hydrolytic enzymes are kept isolated, in an active form, from their substrate by the lysosome membrane and are not functional in the condensed granules. These granules are the *primary lysosomes* of DE DUVE and WATTIAUX (1966). They can coalesce, either with a phagocytotic vacuole (phagosome) into which they transport the enzymes necessary to digest the ingested particles, or with a cytoplasmic region previously bound with plasma membranes and intended for destruction. In the first case, fusion produces a *digestive vacuole*, in the second case, an *autophagic vacuole (cytolysome)*. The enzymes of the latter are able to destroy most structures and most cytoplasmic organelles, such as mitochondria, dictyosomes, and parts of E. R. The autophagic vacuoles therefore form "demolition" systems normally compensated by systems of permanent synthesis of new protoplasm. The whole process effectuates the "turnover" or never-ending renewal of the living matter. The digestive vacuoles and the autophagic vacuoles are part of the "*secondary lysosomes*" of DE DUVE and WATTIAUX (1966). When everything possible is hydrolysed in these two categories of vacuoles, there may be left an unusable material and the enclaves that only contain these fragments form the "residual bodies" that are excreted by the cell.

The discovery of lysosomes and of their properties brings new concepts about the origin and the physiological role of the enclaves previously described as *vacuoles*. Far from passively accumulating metabolic products, storage or waste, the vacuoles appear to be components of first order importance for the biochemical function of the living cell and in the controlled renewal of its structures. An example of this physiological regulation of the cytoplasmic activities is given by the mammotrophic hormone-producing cells of the anterior pituitary gland of the rat (SMITH and FARQUHAR, 1966). In these cells, lysosomes and Golgi vesicles perform an elimination of secretion grains and synthesizing structures (E. R. and ribosomes), when the animal enters a postlactating condition.

Indeed, the vacuoles may contain all the enzymes necessary to the complete lysis of the cell. The aggressivity of these enzymes is checked by one surrounding plasma membrane, that is to say, by the equivalent of the "tonoplast". Any alteration of this 50 Å thick membrane, affecting its tightness may be fatal to the cell. This phenomenon occurs during cell *autolysis*, killing the cells by processes that do not denature the hydrolysing enzymes but destroy the tonoplasts.

3. The Peroxisomes

These cytoplasmic particles are also limited by a single-layered plasma membrane. They were first described in liver cells and called *microbodies*, particularly noticeable in the process of regeneration after partial ablations (ROUILLER and BERNHARD, 1956). Usually they contain a dense, sometimes cristalline mass ("core" or "nucleus") (Fig. 7). Peroxisomes have also been identified in kidney cells and in ciliates (*Tetrahymena*).

These enclaves are unequivocally different from lysosomes by their enzymatic contents (Baudhuin, Beaufay, and de Duve, 1965; de Duve and Baudhuin, 1966). Besides various oxidases, e.g. L-α-hydroxy-acid-oxydase, D-amino-acid oxidase and urate oxidase, the peroxisomes contain *catalase* (Leighton et al., 1968) but practically no acid hydrolases typical of lysosomes.

The origin, the evolution and the functions of the peroxisomes are still poorly understood.

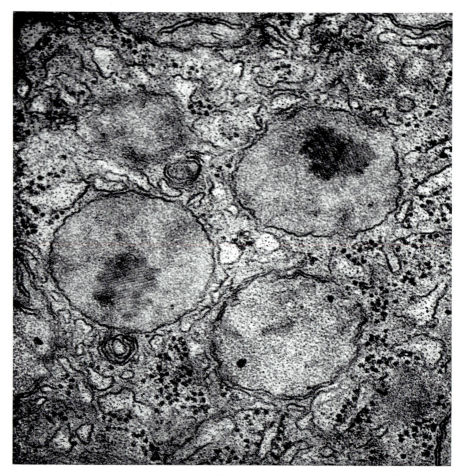

Fig. 7. Infrastructural aspects of microbodies (= peroxisomes) of a rat-liver. Two of them reveal a crystalline "nucleoïd". From De Duve and Baudhuin (1966)

4. The Vacuoles of Pinocytosis

Since Lewis's work (1931, 1937) in microcinematography, it has been recognized that animal cells, as a rule (cf. Holter, 1959) take up liquid droplets from the outer medium by the formation of small cavities at the surface of their plasmalemma and by isolation of these invaginations in the cytoplasm. This process is called pinocytosis. The small liquid enclaves, bound by a plasmalemma-like membrane, proceed through the ectoplasmic regions towards the perinuclear space, getting smaller in volume and it is not impossible that some of them coalesce with other vacuoles. They differ from phagosomes in the lack of solid particles. They are formed after a superficial absorption of soluble or highly dispersed components of the medium, so that these substances seem to be the main object of the pinocytosis. The gradual development of the vacuoles of pinocytosis has been clarified only in a few particular cases. It is known, for instance, that the cells of intestinal villosities can take up lipid microdroplets by pinocytosis (PALAY and KARLIN, 1959). The pinocytotic vacuoles generally meet with the E.R., later the lipid droplets are found again in *Golgi vacuoles* that excrete them into the chyliferous spaces of the villosities.

It is likely that, in many other cases, the useful materials of the vacuoles of pinocytosis are absorbed and metabolized and that the residual content is excreted. The Golgi apparatus seems to play an important role in the reception and distribution of these materials, but the morphological modalities of its intervention will remain a matter of speculations until we have better information on the functional properties of cytoplasmic membranes.

5. The Pulsatile Vacuoles

The pulsatile vacuoles form complex and very particular organelles, doubtless of a nature quite different from the other categories of vacuoles. They are seen in most protists, especially in flagellates (e. g. *Euglena*), in ciliates, in volvocal algae etc. They are liquid enclaves generally located near the base of the flagella or near the cytopharynx (ciliates); they are capable to swell when receiving a liquid secretion from their environment and to contract alternatively in order to excrete their contents.

Electron microscopical work has defined more accurately the complex structures, different according to families, of this system that acts like a cell kidney. The vacuole itself is surrounded with a system of canaliculi, limited by a thin plasma membrane, diversely anastomosed, and in continuity with wider vesicles. The latter ones, located around the central vacuole, receive the excreted liquid and convey it to the vacuole (see CARASSO, FAURÉ-FREMIET, and FAVARD, 1962). The origin of the pulsatile vacuoles does not seem to have been studied in all the systematic groups till now. It seems that the "cytonephros" undergoes a duplication when the protist prepares to divide but the processes underlying this splitting are unknown.

However, in *Vacuolaria virescens* (Chloromonadophyceae), studied by SCHNEPF and KOCH (1966), the pulsatile vacuole arise, in an "excretion plasma", from the fusion of Golgi cisternae. It disappears at each systole; thus it is not a permanent structure.

6. Recapitulation of the Various Morphological or Biochemical Types of Hydrophilous Enclaves of the Cytoplasm

Apart from the pulsatile vacuoles that are very particular excretive systems, the preceding survey leads to identify, at least, eight types of vacuoles in animal cells:

a) *the Golgi vacuoles* resulting from the vesiculation of Golgi saccules at the distal pole of the dictyosomes;

b) *the vacuoles of phagocytosis*, or *phagosomes*, containing in an envelope *coming from the plasmalemma* nutritional particles, meant to be digested;

c) *the vacuoles of pinocytosis* similar to the preceding ones, except that they contain solutions or microdroplets and no solid particles;

d) *the "primary" lysosomes* with concentrated contents (possibly identical with the "secretion grains" or "storage granules" described by earlier authors);

e) *the digestive vacuoles* resulting from the acquisition of acid hydrolysing enzymes through phagosomes;

f) *the autophagic vacuoles* that are active in partial cytoplasmic lysis;

g) *the residual bodies* resulting from the two preceding types after the processes of digestion have been completed; and

h) *the peroxisomes*, characterized by their enzymatic contents uniquely composed of oxidases and catalase.

In the following sections, origin and continuity of these different vacuoles will be discussed.

IV. The Origin of Animal Vacuoles

A. Relations with the Golgi Apparatus

Most authors attribute a primary role to the Golgi apparatus in the formation of various types or animal vacuoles or of secretion grains containing hydrolysing enzymes. Thus, Caro and Palade (1964) have demonstrated by autoradiography that the enzymes of pancreatic zymogen grains are synthesized in the E. R., then gathered and concentrated in the Golgi zone. Novikoff (1963), Goldfischer, Essner, and Novikoff (1964) have identified in electron micrographs, the presence of acid phosphatase in the distal cisternae of dictyosomes and in the Golgi vesicles that they emit, with regard to hepatic cells, kidney cells, and rat neurons. The presence of acid phosphatase in the dictyosomes and in the vesicles which they produce, has been confirmed by many workers (Smith, 1963; Osinchak, 1963) but little evidence was presented on the relationship between Golgi vesicles and vacuoles. However there are many cases where secretion grains or typical lysosomes are constituted from Golgi vesicles. The presence of acid phosphatase in the dictyosomes seems especially obvious when new lysosomes, still little developed, form beside dictyosomes.

The Golgian origin of hydrolysing vacuoles has been shown in a few examples. In the cells of the "fat body" of a Lepidopteran, Locke and Collins (1965) have observed the development of proteinic inclusions of different origin but having all the attributes of Golgi vesicles. Some of them come directly from such vesicles that become rich in proteins and grow through cytoplasmic contributions, integrated in the shape of vesicles of inner pinocytosis. Others result from the the wrapping of cytoplasmic portions, containing rough E. R., by double-layered "isolation membranes", themselves coming from the increase of Golgi vesicles. In these enclaves, either the membranes disappear and not the ribosomes, or the ribosomes are lysed and the inclusions grow like those of the first type and become purely proteinic granules. Lastly, a third kind of sequestration, also by membranes of golgian origin, isolates mitochondria and determines their degradation.

The structures resulting from these processes show the characteristics, either of enclaves with hydrophilous, proteinaceous storage, or of cytolysomes. They are consequently part of the vacuole system in a broad sense and their final bounding membrane is of golgian origin.

In making use of fibroblast cultures (strain L), GORDON, MILLER, and BENSCH (1965) observed the development of phagocytotic vacuoles by labelling the ingested product (protein coacervates + DNA) with colloïdal gold. Thus, they followed up the migration of the "phagosomes" towards the Golgi zone where the process of lysis is always beginning. During intense phagocytosis, the Golgi apparatus becomes very active and its activity is especially discernible with the increasing of the Golgi vesicles that surround the saccules. These vesicles spread around the phagosomes and some figures suggest that fusions occur between both types of enclaves. Anyhow, many Golgi vesicles are soon found within the phagosomes *at the beginning* of the lysing processes.

These latter ones are revealed through a clearer content of the so-formed vacuoles; these vacuoles grow at first by osmotic effect of the products of hydrolysis (intake of water), then they shrink progressively. During this regression they keep on absorbing Golgi vesicles and their content concentrates while becoming opaque. Thus they turn into "dense bodies". These vacuoles reveal acid phosphatase and E-600-resistant esterase activities from the beginning of the lysis, in particular *in* the vesicles that they have enclosed. The dense bodies are equally rich in enzymes (telo-lysosomes of the authors).

By labelling dense bodies with proferrin (sucrosed iron oxide) and by labelling phagosomes with colloïdal gold, GORDON, MILLER and BENSCH showed that the dense bodies coming from the preceding digestion may fuse with new phagosomes to which they bring their enzymatic content.

Finally, the ageing fibroblasts get rich in "autophagic vacuoles" or cytolysomes that develop in the same way as the phagosomes (auto-lysosomes of the authors).

Fig. 8 shows the origin and the evolution of the vacuoles in L-fibroblasts. In this example one can see that the membrane of the phagocytotic vacuoles is at first a fragment of the plasmalemma which is joined by portions of Golgian origin (?) or of dense bodies. In the case of auto-lysosomes, the origin of the sequestration membrane of cytoplasmic portions is less certain but the contribution of Golgian vesicles and of dense bodies is the same. Finally, it should be noted that the telo-lysosomes (dense bodies) from either origin may fuse and thus furnish "residual bodies" meant to be excreted.

The work of GORDON, MILLER, and BENSCH (1965) essentially concerns the phagosomes, but it is likely that pinocytotic vacuoles behave in the same manner. This view is supported by the paper of HIRSCH, FEDORKO and COHN (1968) who show that in macrophages some Golgian vesicles fuse with pinocytotic vacuoles.

Still other examples point to the Golgian origin of hydrolysing enclaves. In this connection BAINTON and FARQUHAR's work (1966) on the rabbit polymorphonuclear leucocytes should be mentioned. Following the genesis of proteinaceous granules (= primary lysosomes) from myeloblasts to the completed polymorphonuclears, the authors showed that the "azurophilous grains" are formed by vesiculations of the saccules in the concave side of the dictyosomes. At first, the isolated vesicles contain a dense inclusion of small size. They fuse, forming larger vacuoles of heterogeneous content, which, after their dispersion out of the Golgi zone, eventually concentrate and become the azurophilous grains. Later on, during the evolution of these leucocytes, another category of enzymatic granules called "specific granules" is formed in a similar way, from saccules of the opposite, convex side, of the same dictyosomes.

Another case of relations between hydrolysing vacuoles and Golgi vesicles is described by Friend and Farquhar (1967) in the epithelial cells of the *vas deferens* of the rat. These cells can absorb a protein, horseradish peroxidase, by a pinocytotic process in the form of *large* coated vesicles, which become smooth and fuse, pro-

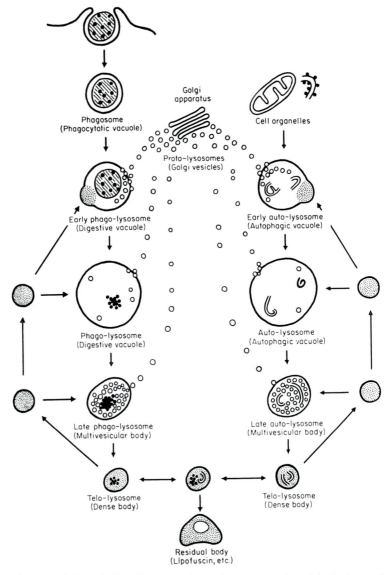

Fig. 8. Diagram of the relations between the various categories of inclusions related to digestive vacuoles in L fibroblasts in tissue cultures. From Gordon, Miller, and Bensch (1965)

ducing multivesicular bodies. As a consequence of this uptake, a doubling of the number of *small* coated vesicles is observed in the Golgi zone. Some of them, which react positively to the tests for acid phosphatase and some other hydrolase activities, are shown to convey these enzymes to the multivesicular bodies in the apical region of the cell. The result is the formation of *heterolysosomes*, in the sense of DE DUVE and WATTIAUX (1966).

B. Relations with the Endoplasmic Reticulum

The characteristic enzymes of the lysosomes and of the related hydrolysing enclaves, such as acid phosphatase, are almost certainly synthesized by the *rough* E.R., although it by itself very rarely shows acid phosphatase activity (NOVIKOFF, ESSNER, and QUINTANA, 1963; GOLDFISCHER, ESSNER, and NOVIKOFF, 1964). The work of PALADE and his associates (PALADE, 1960; CARO and PALADE, 1964) shows, especially by autoradiography, that the pancreatic enzymes are synthesized by the *rough* E.R. of the exocrine cells, gathering in the Golgi vacuoles in which they concentrate into secretion grains (zymogen).

According to ESSNER and NOVIKOFF (1962), the enzymes formed in the *rough* E.R. will be transferred into vesicles coming from the *smooth* E.R. and located in the Golgi zone. These vesicles will then be changed into Golgi membranes and thus incorporated in Golgi apparatus. In fact, Golgi vesicles or vacuoles and profiles of *smooth* E.R. are confusingly mixed in Golgi zones of various animal cells.

Different from the primary lysosomes, the E.R. would thus play an indirect, initiatory part in the genesis of the Golgi vacuoles. With regard to the autophagic vacuoles, NOVIKOFF and SHIN (1964) assign a reticular origin to membranes surrounding cytoplasmic regions and organelles before their destruction. These membranes surround simultaneously Golgi or endoplasmic vacuoles.

We have already seen that in the fat body of the Lepidopteran *Calpodes ethlius*, LOCKE and COLLINS (1965) regard the enclosing double membranes as coming from the Golgi apparatus.

The origin of the membranes bounding the primary lysosomes and the autophagic vacuoles still remains obscure, but we can reasonably assume that their enzymatic content is mainly brought to them by the vesicles of the Golgi apparatus. It is quite likely that the enzymes of these vacuoles are synthetized by the E.R. and then transferred to this apparatus.

C. Relations with the Plasmalemma

The phagosomes and the vacuoles of pinocytosis obviously derive from invaginations of the plasmalemma. It is not certain whether the portion of the membrane thus used, undergoes any alteration very soon after the formation of the vacuole. Moreover we have seen that these vacuoles may fuse with others from Golgi apparatus, or with *dense bodies* resulting from the development of secondary lyosomes.

In various animal cells, vacuoles containing numerous smaller vesicles within themselves (multivesicular bodies) often occur. Recently there is a growing tendency to regard these vesicles as deriving from the dictyosomes and transporting their enzymes to the vacuoles of phagocytosis or pinocytosis (GORDON, MILLER, and BENSCH, 1965; see also Fig. 8).

V. The Problem of the Continuity of Animal Vacuoles

If, with reference to their contents, the vacuoles may be regarded as inert products of elaboration of the living matter, the bounding membrane is a structure strictly related to the living state. Therefore it was logical to consider the continuity of those cell systems, like that of the mitochondria or plastids. The preceding pages abundantly show that vacuoles do not inevitably derive from a preexisting vacuole of the same nature, by simple bipartition or by budding. It is certain that many vacuoles may disappear from the cell, by resorption, by excretion or fusion with other structures, without continuity in any lineage. However, it is most probable that the vacuole membrane does not form *de novo* in the cytoplasm. It usually forms by separation of parts of membrane systems, such as the endoplasmic reticulum. Moreover it may be composed by the union of membrane fragments of various origin.

The problem of the continuity of animal vacuoles is thus transferred to the problem of the origin of the cytoplasmic membranes that constitute the plasmalemma, the E.R., and the dictyosomes. This question is the subject of other chapters in this volume, and the reader is referred to those.

Out of these fundamental functions of cytoplasmic membranes results the notion of a remarkable plasticity. This notion is in contrast with the view that the structures observed in the electron microscope appear inert and rigid. In fact, they are molecular organizations continuously changing and more or less mobilized in a fluid medium constantly in motion. The fine structure of these membranes varies in spite of the fundamental unity of their lipoprotein nature. It is likely that they can only unite by undergoing alterations that allow them to acquire characteristics proper to the new structures resulting from such associations. Independently from any differences in thickness, symmetry or asymmetry, discernible in ultra-thin sections, the appearance of their surfaces, now studied by techniques of freeze etching, are different according to the structure to which they belong and probably varying in time as well as in space.

VI. Recent Developments in Studies of Plant Vacuoles

The recent results obtained on animal cells have triggered much interest and work on the enzymatic contents and the ultrastructure of plant vacuoles.

A. Intravacuolar Lytic Processes

Poux's analyses (1963 *b*) are among the first to focus attention upon the existence of cytoplasmic fragments during their degeneration in vacuoles of apical meristems and of young corn leaves. (Fig. 9 *a* and *b*). These enclaves, infrequent in normal cells, become more frequent when the samples stay a few hours in a sucrose solution before fixation. The fragments of cytoplasm, containing at times mitochondria, dictyosomes, or E.R. membranes are first surrounded by a membrane similar to the tonoplast. This inner membrane disappears during the lysis of the vacuolar content.

A more recent and singular case of lytic vacuolar process has been shown by Gifford and Stewart (1968) in the shoot apices of *Bryophyllum*. The proplastids of this plant form predominantly lipid inclusions which are discharged into the vacuoles, with parts of proplastid and tonoplast membranes. These membranes disappear after the release of the inclusion.

Other examples of enclaves with a degenerating content have been pointed out by several authors in very diverse species: apical cells of rhizoids of *Chara foetida* (SIEVERS, 1966); sporangiophores of *Phycomyces blakesleeanus* (THORNTON, 1967); cells of potato germs (PH. COULOMB, 1968); radicular barley meristems (BUVAT, 1968); and

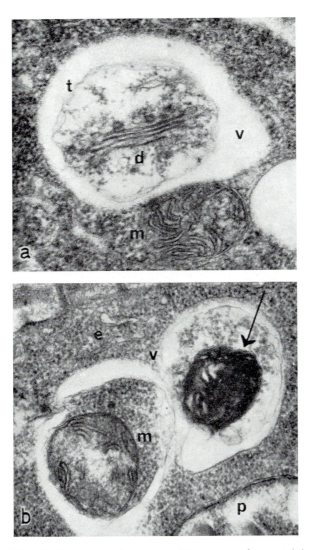

Fig. 9 a and b. Cells of wheat vegetative points. Young vacuoles containing cytoplasmic residues in the process of degeneration. a: very degenerated cytoplasm containing one dictyosome. (*t*) still present tonoplast. b: the protruding of cytoplasm of the lower vacuole is still fairly intact and so is the mitochondrion (*m*) that it contains. The cytoplasm and the mitochondrion of the upper vacuole are, on the contrary, very altered. From POUX (1963*b*)

root meristems of *Cucurbita* (C. Coulomb and Buvat, 1968). The vacuoles and enclaves in question generally show acid phosphatase activity and are reminiscent of *autophagic vacuoles* (Fig. 14d).

B. Particles Similar to the Primary Lysosomes

Plant vacuoles react sometimes positively to the detection criteria of various hydrolases being active at pH 5: acid phosphatase (Poux, 1963a, 1965; Coulomb, 1969), or at pH 7: many neutral phosphatases (Poux 1966, 1967). However, the test of various enzymatic activities, on the same cell types, proves that all vacuoles do not have the same enzymatic content. Thus, plant vacuoles can be fairly dissimilar in the same cell, and their biochemical diversity corresponds with a morphological diversity. More specifically, it is possible to find all the desirable intermediates between typical vacuoles and granules with a concentrated content, furnishing the same hydrolase activities. These granules, seen in meristematic as well as in differentiated cells (Coulomb 1968; 1969) recall closely the primary lysosomes of animal cells. Dense particles revealing acid phosphatase activity have also been shown in maize root cap cells (Berjak, 1968). These particles grow during the ageing of cells and then contain a heterogeneous precipitate similar to that of vacuoles. They meet the conventional criteria of distinction of animal lysosomes. Findings of Matile (1966) and Matile et al. (1965) indicate that maize seedlings contain hydrolysing particles identified as *spherosomes* which would be the equivalent of lysosomes in plant cells. However, it is known that spherosomes contain proteins and phospholipids (Sorokin and Sorokin, 1966), and represent a rather distinct category of enclaves. According to Frey-Wyssling, Grieshaber, and Mühlethaler (1963) they typically develop into lipid granules.

In the single vacuole of the yeast cell, Matile and Wiemken (1967) have characterized, after isolation, four hydrolytic enzymes (RNA-ase, leucylaminopeptidase, and two acid proteases) by which this vacuole resembles to a lysosome equivalent of the cell.

C. The Plant Peroxisomes

Suitable techniques for the detection of peroxidases and catalase (Graham and Karnovsky, 1966) prove some vacuoles of plant cells as active. The same is true for some particles resembling the microbodies of animal cells in many features, especially with regard to their having sometimes an amorphous or crystalline "nucleoïd" concentrated content (Marty, 1969; Czaninski and Catesson, 1969; Poux, 1969; Fig. 10a, b, to be compared with Fig. 7).

In the young foliar primordia of *Euphorbia characias* these dense particles seem to fuse with vacuoles and to bring to them the enzymes responsible for the transformation of the substrate 3,3'-diamino-benzidin (Marty, 1969; cf. Fig. 10c).

D. Possible Origins of the Particles Similar to Lysosomes and Typical Vacuoles

In gourd-roots (*Cucurbita pepo*), lysosome-like particles can be recognized by their morphological characteristics and enzymatic activities. These particles generally form in the proximity of dictyosomes (Coulomb, 1969). Around these are located

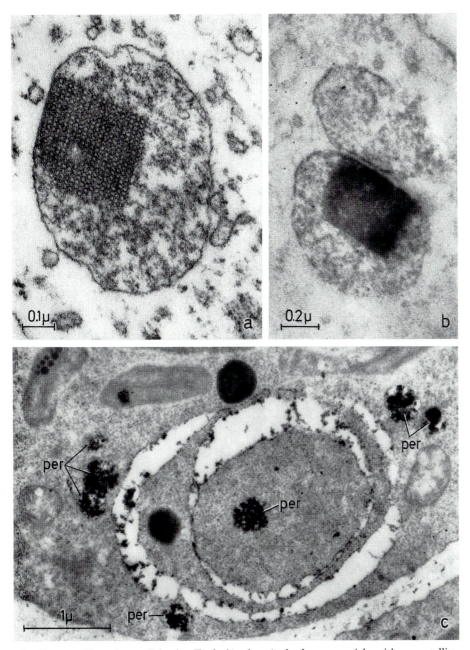

Fig. 10 a—c. "Peroxisomes" in the *Euphorbia characias* bud. a: a particle with a crystalline "nucleoïd". b: a similar particle and its nucleoïd show a positive peroxidase test. c: vacuoles of cytoplasmic sequestration equally positive to the Graham and Karnovsky reaction. In its proximity, many particles similar to typical peroxisomes are found.
a: × 90,000; b: × 50,000; c: × 23,000. From MARTY (1969)

10*

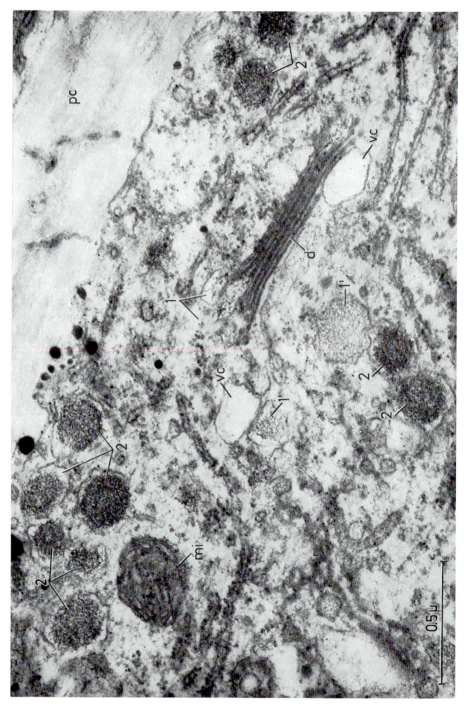

Fig. 11. Radicular cells of *Cucurbita pepo*. Dictyosome (*d*) emitting clear vesicles (*vc*) and vesicles (1) with reticulated contents. These latter ones seem to grow and to concentrate after their freeing into the cytoplasm (1') and then here or elsewhere half

relatively large-sized vesicles (about $0,2 \mu$) with translucent contents, which may derive from dilatation of the Golgi saccules or from the separation of *smooth* swellings of E. R. Beside these vesicles, the dictyosomes form on their periphery much smaller vesicles (about 250 Å) with dense contents. Coulomb (1969) suggests that the small dense vesicles are absorbed by the large ones where they may form

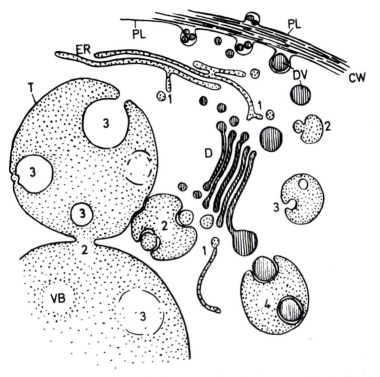

Fig. 12. Diagram of the origin and development of maize-root vacuoles. 1: Formation of provacuoles derived from the endoplasmic reticulum (ER). 2: Fusion between provacuoles and more developed vacuoles. 3: Invaginations of tonoplast (T) from which the incorporation of cytoplasmic material in the vacuole results. 4: Incorporation of the Golgi vesicles (DV) through invaginations of the tonoplast. PL = plasmalemma; CW = cell wall; VB = vacuolar body; D = Dictyosome. From Matile and Moor (1968)

temporary enclaves similar to the *multivesicular bodies* described in animal cells (cf. Novikoff and Shin, 1964 or Gordon, Miller and Bensch, 1965, e. g.) or will be diluted at once. The Golgi zone in question then reveals enclaves with fibrillar or granular contents forming a scale of increasing concentrations very much like the animal primary lysosomes (Fig. 11). In the green alga *Chlorogonium elongatum*, Ueda (1966) showed by micrographs the incorporation of buds of the E. R. in Golgi cisternae, on the one hand, and the entering of Golgi vesicles in provacuoles coming either from Golgi cisternae or from E. R., on the other hand.

It is in fact very difficult to distinguish in the cytoplasm between the fluid enclaves with a smooth membrane coming from the E. R. and those coming from the dictyosomes. The present tendency is to minimize the differences between the enclaves from both origins. Further changes of these plant analogues of primary lysosomes have not been sufficiently followed up in ultrathin sections and with cytochemical techniques.

Studying the ultrastructural effect of vital staining by neutral red, KONČALOVÁ (1965) and AHMADIAN (1969) conclude that the Golgi apparatus probably participates in the transport of the dye which enters the cell by pinocytosis and is led to the vacuoles. Another example of Golgi vesicles which are discharged into vacuoles is given by PICKETT-HEAPS (1967) in the stomatal cells of wheat epidermis.

Thus it seems that the endoplasmic reticulum in its *smooth* type, the dictyosomes and the plasmalemma through vesicles of pinocytosis may associate and confluate to give rise to typical vacuoles or to contribute to their evolution in supplying them with the materials that they either accumulate or hydrolyse.

MATILE and MOOR's freeze-etching work (1968) on the root-cap cells and in the maize radicular meristem led these authors to similar conclusions. The origin of the vacuoles would lie in vesicles of small size, the *provacuoles* coming from the E. R. (Fig. 12,1). These provacuoles amalgamate repeatedly, thus producing more and more voluminous vacuoles characterizing the differentiation of the cells emitted by the meristem (Fig. 12, 2). The vacuoles incorporate Golgi vesicles through tonoplast invaginations. Other invaginations of the vacuole membrane introduce portions of cytoplasm into the vacuoles thus constituted so that they become digestive or autophagic vacuoles (Fig. 12, 3 and 4). According to MATILE and MOOR, the provacuoles are homologous with the animal primary lysosomes, the typical vacuoles being the secondary lysosomes. These observations are confirmed by very precise enzymatic analyses. It still remains to be clarified whether *all* the vacuoles of plant cells undergo this uniform evolution. In meristematic cells, the vacuolar system is sometimes polymorphous and in particular the autophagic vacuoles are distinct from other elements without any content (BUVAT, 1968).

In the meristematic cells of *Hordeum* roots, we are able to grasp the origin of vacuoles that isolate cytoplasmic regions before destroying them (BUVAT, 1968). At the beginning closed *double* membranes are observed forming one or two concentric systems (Fig. 13) with profiles similar to saccules of smooth E. R. These saccules swell and form small vesicles, partially confluating in cupuliform vacuoles, enclosing a part of the cytoplasm (Fig. 14, *a* to *d*). These vacuoles with circular or arc-like profiles are sometimes frequent in various meristematic cells. MARTY (1969) has observed them frequently in the shoot meristem or in foliar primordia of *Euphorbia characias* in which they sometimes reveal intense peroxidase activity with diaminobenzidin as substrate (Fig. 10, *c*).

E. Additional Remarks about the Continuity of Plant Vacuoles on the Infrastructural Scale

From the point of view of the *continuity* of plant vacuoles, it is safe to say that vacuoles may form by mechanisms other than bipartition or budding of preexisting vacuoles. Others, as mentioned before, result from the division of enclaves of the same nature. Anyhow, at the electron microscopic scale, it does not mean much

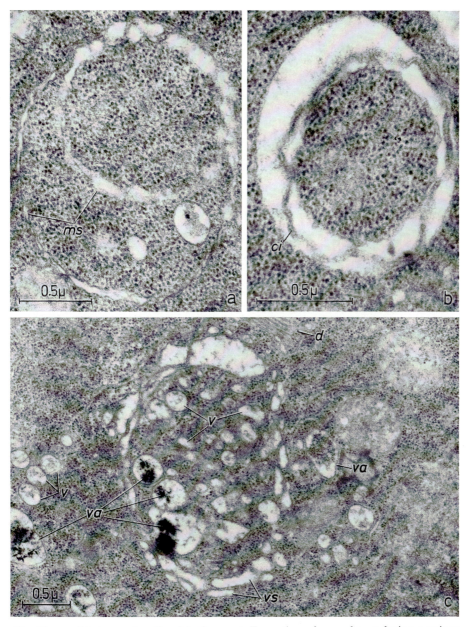

Fig. 13 a—c. Meristematic cells of barley root. Formation of vacuoles enclosing portions of cytoplasm (sequestration vacuoles, *vs*) or autophagic vacuoles. a: beginning of the formation of two systems from double membranes of sequestration *ms*. × 40,000. b: degeneration of the intermediary cytoplasm, *ci*, crushed between two concentric vacuoles. × 52,000. c: formation of membranes and of vacuoles of sequestration (*vs*) surrounding ordinary meristematic (*v*) or autophagic vacuoles (*va*). × 27,000.

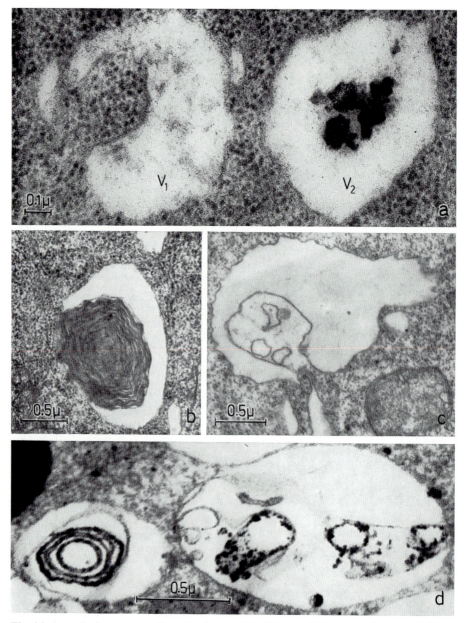

Fig. 14. a—c: Barley root. a: Autophagic vacuoles. V_1: cupuliform vacuole surrounding a protrusion of still intact cytoplasm. V_1: vacuole containing residues of lysis. $\times$ 70,000. b: cupuliform vacuole surrounding a degenerated cytoplasmic protrusion composed of concentric lamellated systems. $\times$ 25,000. c: cupuliform vacuole, surrounding a cytoplasmic protrusion almost emptied of its contents. $\times$ 27,000. d: Gourd root (*Cucurbita pepo*). Autophagic vacuoles; vacuole content gives positive acid phosphatase test. $\times$ 52,000. c and d: from Coulomb and Buvat (1968)

to say that a vacuole forms *de novo*. As far as we know presently, the bounding membrane of a vacuole always comes from parts of preexisting membrane systems of the cytoplasm.

VII. Conclusions

The immense development of plant cell vacuoles bestows upon them important mechanical functions which are not found in animal cells (e. g. turgescence); besides it constitutes a kind of "extracytoplasmic" compartment for waste products (heterosids, alcaloids) but which is also used for storage (for instance inulin). Because of these physiological functions the plant vacuoles are different from the lysosomes or from the vacuoles of animal cells. The combination of infrastructural and cytoenzymological research has exposed the fundamental nature of plant cell vacuoles and has brought them nearer, despite some pronounced morphological differences, to the primary and secondary lysosomes of animal cells. Nevertheless, the differences of morphological development bring about physiological and mechanical consequences, so that plant vacuoles still represent structures that permit a satisfactory distinction between animal and plant cells.

References

AHMADIAN-TEHRANI, P.: Application de la microscopie électronique à l'étude de la pénétration du rouge neutre dans les cellules du méristème radiculaire du Blé (*Triticum vulgare*). Thèse Docteur Ingénieur, Marseille-Luminy 1969.

ARNOLD, J.: Über Plasmastrukturen und ihre funktionelle Bedeutung. Jena: Gustav Fischer 1914.

BAINTON, D. F., FARQUHAR, M. G.: Origin of granules in polymorphonuclear leukocytes. Two types derived from opposite faces of the Golgi complex in developing granulocytes. J. Cell Biol., 28, 277—301 (1966).

BAUDHUIN, P., BEAUFAY, H., DE DUVE, C.: Combined biochemical and morphological study of particulate fractions from rat liver. Analysis of preparations enriched in lysosomes or in particles containing urate-oxidase, d-amino-acid oxidase, and catalase. J. Cell Biol. 26, 219—243 (1965).

BERJAK, P.: A lysosome-like organelle in the root cap of *Zea mays*. J. Ultrastruct. Res. 23, 233—242 (1968).

BERKALOFF, C.: Les cellules méristématiques d'*Himanthalia lorea* (L.) S. F. Gray. Etude au microscope électronique. J. Microscop 2, 213—228 (1963).

BOWES, B. G.: The origin and development of vacuoles in *Glechoma hederacea*, L. La Cellule 65, 357—364 (1965).

BREWER, D. B., HEATH, D.: Lysosomes and vacuolation of the liver cell. Nature (Lond.) 198, 1015—1016 (1963).

BUVAT, R.: Relations entre l'ergastoplasme et l'appareil vacuolaire. C. R. Acad. Sci. (Paris) 245, 350—352 (1957).

— Recherches sur les infrastructures du cytoplasme dans les cellules du méristème apical, des ébauches foliaires et des feuilles développées d'*Elodea canadensis*. Ann. Sci. Nat. Bot. 11e série, 19, 121—161 (1958).

— Le réticulum endoplasmique des cellules végétales. Ber. Dtsch. Bot. Ges. 74, 261—267 (1961).

— Le cytoplasme végétal. In: Travaux dédiés à *L. Plantefol*. 81—124. Masson et Cie 1965.

— Diversité des vacuoles dans les cellules de la racine d'Orge (*Hordeum sativum*). C. R. Acad. Sci. (Paris) 267, 296—298 (1968).

— MOUSSEAU, A.: Origine et évolution du système vacuolaire dans la racine de *Triticum vulgare*; relations avec l'ergastoplasme. C. R. Acad. Sci. (Paris) 251, 3051—3053 (1960).

Carasso, N., Fauré-Fremiet, E., Favard, P.: Ultrastructure de l'appareil excréteur chez quelques ciliés péritriches. J. Microscop. **1**, 455—468 (1962).

Caro, L. G., Palade, G. E.: Protein synthesis, storage and discharge in the pancreatic exocrine cell. An autoradiographic study. *J. Cell Biol.* **20**, 473—495 (1964).

Cassaigne, Y.: Origine et évolution du vacuome chez quelques Champignons. *Rev. Gén. Botan.* **43**, 140—167 (1931).

Coulomb, C., Buvat, R.: Processus de dégénérescence cytoplasmique partielle dans les cellules de jeunes racines de *Cucurbita pepo*. *C. R. Acad. Sci.* (Paris) **267**, 843—844 (1968).

Coulomb, Ph.: Sur la présence de structures lamellisées dans les cellules méristématiques du bourgeon de *Solanum tuberosum*. *C. R. Acad. Sci.* (Paris) **267**, 1373—1374 (1968).

— Mise en évidence de structures analogues aux lysosomes dans le méristème radiculaire de la Courge (*Curcurbita pepo* L. Cucurbitacée). *J. Microscop.* **8**, 123—138 (1969).

Czaninski, Y., Catesson, A. M.: Localisation ultrastructurale d'une activité peroxydasique dans les tissus conducteurs végétaux. (Colloque franco-suisse de micr. électr. Lausanne 1969). J. Microscop. **8**, p. 43a (1969).

Dangeard, P.: Etudes de Biologie cellulaire : évolution du système vacuolaire chez les végétaux. *Le Botaniste* **15**, 1—267 (1923).

Dangeard, P. A.: Nouvelles recherches sur le système vacuolaire. *Bull. Soc. Bot. France* **63**, 179—187 (1916).

Dujardin, F.: Recherches sur les organismes inférieurs. *Ann. Sci. Nat. Zool.*, 2e série, **4**, 343—377 (1835).

De Duve, C.: Lysosomes, a new group of cytoplasmic particles. In: Subcellular Particles. (Ed.: T. Hayashi) 128—159. Ronald Press: New York, 1959.

— The lysosome concept. In: Ciba Foundation Symposium "Lysosomes" (Ed.: A. V. S. De Renck and M. P. Cameron: pp. 1—35. J. & A. Churchill Ltd.: London, 1963.

— Baudhuin, P.: Peroxisomes. Physiol. Rev. **46**, 323—357 (1966).

— Wattiaux, R.: Functions of lysosomes. *Ann. Rev. Physiol.* **28**, 435—492 (1966).

Engleman, E. M.: Ontogeny of aleuron grains in Cotton embryo. *Amer. J. Botany* **53**, 231—237 (1966).

Essner, E., Novikoff, A. B.: Cytological studies on two functional hepatomas. Interrelations of endoplasmic reticulum, Golgi apparatus and lysosomes. *J. Cell Biol.* **15**, 289—312 (1962).

Favard, P., Carasso, N.: Etude de la pinocytose au niveau des vacuoles digestives de Ciliés Péritriches. *J. Microscop.* **3**, 671—696 (1964).

Frey-Wyssling, A., Grieshaber, E, Mühlethaler, K.: Origin of spherosomes in plant cells. *J. Ultrastruct. Res.* **8**, 506—516 (1963).

Friend, D. S., Farquhar, M. G.: Functions of coated vesicles during protein absorption in the rat vas deferens. *J. Cell Biol.* **35**, 357—376 (1967).

Gifford, E. M., Jr. Stewart, K. D.: Ultrastructure of the shoot apex of *Chenopodium album* and certain other seed plants. *J. Cell Biol.* **33**, 131—142 (1967).

— — Inclusions of the proplastids and vacuoles in the shoot apices of *Bryophyllum* and *Kalanchoë*. *Amer. J. Bot.* **53**, 269—279 (1968).

Goldfischer, S., Essner, E., Novikoff, A. B.: The localization of phosphatase activities at the level of ultrastructure. *J. Histochem. Cytochem.* **12**, 72—95 (1964).

Gordon, G. B., Miller, L. R., Bensch, K. G.: Studies on the intracellular digestive process in mammalian tissue culture cells. *J. Cell Biol.*, **25**, 41—55 (1965).

Graham, R. C. and Karnovsky, M. J.: The early stages of absorption of injected horseradish peroxidase in the proximal tubules of Mouse kidney: ultrastructural cytochemistry by a new technique. *J. Histochem. Cytochem.* **14**, 291—302 (1966).

Grassé, P. P.: Ultrastructure, polarité et reproduction de l'appareil de Golgi. *C. R. Acad. Sci.* (Paris) **245**, 1278—1281 (1957).

Guilliermond, A.: Recherches cytologiques sur les Levures et quelques moisissures à formes-levures. *Rev. Gén. Bot.* **15**, 49—66; 104—124; 166—185 (1903).

— Les corpuscules métachromatiques ou grains de volutine. *Bull. Inst. Pasteur*, **4**, 145—151; 193—200 (1906).

— Recherches cytologiques sur la germination des graines de quelques Graminées et contribution à l'étude des grains d'aleurone. *Arch. Anat. Micr.* **10**, 141—226 (1908).

GUILLIERMOND, A.: Recherches cytologiques sur la formation des pigments anthocyaniques. Nouvelle contribution à l'étude des mitochondries. *Rev. Gen. Bot.* **25**, 295—387 (1914).

— Nouvelles recherches sur l'Appareil vacuolaire dans les Végétaux. *C. R. Acad. Sci.* (Paris) **171**, 1071—1074 (1920).

— Nouvelles recherches sur les constituants morphologiques du cytoplasme de la cellule végétale. *Arch. Anat. Micr.* **20**, 1—210 (1924).

— Observations sur l'origine des vacuoles. *La Cellule* **36**, 217—229 (1925).

— Sur le développement d'un *Saprolegnia* en milieu additionné de colorants vitaux et la coloration vitale de son vacuome pendant son développement. *C. R. Acad. Sci.* (Paris) **188**, 1621—1623 (1929).

— Sur la formation des zoosporanges et la germination des spores chez un *Saprolegnia*, en cultures sur milieux nutritifs additionnés de rouge neutre. *C. R. Acad. Sci.* (Paris) **190**, 384—386 (1930).

HIRSCH, J. G., FEDORKO, M. E., COHN, Z. A.: Vesicle fusion and formation at the surface of pinocytic vacuoles in macrophages. *J. Cell Biol.* **38**, 629—632 (1968).

HOLTER, H.: Pinocytosis. *Int. Rev. Cytol.* **8**, 481—504 (1959).

JENSEN, W. A.: Cotton embryogenesis : the zygote. *Planta (Berl.)* **79**, 346—366 (1968).

KONČALOVÁ, M. N.: Vitalfärbung der Meristemzellenvakuolen bei Weizenpflanzen mit Neutralrot. Licht- und elektronenmikroskopische Untersuchungen. *Protoplasma* **60**, 195—210 (1965).

LEIGHTON, F., POOLE, B., BEAUFAY, H., BAUDHUIN, P., COFFEY, J. W., FOWLER, S., DE DUVE, C.: The large scale separation of peroxisomes mitochondria and lysomomes from the livers of rats injected with triton WR-1339. *J. Cell Biol.* **37**, 482—513 (1968).

LEWIS, W. H.: Pinocytosis. *Bull. John Hopk. Hosp.* **49**, 17—27 (1931).

— Pinocytosis by malignant cells. *Amer. J. Cancer* **29**, 666—679 (1937).

LOCKE, M., COLLINS, J. V.: The structure and formation of protein granules in the fat body of an insect. *J. Cell. Biol.* **26**, 857—884 (1965).

MANTON, I.: Observations on stellate vacuoles in the meristem of *Anthoceros*. *J. exptl. Bot.* **13**, 161—167 (1962).

MARTY, F.: Caractérisation cytochimique infrastructurale de peroxysomes (="microbodies" *sensu stricto*) chez *Euphorbia characias*. *C. R. Acad. Sci.* **268**, 1388—1391 (1969).

MATILE, PH.: Enzyme der Vakuolen aus Wurzelzellen von Maiskeimlingen. Ein Beitrag zur funktionellen Bedeutung der Vakuole bei der intrazellulären Verdauung. *Z. Naturforsch.* **21b**, 871—878 (1966).

— Lysosomes of root tip cells in corn seedling. *Planta* (Berl.) **79**, 181—196 (1968).

— BALZ, J. P., SEMADENI, E., JOST, M.: Isolation of spherosomes with lysosome characteristics from seedlings. *Z. Naturforsch.* **20b**, 693—698 (1965).

— MOOR, H.: Vacuolation: origin and development of the lysosomal apparatus in root-tip cells. *Planta* (Berl.) **80**, 159—175 (1968).

— WIEMKEN, A.: The vacuole as the lysosome of the yeast cell. *Arch. Mikrobiol.* **56**, 148—155 (1967).

MESQUITA, J. F.: Electron microscope study of the origin and development of the vacuoles in root-tip cells of *Lupinus albus L. J. Ultrastruct. Res.* **26**, 242—250 (1969).

MÖLLENDORFF, W.: Zur Morphologie der vitalen Granulfärbung. *Arch. Mikr. Anat.* **90**, 463—502 (1918).

NAEGELI, C.: Zellkern, Zellbildung und Zellenwachstum bei den Pflanzen. *Z. Wiss. Bot.* **1**, 34—133 (1844).

NOVIKOFF, A. B.: Lysosomes in the physiology and pathology of cells: contributions of staining methods. In: Lysosomes, Ciba Foundation Symposium "Lysosomes" (Ed.: A. V. S. de RENCK and CAMERON, M. P. J. & A. Churchill Ltd, London, pp. 36—73, 1963.

— BEAUFAY, H., DE DUVE, C.: Electron microscopy of lysosome-rich fractions from rat liver. *J. Biophys. Biochem. Cytol.* **2**, 179—184 (1956).

— ESSNER, E., QUINTANA, N.: Relations of endoplasmic reticulum, Golgi apparatus and lysosomes. *J. Microscop.* **2**, 3 (1963).

Novikoff, A. B., Shin, W. Y.: The endoplasmic reticulum in the Golgi zone and its relations to microbodies, Golgi apparatus and autophagic vacuoles in rat liver cells. *J. Microscop.* **3**, 187—206 (1964).

Osinchak, J.: Acid phosphatase activity and the identity of intracellular granules in neurosecretory cells of the rat. *J. Cell Biol.* **19**, 54 A (1963).

Palade, G. E.: The endoplasmic reticulum. *J. Biophys. Biochem. Cytol.* **2**, suppl., 85—98 (1956).

— The secretory process of the pancreatic exocrine cell. Symposium "Electron Microscopy in Anatomy". *Anat. Soc. of Great Britain*, 176—206. Arnold, London 1960.

Palay, S. L., Karlin, L. J.: An electron microscopic study of the intestinal villus. II. The pathway of fat absorption. *J. Biophys. Biochem. Cytol.* **5**, 373—384 (1959).

Parat, M.: Contribution à l'étude morphologique et cytologique du cytoplasme. *Arch. Anat. Micr.* **24**, 73—357 (1928).

— Painlevé, J.: Appareil réticulaire interne de Golgi, trophosponge de Holmgren et vacuome. *C. R. Acad. Sci.* (Paris) **179**, 844—846 (1924).

— — Sur l'exacte concordance des charactères du vacuome et de l'appareil de Golgi classique. *C. R. Acad. Sci.* (Paris) **180**, 1134—1136 (1925).

Pfeffer, W.: Zur Kenntnis der Plasmahaut und der Vakuolen, nebst Bemerkungen über den Aggregatzustand des Protoplasmas und über osmotische Vorgänge. *Abh. math. phys. Kgl. Sachs. Ges. Wiss.* **16**, 185—344 (1890).

Pikett-Heaps, J. D.: Further observations on the Golgi apparatus and its functions in cells of the wheat seedling. *J. Ultrastruct. Res.* **18**, 287—303 (1967).

Poux, N.: Nouvelles observations sur la nature et l'origine de la membrane vacuolaire des cellules végétales. *J. Microscop.* **1**, 55—66 (1962a).

— Sur l'origine et l'évolution de l'appareil vacuolaire dans les organes aériens des Graminées. *5th Intern. Congr. f. Electron Micr. Proc.*, (Ed.: S. S. Breeze), II, W2. Academic Press. New York 1962b.

— Localisation de la phosphate acide dans les cellules méristématiques de Blé (*Triticum vulgare* Vill.). *J. Microscop.* **2**, 485—490 (1963a).

— Sur la présence d'enclaves cytoplasmiques en voie de dégénérescence dans les vacuoles des cellules végétales. *C. R. Acad. Sci.* (Paris) **257**, 736—738 (1963b).

— Localisation de l'activité phosphatasique acide et des phosphates dans les grains d'aleurone. I. Grains renfermant à la fois globoïdes et cristalloïdes. *J. Microscop.* **4**, 771—782 (1965).

— Localisation d'activités phosphatasiques à pH 7 dans les cellules du méristème radiculaire de *Cucumis sativus* L. *6th Int. Congr. f. Electron Microscopy, Kyoto.* Proc. II, 95—96. Tokyo: Maruzen and C°, 1966.

— Localisation d'activités enzymatiques dans les cellules du méristème radiculaire de *Cucumis sativus* L. I. Activités phosphatasiques neutres dans les cellules du protoderme. *J. Microscop.* **6**, 1043—1058 (1967).

— Premier essai de localisation ultrastructurale de l'activité peroxydasique dans le méristème radiculaire de *Cucumis sativus* L. *J. Microscop.* **8**, 78a (1969).

Rouiller, C., Bernhard, W.: "Microbodies", and the problem of mitochondrial regeneration in liver cells. *J. Biophys. Biochem. Cytol.* **2**, suppl., 355—360 (1965).

Schnepf, E., Koch, W.: Über die Entstehung der pulsierenden Vakuolen von *Vacuolaria virescens* (Chloromonadophyceae) aus dem Golgi-Apparat. *Arch. Mikrobiol.* **54**, 229—236 (1966).

Schulz, Sister R., Jensen, W. A., *Capsella* embryogenesis: the synergids before and after fertilization. *Amer. J. Botany* **55**, 541—552 (1968).

— — *Capsella* embryogenesis: the egg, zygote and young embryo. *Amer. J. Botany* **55**, 807—819 (1968).

Sievers, A.: Lysosomen — ähnliche Kompartimente in Pflanzenzellen. *Naturwissenschaften* **13**, 334—335 (1966).

Smith, R. E.: Acid phosphatase activity of rat adenohypophysis during secretion. *J. Cell Biol.* **19**, 66 A — 67 A (1963).

— Farquhar, M. G.: Lysosome function in the regulation of the secretory process in cells of the anterior pituitary gland. *J. Cell Biol.* **31**, 319—347 (1966).

SOROKIN, H. P., SOROKIN, S.: The spherosomes of *Campanula persicifolia* L. A light and electron microscope study. *Protoplasma* **62**, 216—236 (1966).

THORNTON, R. M.: The fine structure of *Phycomyces*. I. Autophagic vesicles. *J. Ultrastruct. Res.* **21**, 269—280 (1967).

DE VRIES, H.: Plasmolytische Studien über die Wand der Vakuolen. *Jahrb. wiss. Bot.* **16**, 465—598 (1885).

WENT, F. A. F. C.: Die Vermehrung der normalen Vakuolen durch Teilung. *Jahrb. wiss. Bot.* **19**, 295—356 (1888).

WHALEY, W. G., KEPHART, J. E., MOLLENHAUER, H. H.: The dynamics of cytoplasmic membranes during development. In: "Cellular membranes in development" (Ed.: M. LOCKE), 135—173. Academic Press New York-London 1964.

Origin and Continuity of Polar Granules *

ANTHONY P. MAHOWALD **

Biology Department, Marquette University, Milwaukee, Wisconsin

I. Introduction

Embryonic development is not totally an epigenetic phenomenon but is clearly dependent upon the cytoplasmic organization present in the egg at the inception of development. While the general acceptance of the theory of differential gene activity as the basic epigenetic mechanism is an important advance in our understanding of development, it is only more recently that the presence of organized cytoplasmic "information" in the egg is recognized to be one of the key mechanisms for establishing the sequence of differential gene action that characterizes development (the argument for this understanding of development has been thoroughly and brilliantly explained by DAVIDSON, 1968). The best example of such developmentally significant localizations of the egg cytoplasm are the polar granules or "germ cell determinants" of insects. These organelles are found in a variety of species principally within the insect orders Diptera, Hymenoptera, and Coleoptera. Their developmental significance was indicated morphologically, in that they were incorporated into one or more cells which then gave rise to the primordial germ cells of the embryo, and then experimentally, by HEGNER (1911) who found that if the posterior pole plasm was destroyed by cauterization, normal embryos were formed except that they lacked germ cells. HEGNER termed these organelles "germ cell determinants" and the search was begun to find other examples of organ determinants in other embryonic systems. Similar germ cell determinants have been found in other organisms, most clearly in amphibians where the most elegant experiments have been performed by SMITH (1966), but in no other instance has the existence of cytoplasmic localizations of information for embryonic development been related to the existence of organelles that can be directly studied.

In some organisms the posterior ooplasm appears to have further roles. MENG (1968) has summarized these roles into three categories: first, pole cell formation as already mentioned; secondly maintenance of the germ line number of chromosomes in those species in which chromosomes are eliminated in the somatic line. This role was clearly shown by GEYER-DUSZYNSKA (1959) in *Wachtliella*, a cecidomyiid. Normally, the cleavage nucleus which reaches the posterior polar plasm retains all its chromosomes while the remaining nuclei of the embryo go through

---
* This work has been supported by research grants from the National Science Foundation, GB-5155, GB-5780 and GB-7980.

** Present address: Institute for Cancer Research, Philadelphia, Pennsylvania.

an abnormal mitosis in which many of the chromosomes are eliminated. When she prevented the cleavage nuclei from reaching the posterior polar plasm until after elimination of the usual chromosomes from the somatic cells, and only then allowed a nucleus, now with a somatic complement of chromosomes, to reach the polar plasm, pole cells still formed but were incapable of producing mature eggs. Thus pole cell formation is separable from chromosome retention. Thirdly, BIER (1954) has shown in *Formica* that the polar plasm is important in caste determination.

From these classic studies the developmental significance of these organelles is clear. Even today "germ-cell determinants", in whatever organism studied, remain the only case known where developmentally significant information is localized in organelles that can be seen and followed during development. Hence it becomes especially important to increase our understanding of how they function in producing pole cells and moreover to learn how they originate and what is their fate. The additional information that is available is derived almost solely from electron microscopy but hopefully these studies will lead to analyses of the molecules which compose the polar granules, and of their function.

II. Polar Granules during Pole Cell Formation

A. Morphology

The original cytological observations of polar granules during early development have been repeated many times (cf. HEGNER, 1914, for review of early literature). In every instance the organelles are restricted to the posterior portion of the egg, either in a spherical mass called the "oosome" or as a thin layer of densely staining granules called the polar disk or simply polar granules. As the first cleavage nuclei reach the polar plasm, this region immediately pinches off from the egg to form the pole cells. In only a few instances have workers attempted to follow the fate of the polar granules during subsequent stages of embryogenesis. Most recently, COUNCE (1963) has shown that polar granules of *Drosophila* usually assume a juxtanuclear arrangement at the time that the pole cells are moving to the gonadal region of the embryo, but she did not try to follow this localization in later stages.

Cytological studies of polar granule morphology during these early embryonic stages have frequently demonstrated that the polar granule region of the embryo goes through a regular sequence of changes. Before fertilization the polar granule material is restricted to a spherical body or to a flat disc at the posterior tip of the egg, and during the early cleavage divisions the polar granule material disperses in the posterior ooplasm. Recently the greater resolution of the electron microscope has made it possible to follow the morphological changes not only during the early stages of the embryo, but throughout their life cycle in the germinal tissue. The fine structure of polar granules was first described by MAHOWALD (1962) in *Drosophila melanogaster* and subsequently in other *Drosophila* species (ULLMANN, 1965; MAHOWALD, 1968), in *Wachtliella persicariae* (WOLF, 1967), and in *Miastor* (MAHOWALD and STOIBER, 1970). In each case the fine structure is basically the same (Fig. 1, 2). Polar granules are meshes of interwoven fibrils, approximately 15 mμ in thickness in the early embryo, and are not bounded by any membrane.

Two different types of organizations are present: in *Drosophila*, the polar granules are discrete organelles varying in size from under 0.5 μ in *D. melanogaster* and *D. hydei* to over 1 μ in *D. willistoni*. In the mature egg they are found attached end to end to varying extents depending upon the species (Fig. 3), but in each instance the approximate outlines of the individual granules can be discerned. In the mature

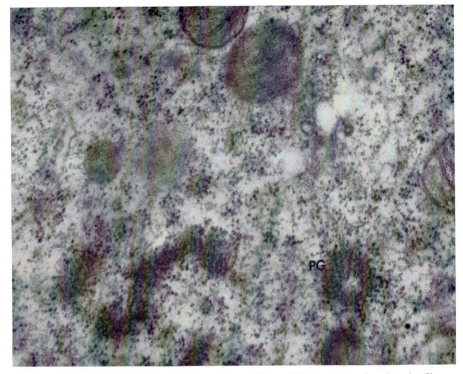

Fig. 1. Polar granule (PG) of an early embryo of *Drosophila immigrans*, showing the fibrous structure and the typical accumulation of ribosomes at the periphery (47,000 ×)

egg of *Miastor* (Fig. 2), and apparently in *Wachtliella* (WOLF, 1967), the polar granules are present as strands of loosely woven fibrous material which appear to be continuous throughout the posterior region of the egg. Between these strands free ribosomes, endoplasmic reticulum, and occasionally other organelles are found. In the mature egg of every species of *Drosophila* examined, except *D. hydei*, and in *Miastor* polar granules are attached to mitochondria. Shortly after fertilization a number of changes are observed which are constant for each species: the granules become detached from mitochondria; they fragment into small spherical or rod-shaped structures; clusters of ribosomes become closely associated with the periphery of the granule, and the polar granules spread deeper into the surrounding ooplasm. This spreading of the polar disc of the unfertilized egg after fertilization (COUNCE, 1963; MAHOWALD, 1968) can be attributed to the type of cortical changes that accompany fertilization in other organisms and remains today as our clearest example in insects of a cortical reaction produced at fertilization.

At the time the cleavage nuclei reach the polar plasm the polar granules have reached their most fragmented state and tend to cluster around the centrioles especially during the pole cell mitoses (COUNCE, 1963). After the pole cells have ceased dividing the polar granules reaggregate either into a few large granules (e.g. in *Miastor* and *D. melanogaster*, *D. hydei*, and *D. willistoni*) or into a chain or

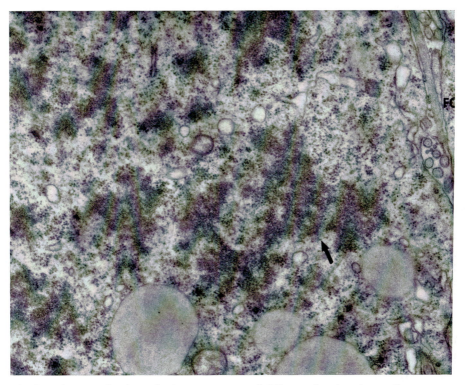

Fig. 2. Polar granules (arrow) of a mature egg of *Miastor*, showing the confluent nature of the polar granule material. The follicle cells (FC) are at the right border (32,000 ×)

plaque of attached granules *(D. immigrans)* (MAHOWALD, 1968). In the latter species, the fibrous matrix becomes reorganized into rods about 25 mμ in thickness and usually about 0.5 μ long.

B. An Hypothesis on the Mechanism of Polar Granule Function

There has been little questioning of the role of the polar plasm in pole cell formation since the experimental work of HEGNER (1911) and especially the studies using ultraviolet irradiation of GEIGY (1931). These latter studies are especially convincing because he obtained adult flies after UV irradiation of the polar plasm of *Drosophila melanogaster* which were normal in all respects except that they lacked germ cells. What has not been clear is whether the polar granules themselves are the effective agents in the formation of pole cells or whether other constituents of the polar plasm, also sensitive to UV irradiation, may actually be responsible for the formation of pole cells. Jazdowska-Zagrodzinska (1966) has recently approached this question. She found that after polar granules had been dislocated from the polar plasm by centrifugation in an anterior direction, no pole cells formed either at the posterior tip in the remaining polar plasm or elsewhere in the centrifuged egg. When centrifugation was in a lateral direction, the polar granules

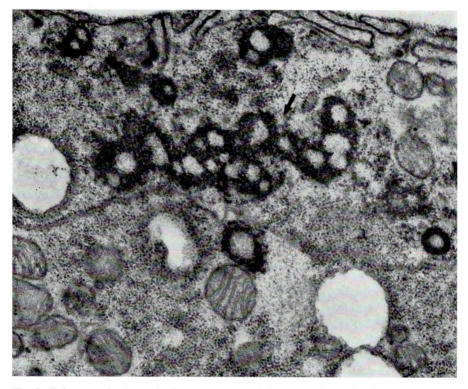

Fig. 3. Polar granules (arrow) of a mature egg of *Drosophila hydei*, showing the characteristic fusing of individual polar granules into chains of polar granule material. The many cell membrane folds at the edge of the egg are typical of mature *Drosophila* eggs (32,000 ×)

remained within the posterior ooplasm and pole cells formed. Since she was unable to detect cytologically the new location of the polar granules in the first type of centrifugation, this is not a convincing proof that the polar granules must be present in the polar plasm in order to produce pole cells. Probably after centrifugation, the polar granules are so dispersed that they are no longer present in sufficient concentration in a localized cortical region of the egg for pole cell formation to result.

A number of further observations form the basis of an hypothesis of polar granule function during these early stages when they are responsible for pole cell formation. If the clusters of ribosomes are true polysomes, then presumably protein synthesis is occurring in conjunction with the polar granules. Many cytochemical studies of polar granules have indicated the presence of RNA and protein in polar granules of the egg and early embryo (e.g., NICKLAS, 1959), but RNA is no longer detectable after pole cell formation even though the granules are larger and more compact (MAHOWALD, 1970b). These observations have led to the postulate that the mode of function of the granule is probably through proteins synthesized by a messenger RNA stored in the granule during oogenesis (MAHOWALD, 1968). This hypothesis also accounts for the experimental results of JAZDOWSKA-ZAGROD-ZINSKA (1966) who found that after polar granules had been moved to other portions of the embryo by centrifugation, pole cells no longer formed. If the mode of function of the polar plasm is through an accumulation of specific proteins which will interact with the cleavage nuclei which penetrate the posterior ooplasm, then the disruption of this region of the egg will also spread these specific proteins (or whatever is ultimately the determining factors) throughout the egg so that an effective concentration of these specific factors is lost. Definitive experimental verification of the manner in which the polar plasm functions to produce pole cells will probably have to wait until it is possible to isolate the organelles and characterize their composition in more specific terms.

III. Continuity of Polar Granules in the Germ Cells

No further significant change occurs in the ultrastructure of the polar granule until the pole cells become transformed into primordial germ cells. At this time in *Drosophila* (this has not yet been studied in other species) the polar granules again fragment and become associated with the outer nuclear membrane (COUNCE, 1963; MAHOWALD, 1970a). At first the granules still retain their characteristic 15 mμ fibrils but these become quickly transformed into fine fibrils approximately 5 mμ in thickness. In *D. immigrans* this process of change can be more clearly

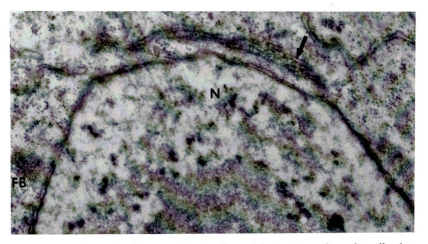

Fig. 4. Fragmentation of polar granules in *Drosophila immigrans* in pole cells that are differentiating into primordial germ cells. Parts of the polar granule still retain the rod-like substructure (arrow) characteristic of this species while in other regions (FB) the polar granules are adjacent to the nuclear membrane and are composed totally of fine fibrillar material (33,000 ×)

seen because stages in the transition from rod-shaped components to the inter-woven mesh of fibrils can be found (Fig. 4; Mahowald, 1970a). From this time on these fibrous bodies can be found on the nuclear membrane of the primordial germ cells (Fig. 5) throughout the larval and prepupal stages of the fly. In no case have these fibrous bodies been found in other cell types of the fly. Since the pri-mordial germ cells are increasing in number throughout larval development, it is

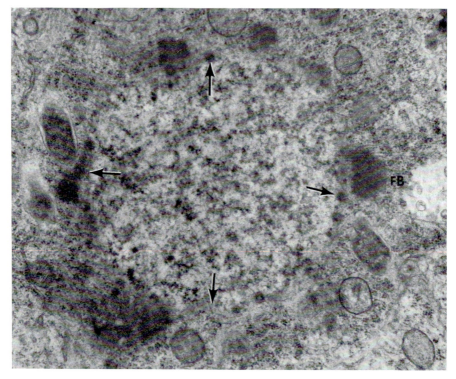

Fig. 5. Fibrous bodies (FB) attached to the outer nuclear membrane in ovary of early prepupal stage of *Drosophila willistoni*. The limits of the nucleus are indicated by the arrows (20,000 ×)

apparent that more of this material is being produced. It is presumed that this material found in great abundance at later stages is identical to that derived from the polar granules in the embryo since no change in fine structure has been detected and the localization on the nuclear envelope is unique.

The continuity of these fibrous bodies during the formation of the adult ovary and the process of oogenesis is especially intriguing. During dipteran ooge-nesis, the oogonium first divides four times to give rise to a cluster of 16 cells interconnected by intercellular bridges or ring canals in a definite pattern: two cells have four ring canals, two cells have three canals, four cells have two canals and eight cells have one canal (Brown and King, 1964). At first the nuclear membranes of all 16 cells possess fibrous bodies. The two cells with four ring canals both start meiosis as judged by the presence of synaptonemal complexes and consequently have been termed pro-oocytes by KOCH, SMITH, and KING (1967). One of these pro-oocytes redifferentiates into a nurse cell and subsequently is indistinguishable from the 14 nurse cells originally derived from the oogonium. At the time one pro-oocyte is becoming a nurse cell, the fibrous bodies on the nuclear membrane of the true oocyte disappear. Some fibrous bodies are found in the oocyte cytoplasm at this time, but it is not certain that these originate from the fibrous bodies that had been localized on the oocyte nuclear envelope (this will be treated again in the next section). Occasionally, single polar granules have been found free in the oocyte cytoplasm prior to the time of vitellogenesis (MAHOWALD, 1970a). It is reasonable to postulate that these arose from the fibrous material released from the oocyte or nurse cell nuclear membrane. Throughout the rest of oogenesis the

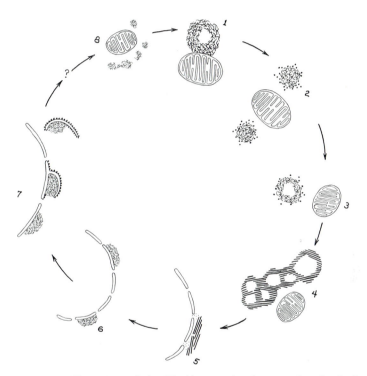

Fig. 6. Diagrammatic illustration of the life history of polar granules. Beginning in the mature oocyte (1), the polar granules are typically attached to mitochondria, then after fertilization (2) they become free, fragment and have ribosomes attached to their periphery. After pole cells have formed the number of ribosomes attached to the granules decreases, and reaggregation of the fragments occurs (3) until the completion of the blastoderm stage when large aggregates have formed and ribosomes are not associated with the granules. (4). Subsequently, during the formation of the embryonic gonad, the large aggregates again fragment and become localized along the outer nuclear envelope (5 and 6) and remain in this location until the next cycle of oogenesis (7). At this time they are frequently associated with rough endoplasmic reticulum. Although the formation of polar granules during oogenesis is not clear (cf. text), it is possible that this same fibrous material becomes localized in the posterior region of the egg (8) and becomes organized into the typical polar granules of the mature egg (1)

nuclear envelope of the oocyte is free of fibrous bodies while the amount of this material on the nurse cells' nuclear membranes continues to increase and can be detected even with the light microscope (King, 1960). No change in fine structure can be seen from that originally present after the polar granules fragmented in the embryo and became attached to the nuclear membrane.

Thus the morphological evidence is good that the protein portion of the polar granule and presumably material of similar properties made by the primordial germ cells is continuous in the germ line until the next cycle of oogenesis. After the oocyte has become differentiated from the nurse cells, we no longer find these fibrous bodies in the oocyte, but they are present in abundance in the nurse cells whose cytoplasm is continuous with the ooplasm and is the source of most of the RNA and other organelles of the egg. In Fig. 6 this life cycle of the polar granule is diagrammatically summarized.

IV. Origin of Polar Granules during Oogenesis

The origin of the polar granule material has been an especially intriguing question in insect embryology. Early workers were at a loss to know the source of the polar plasm. In dipterans and coleopterans the material seemed to appear at the end of vitellogenesis with no appreciable intermediate stages. Because of the suddenness of its appearance, there were only speculations concerning the origin. Nearly every cellular organelle was at some time supposed to give rise to the polar granule. With the advent of clearer cytochemical evidence for the presence of RNA and protein and the absence of DNA or lipid, speculation on the origin became more restrained. At the light microscope level, the best hypothesis concerning the origin of the polar granule material, or oosome as it is usually called in hymenopteran species, was already proposed by Buchner (1918) who first suggested the nurse cells as the origin of the polar granule material. Bier (1954) reasoned that if this is true, then the nurse cell nuclei must be the source of the nucleic acid of the oosome. The eggs of *Formica rufa* show seasonal dimorphism; winter eggs are twice the size of summer eggs with an oosome eight times the volume of summer eggs. Bier was also able to note the greatly increased size of nurse cell nuclei in the winter egg which correlated with his hypothesis of a nurse cell origin of oosomal material.

Meng (1968) has recently described the formation of the oosome in the hymenopteran *Pimpla turionellae*. Just prior to yolk formation she finds the first appearance of this material in the shape of a thin stream starting midway in the egg and stretching to the oosome region at the posterior tip. Later all of the oosome material accumulates in the oosome of the mature egg. Because of this sequence, it is reasonable to assume that the oosome or polar granule material is made in the nurse chamber.

Electron microscopy has not yet clarified the origin of polar granules to any great extent. Nurse cell nuclear membranes have large accumulations of fibrous material at their surface in *Drosophila*, and, as already mentioned, its fine structure is similar to the fibrous bodies derived from polar granules in the embryo. A reasonable hypothesis is that this proteinaceous material acquires the RNA component of the polar granule and then accumulates at the posterior tip of the egg.

In *Drosophila* there is no direct evidence for this transport. Polar granules appear during the process of vitellogenesis as small fibrous bodies which become attached to mitochondria in the maturing egg (MAHOWALD, 1962). In *Miastor* (MAHOWALD and STOIBER, 1970) there is more evidence for such a transport. This species is especially suited for such a study because very little yolk is produced and there are, moreover, no protective egg coverings, thus making fixation for electron microscopy easier. The fibrous material found on the nurse cell nuclear envelopes becomes attached to mitochondria principally at two stages: during the early stages of formation of the egg chamber and the end of oogenesis when the nurse cell chamber is breaking down. At both times there is a movement of cellular organelles from the nurse chamber into the oocyte so that these fibrous bodies probably also move into the egg. The movement at the end of oogenesis can have no effect on polar granule formation because the granules have already appeared at the posterior tip. But the first movement could be related to the formation of polar granules. Recent analysis of serial sections of the germarium in *Drosophila melanogaster* (MAHOWALD, 1970a) has also indicated that there may be a similar movement of fibrous material into the oocyte at the time that oocyte nucleus in this species is losing the fibrous material attached to its membrane. However, it is clear that in both species there is no evident formation of polar granule material at the posterior tip until late in oogenesis.

It would seem that autoradiography after pulses of RNA and protein precursors might tell us something about the origin of polar granules during oogenesis. In studies with ³H-thymidine (MUCKENTHALER and MAHOWALD, 1965) no significant incorporation of thymidine into the polar granule region was found even though it was possible to detect mitochondrial DNA synthesis by autoradiography. This result confirms the reports of negative Feulgen reactions for polar granules (e.g. MUKERJI, 1930). BIER (1963 a, b) has made extensive studies of the incorporation of protein and RNA precursors during dipteran oogenesis and has not found any indication of labelled RNA or protein in the pole plasm region. Thus, instead of discovering that the RNA and protein components of the polar granule are made at the posterior tip of the egg, the incorporation data support the argument presented above that polar granule material must be made elsewhere, or earlier. Moreover, since nearly all of the RNA of the egg is produced by the nurse cells in dipterans (BIER, 1963a; MAHOWALD and TIEFERT, 1970), the most reasonable site for the production of polar granule precursors is the nurse cell chamber.

An origin of the polar granule material from the follicle cells bordering the posterior tip cannot be absolutely excluded (e.g., MENG, 1968). But neither ultrastructural (MAHOWALD, 1962) nor autoradiographic studies give any support to the possible follicular origin of polar granules.

V. Concluding Remarks

During the second half of the 19th century the discovery of polar granules or germ plasm led to the theory of the continuity of the germ plasm from one generation to the next. From the early work it was not at all clear whether the uniqueness of the germ line pertained to the unique cytoplasm that the germ line cells possessed, or to the nuclei. Today we know much about the continuity of information from generation to generation through the genetic material of the nucleus even though many details of the process remain to be worked out. But we are just beginning to appreciate the importance of other types of continuity of cellular organelles from generation to generation. It is not difficult conceptually to cope with the problem when the organelles appear to have their own proper genetic material, e.g. the mitochondria and chloroplasts and possibly the centrioles. But, it is equally certain that the egg becomes the medium through which many other types of cytoplasmic organelles are passed on from the mother generation to the egg. In some cases we also know there is no question of true cytoplasmic continuity in as much as the organelle is always dependent upon the nucleus of the embryonic cells. For example, even though ribosomes are provided to the egg in great abundance by the mother, the work on the anucleolate mutant in *Xenopus laevis* clearly demonstrates that unless the embryonic cells themselves produce more ribosomes, the embryo will die before reaching maturity (Brown and Gurdon, 1964). The problem of the origin and continuity of polar granules may fit somewhere in between these two extremes. There is no question that polar granules are formed during oogenesis and become localized at the posterior tip of the insect egg, and, possibly, analogous organelles may be present in many other species throughout the animal kingdom. But it also seems evident that the organelle, or at least part of it, is always present in the germ line cells of the insects possessing them. There is no evidence available to indicate an autonomous life cycle for the organelle, but the same is certainly true for mitochondria even though it is certain that this organelle possesses its own genetic reservoir of information. But there is suggestive evidence that the polar granule has a life cycle and that at least part of it has a continuity throughout the life of the organism.

One final observation is necessary concerning the unique properties of polar granules. It is clear that the egg cytoplasm is organized so that information for development is specifically localized in the egg (cf. Davidson, 1968). Thus, an important property of the polar granule is to localize the information for forming pole cells and ultimately germ cells in one specific region of the embryo. It is reasonable to suppose that this is part of the specific role of the protein component of the granule which is always present in the germ cell. Thus, further studies on the molecular properties of polar granules may, beyond aiding us to understand the process of cellular determination, lead to an understanding of the mechanisms whereby information for development becomes localized in the egg during oogenesis.

References

Bier, K.: Über den Saisondimorphismus der Oogenese von *Formica rufa rufo-pratensis mino* Gössw. und dessen Bedeutung für die Kastendetermination. Biol. Zbl. **73**, 170—190 (1954).
— Synthese, interzellulärer Transport und Abbau von Ribonukleinsäure im Ovar der Stubenfliege *Musca domestica*. J. Cell Biol. **16**, 436—440 (1963a).
— Autoradiographische Untersuchungen über die Leistungen des Follikelepithels und der Nährzellen bei der Dotterbildung und Eiweißsynthese im Fliegenovar. Wilhelm Roux' Arch. **154**, 552—575 (1963b).
Brown, D. D., Gurdon, J. B.: Absence of ribosomal RNA synthesis in the anucleolate mutant of *Xenopus laevis*. Proc. nat. Acad. Sci. (Wash.) **51**, 139—146 (1964).

BROWN, E. H., KING, R. C.: Studies on the events resulting in the formation of an egg chamber in *Drosophila melanogaster*. Growth **28**, 41—81 (1964).

BUCHNER, P.: Vergleichende Eistudien. I. Die akzessorischem Kerne des Hymenoptereneies. Arch. mikrosk. Anat. **91**, Abt. II, 1 (1918).

COUNCE, S. J.: Developmental morphology of polar granules in *Drosophila* including observations on pole cell behavior and distribution during embryogenesis. J. Morph. **112**, 129—145 (1963).

DAVIDSON, E. H.: Gene activity in early development, 375 pp. New York-London: Academic Press 1968.

GEYER-DUSZYŃSKA, I.: Experimental research on chromosome elimination in Cecidomyidae (Diptera). J. exp. Zool. **141**, 391—488 (1959).

HEGNER, R. W.: Experiments with Chrysomelid beetles. III. The effects of killing parts of the eggs of *Leptinotarsa decemlineata*. Biol. Bull. **20**, 237—251 (1911).

— The germ-cell cycle in animals. 346 pp. New York: Macmillan Company 1914.

JAZDOWSKA-ZADGRODZINSKA, B.: Experimental studies on the role of "polar granules" in the segregation of pole cells in *Drosophila melanogaster*. J. Embryol. Exp. Morph. **16**, 391—399 (1966).

KING, R. C.: Oogenesis in adult *Drosophila melanogaster*. IX. Studies on the cytochemistry and ultrastructure of developing oocytes. Growth **24**, 265—323 (1960).

KOCH, E. A., SMITH, P. A., KING, R. C.: The division and differentiation of *Drosophila* cystocytes. J. Morph. **121**, 55—70 (1967).

MAHOWALD, A. P.: Fine structure of pole cells and polar granules in *Drosophila melanogaster*. J. Exp. Zool. **151**, 201—215 (1962).

— Polar granules of *Drosophila* II. Ultrastructural changes during early embryogenesis. J. Exp. Zool. **167**, 237—261 (1968).

— Polar granules of *Drosophila* III. Continuity of polar granule material throughout the life cycle. J. Exp. Zool. (in press) (1970a).

— Polar granules of *Drosophila* IV. Cytochemical studies showing loss of RNA from polar granules during early stages of embryogenesis. J. Exp. Zool. (in press) (1970b).

— STOIBER, D. L.: Polar granules of *Miastor* (in preparation) (1970).

— TIEFERT, M.: Fine structural changes in the *Drosophila* oocyte nucleus during a short period of RNA synthesis. Wilhelm Roux' Arch. **165**, 8—25 (1970).

MENG, C.: Strukturwandel und histochemische Befunde insbesondere am Oosom während der Oogenese und nach der Ablage des Eies von *Pimpla turionellae* L. (Hymenoptera, Ishneumonidae). Wilhelm Roux' Arch. **161**, 162—208 (1968).

MUCKENTHALER, F. A., MAHOWALD, A. P.: DNA synthesis in the ooplasm of *Drosophila melanogaster*. J. Cell Biol. **28**, 199—208 (1966).

MUKERJI, R. N.: The "nucleal-reaction" in *Apanteles* sp., with special reference to the secondary nuclei and the germ-cell determinant of the egg. Proc. roy. Soc. B **106**, 131—139 (1930).

NICKLAS, R. B.: An experimental and descriptive study of chromosome elimination in *Miastor* spec. (Cecidomyidae; Diptera). Chromosoma (Bul.) **10**, 310—336 (1959).

SMITH, L. D.: The role of a "germinal plasm" in the formation of primordial germs cells in *Rana pipiens*. Devel. Biol. **14**, 330—347 (1966).

ULLMANN, S. L.: Epsilon granules in *Drosophila* pole cells and oocytes. J. Embryol. exp. Morph. **13**, 73—81 (1965).

WOLF, R.: Der Feinbau des Oosoms normaler und zentrifugierter Eier der Gallmücke *Wachtliella persicariae* L. (Diptera). Wilhelm Roux' Arch. **158**, 459—462 (1967).

Centrioles

CHANDLER FULTON

Department of Biology, Brandeis University
Waltham, Massachusetts

What we need we do not know
And what we know we do not need.
GOETHE

If one wandered about asking biologists to complete the sentence "Centrioles are..." the answers might well range from "I don't know" to "Centrioles are self-replicating organelles responsible for the synthesis and assembly of microtubules." Although it is conceivable that the latter reply contains a little truth, the "I don't know" is more likely to be the reply of an expert.

What do we really know about centrioles? This review is a selective attempt to seek both valid generalizations and the frontiers of our knowledge about centrioles. Though statements are documented, no attempt has been made to give all relevant references or, in general, to "review" the voluminous, diffuse, and often conflicting literature. WILSON's definitive "Cell" (1925) provides a thorough and thoughtful discussion of classical studies, and four recent reviews are available: the general reviews of DALCQ (1964) and WENT (1966a), and those of STUBBLEFIELD and BRINKLEY (1967) and DE HARVEN (1968) which emphasize the ultrastructure of centrioles.

I. Genesis of Ideas about Centrioles

The discovery of centrioles is usually credited to VAN BENEDEN and BOVERI[1]. VAN BENEDEN first saw structures at the poles of mitotic spindles, and named them polar corpuscles. In 1887 VAN BENEDEN and BOVERI independently discovered that the polar corpuscles do not disappear at the completion of mitosis, but instead persist throughout the life of the cell. VAN BENEDEN renamed the granules central corpuscles or central bodies. BOVERI suggested the term centrosome, and later centriole. The two men gave similar descriptions of centriole behavior. During interphase a pair of centrioles lie side-by-side next to the nucleus, apparently doing nothing. They separate at the onset of division, and as they move apart asters form around each and a spindle is spun out between them. The nuclear envelope breaks down; the chromosomes mingle with the spindle, organize into the equato-

---

1 I found the first edition of WILSON's "Cell" (1896) an invaluable source of historical perspective as well as good reading. The VAN BENEDEN and BOVERI quotes were found there.

rial plate, and separate. During mitosis the centriole at each pole duplicates, so each daughter cell has a pair of centrioles.

These observations led both men to view the centriole as the organelle primarily responsible for cell division; BOVERI called it "the dynamic center of the cell." Both emphasized the permanence of centrioles. VAN BENEDEN concluded "that every central corpuscle is derived from a pre-existing corpuscle," and BOVERI wrote: "The centrosome is an independent permanent cell-organ, which, exactly like the chromatic elements, is transmitted by division to the daughter cells."

BOVERI and VAN BENEDEN made their first clear observations while studying early embryonic development. The resulting emphasis on the role of the centriole in fertilization led them to a remarkable conclusion. As BOVERI expressed it,

> The ripe egg possesses all of the organs and qualities necessary for division excepting the centrosome, by which division is initiated. The spermatozoon, on the other hand, is provided with a centrosome, but lacks the substance in which this organ of division may exert its activity. Through the union of the two cells in fertilization all of the essential organs necessary for division are brought together.

The concept of the centriole as the "division organ" gave centrioles great importance in the life of cells. VAN BENEDEN's and BOVERI's descriptions of the centriole cycle, first made in the nematode *Ascaris*, were rapidly confirmed in many different organisms, and both the cycle and the theories soon became accepted as general truths (see WILSON, 1896). Then the theories began to be undermined. The first challenge came with descriptions of mitosis in higher plants, where the poles of the mitotic figures are rounded and no centrioles can be found at any time. Though a few were reluctant to admit that some cells never have centrioles, this was certainly as clear as the observation that other kinds of cells seem always to have centrioles. It was not long before evidence accumulated that certain cells have centrioles only sometimes – the centrioles come and go. This provoked controversy that has persisted, in some cases, to the present. Particularly upsetting for the idea of continuity of centrioles, and their role as division centers, was the discovery of parthenogenesis – cleavage without fertilization – which clearly made untenable the concept of the sperm centriole as an essential "division organ." Within a few years the two main ideas about centrioles – continuity and function as division centers – were in serious question.

Centrioles early were given another possible role, that of precursors for the basal bodies of flagella. LENHOSSÉK (1898) was looking at ciliated intestinal epithelium, and noticed that whereas ciliated cells had basal bodies at the base of each cilium, they had no centrioles and did not divide. Intermingled in the epithelium were non-ciliated cells that had a pair of centrioles and divided. He concluded that centrioles gave rise to basal bodies, presumably by multiplication. HENNEGUY (1898) independently noticed an interconversion of mitotic centrioles and flagellar basal bodies in *Bombyx* spermatocytes. Thus came into being the Henneguy-Lenhossék theory that basal bodies are derivatives of centrioles. That centrioles can become or give rise to basal bodies has been demonstrated convincingly many times, with endless variety (see WILSON, 1925; POLLISTER, 1933).

Around the turn of the century doubters appeared who began to question whether, even in those cases where centrosomes were clearly seen, as in *Ascaris* (Fig. 1), they were not coagulation artifacts, optical illusions, or something else.

172   C. FULTON:

The first of these doubters may have been FÜRST (1898), and the number increased
steadily into the 1920s, the best-known being FRY (1929 and other papers). Often
centrioles, even if present, were simply too small to be seen clearly. This difficulty
enhanced the confusion and argument, but led to a declining interest in centrioles.

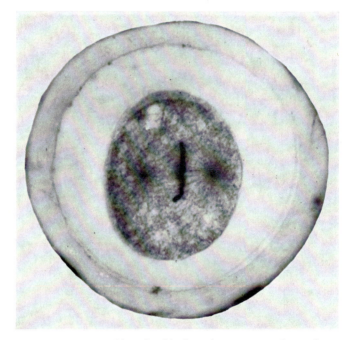

Fig. 1. Centrosomes are very evident in this first cleavage metaphase of an *Ascaris* egg.
These centrosomes are similar to those drawn by BOVERI (1901) and others. Photograph,
from Turtox slide no. E 6.27, provided by General Biological, Inc.

From about 1920 to 1960 few studied centrioles, though WILSON (1925) and some
others discussed them seriously. Centrioles barely received mention in many texts
of general biology or even of cytology.

   With the advent of electron microscopy centrioles were no longer too small
to see. At first centrioles were described as cylindrical forms with a low-density
core and a dense wall made of approximately 9 parallel tubules (esp. BURGOS and
FAWCETT, 1955; DE HARVEN and BERNHARD, 1956; PORTER, 1957; BESSIS, BRETON-
GORIUS, and THIÉRY, 1958; BERNHARD and DE HARVEN, 1960). As methods were
refined, it gradually became clear that the wall is composed of 9 triplet tubules
in pinwheel array (GIBBONS and GRIMSTONE, 1960). Many of these authors emphasized the homology in structure between basal bodies and centrioles.

   During the early 1960s it became clear, through the use of improved fixatives,
that microtubules were components of many structures associated with centrioles
and basal bodies, including the spindle fibers (e.g., DE HARVEN and BERNHARD,
1956; HARRIS, 1962; KANE, 1962; ROTH and DANIELS, 1962; SLAUTTERBACK, 1963;
LEDBETTER and PORTER, 1963).

The ability to see centrioles clearly led to a renewed interest in their continuity and reproduction. However, recent studies have focused on problems in a way that seems largely the result of historical accident. A brief chronology makes this stand out. After VAN BENEDEN and BOVERI argued that centrioles are permanent, self-reproducing organelles, this idea continued to develop (with many pros and cons). WILSON (1925, p. 1127), who knew as much about centrioles as anyone, defined a centriole as an "autonomous cell-organ arising only by the growth and division of a preexisting centriole." During the 1930s CHATTON and LWOFF made a precise description of the reproduction of basal bodies in ciliates. The ideas resulting from this description were discussed in a widely read and charming book by LWOFF (1950), who emphasized throughout that "the kinetosome [basal body] is always formed by the division of a pre-existing kinetosome. It is endowed with genetic continuity." The notion that centrioles and basal bodies were self-reproducing (by "division") became accepted as fact. This made reports of the temporary appearance of centrioles all the more interesting — and implausible. Electron microscopists began to look for a mechanism of reproduction, with the expectation that centrioles would serve as "rubber stamp" templates for the production of new centrioles (see DE HARVEN and BERNHARD, 1956; MANTON, 1959). When the WATSON-CRICK model of DNA "explained" self-reproduction — or replication as it became known — an autonomous, self-replicating entity became one that contained DNA. Soon thereafter people began seeking DNA in basal bodies and centrioles.

Terminology

Centriole, the term chosen by almost all electron microscopists, is used here to include central body, centrosome, cell center, and other terms used by light microscopists. Though these words often have been used loosely and interchangeably, centriole has most often been used to refer to the minute dot, and the other words have acquired various shades of meaning (see WILSON's glossary, 1925; WENT, 1966a). Certainly a contemporary definition of a centriole should be simply structural and should not include or imply a function.

Centriolar pinwheel is here defined specifically as any structure which, by electron microscopy, looks like a centriole: a cylinder about $0.2\ \mu$ in diameter with a wall composed of 9 equally spaced groups of microtubules, usually triplets. Where fine structure is known, the term centriole is here reserved for organelles that have centriolar pinwheels, but this has not always been done in the literature. In studies that depend on light microscopy, often we simply do not know if a centriole is a centriolar pinwheel (Section II, B).

Diplosome is a pair of centrioles.

Basal body is a centriolar pinwheel that bears a flagellum or cilium. There are many synonyms, including *kinetosome*, blepharoplast, basal granule, basal corpuscle, and proximal centriole. In the past some have confused the *kinetoplast* of trypanosomes (see TRAGER, 1965) with the kinetosome, though these are totally different structures.

In this review the term centriole is used both in its specific sense and as an inclusive term for the homologous structure that can serve either as a centriole or a basal body. Thus "centriolar pinwheel" or "centriolar DNA" refers to the pinwheel and the DNA of basal bodies as well.

Centrosphere is the area around one or more centrioles that by light microscopy stains differently and by electron microscopy has a different appearance than the surrounding cytoplasm. Often centrosome is used instead of centrosphere.

Centromere is a synonym for the chromosomal *kinetochore*, not for centrosome or centrosphere, and since this frequently causes confusion only kinetochore will be used here.

Procentriole is an immature centriole (or basal body) which has or is acquiring the centriolar pinwheel but is still developing. This definition is a little broader than the one proposed by GALL (1961), but does not include short centrioles (e.g., RENAUD and SWIFT, 1964). It certainly is a more useful definition than the one suggested in a recent review (GRANICK and GIBOR, 1967) and then in a biological dictionary (RIEGER, MICHAELIS, and GREEN, 1968): "an organelle that is presumed to give rise to all cytoplasmic microtubules."

In describing position in a cell, *proximal* is toward and *distal* is away from the center. These positions are obvious in basal bodies. For centrioles themselves the positions are defined by resemblance to comparable positions in basal bodies, no matter in which direction a centriole is oriented in a cell.

II. Structure and Composition

A. Morphology

The fine structure of centrioles has recently been reviewed in detail by DE HARVEN (1968) and STUBBLEFIELD and BRINKLEY (1967), who also have contributed new observations. FAWCETT's (1966) text has excellent, informative illustrations.

Cylinder Wall. The most striking and regular ultrastructural feature of centrioles and basal bodies is the array of nine triplet microtubules equally spaced around the perimeter of an imaginary cylinder (Fig. 2). The space between and immediately around the triplets is filled with an amorphous, electron-dense material. In transverse section the triplets are arranged like the vanes of a pinwheel. The vanes are tilted so they form an angle of roughly 40° to the radius of the cylinder. Centrioles are about $0.15-0.25\,\mu$ in diameter and usually $0.3-0.7\,\mu$ long, though some are as short as $0.16\,\mu$ (RENAUD and SWIFT, 1964) and others as long as $8\,\mu$ (FRIEDLÄNDER and WAHRMAN, 1966).

Since centrioles have no outer membrane, the triplets are considered to form the wall of the cylinder, and arbitrarily define the inside and outside of the centriole. The centriole looks like a "squirrel-cage" (RANDALL and HOPKINS, 1963), so the ultrastructure does not provide a *visible* boundary that would suggest any restriction on free exchange with the cytoplasm.

Triplets. The nine triplets that make up the wall are indistinguishable from one another, and seem basically similar in centrioles and basal bodies. The three subunit microtubules have been designated A, B, and C, with the innermost tubule being A (GIBBONS and GRIMSTONE, 1960). Individual tubules are $200-260$ Å in diameter. Only the A tubule is round; the others are partial and share their wall with the preceding tubule. Because of this it is misleading to refer to a centriole as having 9 × 3, or 27, microtubules, as is often done. At both ends the C tubules often terminate before the A and B tubules (e.g., GIBBONS and GRIMSTONE, 1960; WOLFE, 1970).

The substructure of basal body tubules has been examined by RINGO (1967b) and WOLFE (1970), and seems to be similar to the structure of other microtubules (reviewed by STEPHENS, 1970b). The A tubule has 12 or 13 40–45 Å globular subunits around its perimeter. Three or four of these subunits are shared with the B tubule, which in turn shares several of its subunits with the C tubule.

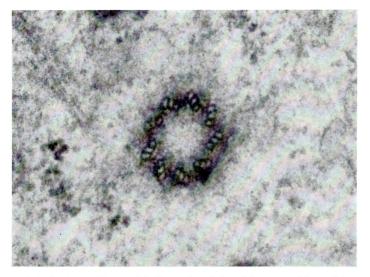

Fig. 2. Centriole in the oviduct epithelium of a 5-day-old mouse. The triplets are tilted inward in a clockwise direction, which indicates that the centriole is viewed from the proximal end. Courtesy of E. R. DIRKSEN. 90,000 ×

Arrangement of Triplets in the Wall. Often the triplets are thought to run parallel to one another and to the long axis of the cylinder, but this is not always the case. In the basal bodies of some organisms, the triplets get closer toward the proximal end, so the diameter of the cylinder gets smaller (GIBBONS, 1961; DINGLE and FULTON, 1966). In some centrioles the triplets are parallel to one another but turn in a long-pitched helix with respect to the cylinder axis (ANDRÉ, 1964;[2] FAWCETT, 1966). In addition, STUBBLEFIELD and BRINKLEY (1967) have found that the angle of the triplet vanes with respect to the cylinder radius can vary within a centriole – a change of 10° in 3 successive sections – and thus that in some centrioles there is a twist in the individual triplets. They proposed that the triplets are hinged along the C tubule so the A tubule can move toward or away from the cylinder axis. It remains to be seen how general these changes of triplet arrangement within a centriole are, and whether they vary during the life cycle of a centriole. The

2 Examples of the excellent micrographs on which ANDRÉ based his conclusions may be found in DALCQ (1964) and in GACHET and THIÉRY (1964).

changes, and especially the helical twist, may explain the frequent observation of centrioles in which the triplets on one side are sectioned transversely whereas those on the other side appear obliquely sectioned (e.g., Fig. 3). Relatively few micrographs of centrioles have been published which show all 9 triplets clearly (see Fawcett, 1966). It is not uncommon to find such images of basal bodies, suggesting the possibility that basal bodies may have less twist than centrioles.

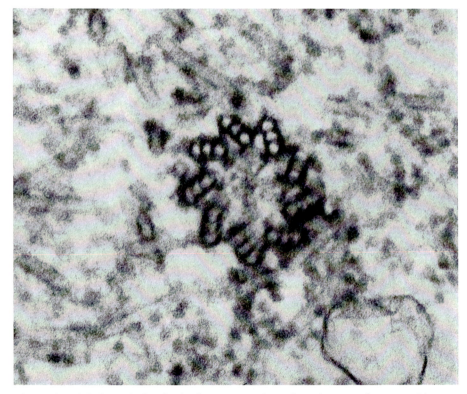

Fig. 3. Centriole in an isolated mitotic apparatus from first cleavage of a sea urchin egg. Courtesy of R. E. Kane. 150,000 ×

This leads to an important point. In discussing the morphology of centrioles, much is sometimes made of one or a few pictures. It is known that most vertebrate cells occasionally make a single flagellum, and thus use their centrioles as basal bodies (see Section V, A). A single section through such a cell might include either a centriole or a basal body; often, unless a flagellum is included in the section, there is no way to be sure which one. As de Harven (1968) emphasized, some of the information about centriole morphology may actually be derived from centrioles serving as basal bodies. Certainly this is not true of all micrographs of centrioles, and probably not of most of them, but indicates a need for extra caution in building models based on single micrographs.

Linkers. The major observed connection between tubules is a dense link between the A and C tubules. These A–C linkers have been seen many times, and show clearly in the micrographs included here (esp. Fig. 3). It has been suggested that the A—C linkers are responsible for the radial tilt of the triplets (e.g., GIBBONS and GRIMSTONE, 1960; DIPPELL, 1968). The A—C linkers do not seem to run the full length of centrioles or basal bodies.

Feet. There are dense condensations from the A tubules toward the center of the cylinder, forming a series of "feet" (Figs. 2 and 3). The probable structures formed by these feet have been considered by GIBBONS and GRIMSTONE (1960), ANDRÉ (1964), STUBBLEFIELD and BRINKLEY (1967), and DE HARVEN (1968).

Cartwheel. One of the most prominent connections between tubules is the hub-and-spokes cartwheel, which is found at the proximal end of centrioles and basal bodies (GIBBONS and GRIMSTONE, 1960; GALL, 1961; and many subsequent observers). The cartwheel consists of a central hub and 9 radial spokes which run toward the A tubules (Fig. 6E). These spokes end in the dense "feet" mentioned above.

Pinwheel is used to refer to the arrangement of triplets, and cartwheel to the hub-and-spokes.

Vacuole. A small, membrane-bounded vacuole or vesicle has been found inside the centriolar pinwheel several times (e.g., Fig. 4), though the regularity of its occurrence is unknown (ANDRÉ, 1964; STUBBLEFIELD and BRINKLEY, 1967).

Other Internal Structures. Several other structures have been observed in centrioles and basal bodies but in general these are less clearly defined and there is less evidence that they occur regularly. For example, both STUBBLEFIELD and BRINKLEY (1967) and DIPPELL (1968) present evidence that a fine strand is found inside the pinwheel, possibly wound as a helix — but the two descriptions are quite different. Basal bodies in general have a pronounced proximal-to-distal structural differentiation, which has been described in detail in several ciliates and flagellates (esp. GIBBONS and GRIMSTONE, 1960; GIBBONS, 1961; DINGLE and FULTON, 1966; DIPPELL, 1968). In most basal bodies considerable dense material has been found in the lumen, but the basal body of *Chlamydomonas* lacks this dense material (RINGO, 1967a). *Chlamydomonas*, as well as other algae and lower plants, have an interconnection of tubules which resembles a 9-pointed star at the distal end of the basal body (LANG, 1963; MANTON, 1964a; RINGO, 1967a).

B. Light Microscopists' Centrioles

It is difficult in many cases to relate what light microscopists called centrioles, by whatever name, to the centriolar pinwheel of electron microscopists. The structures drawn or photographed by light microscopists are often much larger than can be accounted for by a $0.2\,\mu$ diameter centriole. The larger structure seen with the light microscope is frequently the centrosphere, and not the small centrioles it contains. The centrosphere is undoubtedly what is seen in the photograph of *Ascaris* mitosis (Fig. 1), though hidden in the centrosphere are normal centriolar pinwheels (P. FAVARD, cited in DALCQ, 1964). Light microscopists have described an impressive diversity of different structures as centrioles, even within a single species (see WILSON, 1925, p. 673; FRY, 1929, p. 144). *Ascaris*, for example, shows variation in the appearance of the centrosphere region during development (see BOVERI, 1901; WENT, 1966a).

Cleveland (e.g., 1957, 1963) has made elegant light microscope descriptions of the behavior of giant centrioles found in certain flagellates that live in termite guts. These giant centrioles have provided the basis for many discussions of centriolar behavior (e.g., Mazia, 1961; Grell, 1964; Went, 1966a). However, Grimstone and Gibbons (1966) have since found that in at least two species these "giant centrioles" neither resemble nor contain centriolar pinwheels, which these flagellates also have. For this reason it is uncertain what relationship, if any, Cleveland's famous "centrioles" have to the 9-triplet centrioles of most other organisms (see Section VI, A), and so these "giant centrioles" are not considered further in this review.

Many early cytologists reported intranuclear centrioles, especially in cells that accomplish karyokinesis within an intact nuclear envelope (reviewed by Wilson, 1925; Kater, 1929). The repeated finding of intranuclear centrioles seems to have reflected the expectation that cells could not accomplish karyokinesis unless they had division centers *someplace*. No one has ever seen an intranuclear centriolar pinwheel, although recently intranuclear engulfment and apparently destruction of centrioles during spermatogenesis has been reported (Reger, 1969). At present it seems unlikely that any of the early reports of intranuclear centrioles will be confirmed.

Most of the classical work on basal body reproduction in ciliates has depended on silver staining to visualize the basal bodies (see Lwoff, 1950). Dippell (1962 and personal communication) has studied this staining reaction by electron microscopy, and has found that the silver stain is deposited on the outer surface of the cell around the base of the cilium, and also at the similar juncture of a nonciliated basal body and cell surface. The staining is correlated with the presence of basal bodies but the basal bodies themselves are not stained.

The dichotomy between light and electron microscope observations of centrioles and basal bodies makes caution necessary in comparing results. Often the electron microscope studies have confirmed and extended the light microscope studies (e.g., Sharp, 1914, and Mizukami and Gall, 1966). In other cases, such as intranuclear centrioles, there does not seem any likelihood of confirming the classical studies. In some cases where centriolar pinwheels are known to be present, centrioles cannot be seen with the light microscope. For example, Friedländer and Wahrman (1966) reported that they could not see centrioles with the light microscope except when a centrosphere was present. Usually we have no way of knowing whether our extrapolation of light microscopy is valuable or fanciful.

C. Associated Structures

Light microscopists found centrioles associated with the mitotic apparatus and basal bodies with the flagellar apparatus. The nature of these associations is examined in relation to the possible functions of centrioles in Section V; here we examine the ultrastructure of components associated with the electron microscopists' centrioles.

1. Associates of Centrioles

Centrioles are most often found associated in pairs (the diplosome). Usually during interphase these are next to the nuclear envelope, though in some cell types they are near the cell surface (Fawcett, 1966). Sometimes these pairs are arranged end-to-end (see especially Moser and Kreitner, 1970), but more commonly they come together to form a right angle, seen early by light microscopists and subsequently by many electron microscopists (see Gall, 1961, and Section III, A). Often the area in which a centriole or a diplosome lies is relatively devoid of other cellular constituents, including ribosomes. This area, the centrosphere, contains a fine fibrous material (e.g., Robbins, Jentzsch, and Micali, 1968; Kalnins and Porter, 1969).

Amorphous plaques or spheres of electron-dense material, with poorly defined outer limits, are often found near centrioles. These were described as "massules" by Bessis, Breton-Gorius, and Thiéry (1958) and as "pericentriolar satellites" by Bernhard and de Harven (1960). There has been some debate about how regular these are, and in particular whether there is an arrangement of exactly 9 satellites around the centriole (see de Harven, 1968). Szollosi (1964) shows a case where there definitely are 9 dense satellites, which appear temporarily during spermatogenesis in a jellyfish. André (1964) and others have reported that in other cells the number and location of satellites vary, and several have described regular changes in the satellites during mitosis (Section V, A).

2. Associates of Basal Bodies

Basal bodies have a remarkable variety of associated dense plaques and masses, and of rootlets, rhizoplasts, and other filaments (reviewed by Pitelka, 1969). (There are also occasional rootlet-like structures associated with centrioles, described by Sakaguchi, 1965.) Often the arrangement of accessories relative to the basal bodies is very specific. This is especially clear in ciliates, where the accessories are predictably oriented in the kinetosomal unit (see Dippell, 1968; Allen, 1969; Hufnagel, 1969a).

3. Microtubules

Since microtubules were first observed they have been found in association with centrioles, though orderly arrays of microtubules can also form in the absence of centrioles. Many aspects of microtubules, including places where they are found, their structure, and their chemistry, have been reviewed by Porter (1966) and Stephens (1970b).

Cytoplasmic Microtubules in Interphase Cells. In cells that are not dividing, centrioles are often surrounded by a small number of microtubules. In such cells, the diplosome often forms a focal point from which a small interphase aster composed of microtubules radiates (de-Thé, 1964; Stubblefield and Brinkley, 1967).

Centriole-associated microtubules rarely seem actually to contact the centriolar tubules, though there are a few cases described in which microtubules come close to or even into the centriolar pinwheel (see Krishan and Buck, 1965; Stubblefield and Brinkley, 1967). More often the microtubules seem to end in dense material, such as the pericentriolar satellites (e.g., de-Thé, 1964; Szollosi, 1964; Fawcett, 1966; Bilke, Tilney, and Porter, 1966; de Harven, 1968). Even though the microtubules usually do not touch the triplet tubules, the centriole provides a "center" from which the microtubules emanate as rays from a star.

12*

The little interphase asters have no obvious function, but there are other groups of cytoplasmic microtubules that are associated with centrioles that have likely functions, such as serving as an apparent cytoskeleton (Porter, 1966). Sometimes these microtubules radiate from centrioles (e.g., de Harven and Bernhard, 1956; Gall, 1961; Slautterback, 1963; Bilke, Tilney, and Porter, 1966; Gibbins, Tilney, and Porter, 1969), but just as often they are not associated with centrioles (examples in Porter, 1966). The microtubules regularly end in dense material whether or not they are focused toward a centriole (Porter, 1966).

Neurotubules. There is a special class of cytoplasmic microtubules which run along nerve cell processes, the neurotubules. In at least some cases, neurotubules emanate from dense masses around centrioles (Gonatas and Robbins, 1964).

Mitotic Microtubules. The spindle fibers, which are composed of microtubules, are regularly associated with centrioles, but also form in mitotic cells in organisms that do not have centrioles. The microtubules of the asters, the amphiaster, and the continuous and chromosomal fibers of the mitotic apparatus all come to focus toward the centrioles. As in the previous cases, these microtubules usually do not directly contact the centrioles.

Microtubules Associated with Basal Bodies. Basal bodies have several distinctly different arrays of associated microtubules. In marked contrast to the centrioles, one type of microtubule issues directly from the basal-body microtubules. The A and B tubules of the basal body are directly continuous with the A and B tubules of the flagellar or ciliary outer fibers, whereas the basal-body C tubule terminates (Gibbons and Grimstone, 1960, and subsequent observers). This continuity of microtubules provides an important distinction between the behavior of basal bodies and of centrioles. Though the outer fibers are a direct continuation of the centriolar tubules, there are usually complex structural changes in the transition zone between flagellum and basal body (emphasized by Gibbons, 1961), so a basal body is more than simply a centriole with a flagellum.

A flagellum also usually contains a central pair of microtubules, which originate at the transition zone between basal body and flagellum, where there is often a dense structure. The central pair establishes a 2-fold symmetry on the 9-fold rotational symmetry of the centriolar pinwheel; the 2-fold symmetry of cilia is oriented specifically with reference to the direction of beat (Gibbons, 1961).

Basal bodies are also associated with the microtubules of diverse forms of modified cilia, including various sensory structures (reviewed by Fawcett, 1961).

Subsurface or subpellicular microtubules often surround a basal body (Pitelka, 1969). These tubules originate from the proximal end of the basal body, angle sharply to the surface, and there form a regular array. Though the basal body serves as an obvious focal point of origin for these, there is no morphological continuity, so they are like the cytoplasmic microtubules that emanate from centrioles. Still other groups of microtubules often are associated with the proximal end of a basal body without going to the cell surface. The arrangement of the basal-body-associated tubules is usually very precise.

Fuller and Calhoun (1968) described a remarkable case in a water mold where 9 groups of 3 microtubules emanate from dense material surrounding the basal body. There is no continuity between the basal-body tubules and the cytoplasmic tubules, and it is not known whether the numerical correspondence in number of tubules is coincidence or related to the 9 triplets of the basal body.

Complicated accessory filaments made up of microtubules, such as the axostyles of certain flagellates (GRIMSTONE and CLEVELAND, 1965), often form in association with the basal bodies, when basal bodies are present. On the other hand, filaments of similar complexity can form without any association with centrioles, as do the axopodia of *Echinosphaerium* (TILNEY and PORTER, 1965).

D. Isolation and Gross Chemical Composition

Almost all chemical analysis has been done with basal bodies isolated from *Tetrahymena* using variations of the ethanol-digitonin procedure developed by CHILD and MAZIA (1956). Ciliates are obviously advantageous for such studies since each cell has thousands of basal bodies. However, even with this starting material it has not proved easy to isolate basal bodies in a form suitable for analytical study.

There have been some beginnings toward isolation of basal bodies from the ciliate *Euplotes* (GRIM, 1966) and from trout sperm (FISCHER, HUG, and LIPPERT, 1952). Purification of centrioles has been aptly described as "a desperate endeavor" (DE HARVEN, 1968).

The ethanol-digitonin isolated basal bodies of *Tetrahymena* were first studied by SEAMAN (1960), whose study has been frequently and justifiably criticized because only light microscopic examination was used to evaluate the composition of the basal body fraction. SEAMAN found that his fraction was mostly protein but contained a little of everything, including enzymes that made the basal bodies "well equipped to carry out the role of self-duplication and synthesis which has been assigned to them." SEAMAN's interest in synthesis led him (1962) to test the ability of his fraction to synthesize proteins, which apparently it could do if conditions were right.

ARGETSINGER (1965) and HOFFMAN (1965) independently isolated fractions they characterized as mainly basal bodies by electron microscopy. But these fractions were grossly contaminated. For example, ARGETSINGER's final fraction consisted of basal bodies with attached kinetodesmal fibers, a tab of pellicle, and some debris. This fraction had only one-tenth the DNA of SEAMAN's fraction, but about the same amount of RNA (both found about 4% as much RNA as protein, which is about a fifth as much as is found in the ciliates). HOFFMAN also found some RNA but little DNA (see below).

In these studies much hinges on the nature of the fraction analyzed. SATIR and ROSENBAUM (1965) showed that the final fractions, after digitonin treatment, have morphologically altered basal bodies, with the lumen empty of dense material and only doublet instead of triplet tubules left in the wall (HOFFMAN, 1965, also found doublets). They found their purest basal bodies far from pure. They concluded, and I heartily concur, that chemical analysis "still seems to us to be premature."

A number of people have tried to learn something about the composition of centrioles by cytochemical procedures. For example, STICH (1954), ACKERMAN (1961), and DAVID-FERREIRA (1962) all suggested the presence of RNA in centrioles. These studies are inconclusive, especially since with the light microscope probably much of the cytochemistry is of the centrosphere rather than the centriole (see Dalcq, 1964; de Harven, 1968).

Hartmann (1964) studied the cytochemical localization of ATPase during mitosis in tissue culture cells, and "frequently" observed a high concentration toward the poles, suggesting that ATPase "may originate in the centrosome." Abel (1969) demonstrated ATPase by electron microscopic cytochemistry, and found a marked ATPase staining reaction localized on or close to the centriolar tubules in several cell types. The centriolar ATPase might be similar to the dynein associated with flagellar outer fibers (Gibbons, 1965).

Other microtubules that have been studied, and probably by structural homology the triplet tubules of centrioles, are made of one or more species of microtubule structural protein, tubulin (see Section V, B). The marked osmiophilia of the centriolar tubules (greater than most other microtubules) suggests the presence of lipid or other osmiophilic material in the tubules.

E. Nucleic Acids

1. DNA

Searchers for nucleic acid in centrioles — in this case specifically in basal bodies — have been influenced by the idea of autonomous self-reproduction. The question of whether there is DNA in basal bodies has usually been asked using experiments designed to obtain evidence that there *is* DNA in basal bodies, since it seemed reasonable to think that a self-replicating entity would contain its own DNA. It is relatively easy to obtain evidence for something, even something erroneus, especially if one is allowed to disregard experiments that "don't work." In contrast, it is difficult if not impossible to prove that there is *no* DNA in basal bodies. Certainly simple failures to find DNA are not evidence that there is none. In terms of number of reports of DNA in basal bodies, the evidence might be labeled impressive. Many have accepted the conclusion that basal bodies contain DNA. Because of this, I will be a devil's advocate and deliberately lean in the other direction, critically examining the evidence for the presence of DNA. A review based on the premise that basal bodies contain DNA has been published (Granick and Gibor, 1967).

Several distinct types of evidence are available.

Isolated Basal Bodies. The evidence from isolated basal bodies (all in *Tetrahymena*) is unsatisfactory, because on the one hand the basal bodies were contaminated with other cell fractions, and on the other hand much material was leached out of basal bodies by the method of preparation used. It is noteworthy that in these studies the amount of DNA found declined, with the increasing effort to purify basal bodies, from 6% as much DNA as protein (Seaman, 1960), to 0.6% with a range of 0.0 to 0.9% (Argetsinger, 1965), to such a low amount that in many cases Hoffman (1965) was uncertain whether there was any DNA.

Isolated Pellicles. Hufnagel (1969b) made a careful study of the DNA associated with isolated *Paramecium* pellicles, which are rich in basal bodies as well as other cortical components. The yields of pellicular DNA were variable, and the only consistent DNA isolated resembled nuclear DNA in its buoyant density. All the evidence suggested that this DNA was contaminating nuclear DNA. Hufnagel concluded that if there was any basal body DNA it was either (a) lost during isolation of pellicles, (b) similar in buoyant density to nuclear DNA, or (c) present in a quantity not greater than 3.2×10^{-17} g of DNA per basal body.

Feulgen Staining. RANDALL and FITTON-JACKSON (1958) and RANDALL (1959) reported that the basal bodies of *Stentor* and *Tetrahymena* stained clearly with the Feulgen reagent. As was noted in both papers, "stringent ... criteria" were used to make sure that the reaction was real, in view of the small size of the basal body (for *Stentor*, they estimated the size to be $0.2\,\mu$ in diameter by $0.75\,\mu$ long, and saw a "deep magenta" stain). Later RANDALL and DISBREY (1965) noted that these studies "proved difficult to corroborate." McDONALD and WEIJER (1966) reported finding DNA in *Neurospora* centrioles by Feulgen staining, a finding which provided this organelle "with the necessary genetic information for its independent division." The size of the Feulgen-positive bodies varied greatly, ranging up to $1.2 \times 2.4\,\mu$, and the staining intensity indicated "a surprisingly large amount of DNA." It is not clear what these bodies in *Neurospora* are; they do not seem to be equivalent to centriolar pinwheels. Many cytologists have studied cells with Feulgen staining, and there seem to be no other reports of Feulgen-positive centrioles.

Acridine-orange Fluorescence. It is also possible to stain nucleic acids with acridine orange, after which single-stranded nucleic acids fluoresce red and double-stranded fluoresce yellow-green. RANDALL and DISBREY (1965) first used this approach to seek DNA in basal bodies. They isolated *Tetrahymena* pellicles, using an ethanol procedure, stained the pellicles with acridine orange, and examined their fluorescence. They reported a yellow-green fluorescence corresponding in position to rows of basal bodies, removable by DNase, and thus indicating double-stranded DNA. Similarly they reported a red fluorescence, removable by RNase, and thus indicating single-stranded RNA. By visually comparing the yellow-green fluorescence with the amount produced by bacteriophage T2, they concluded that basal bodies had as much DNA as phage T2, or about 2×10^{-16} g of DNA per basal body.

Study of RANDALL and DISBREY's paper leaves many questions unanswered. For example, the photographs which illustrate the paper do not all convincingly support the stated results. Their Figs. 14 and 15 both show pellicles with many red and scattered yellow dots, but their Fig. 14 illustrates the appearance of a pellicle before and Fig. 15 the appearance of one after treatment with DNase.

SMITH-SONNEBORN and PLAUT (1967) used the same approach to study *Paramecium* pellicles, but they gave more attention to detail. The photographs which illustrate their paper are striking, especially one which shows a large area of pellicle with regularly spaced dots of yellow-green fluorescence. Their evidence from acridine-orange fluorescence is the strongest argument for DNA in basal bodies, and so deserves especially critical scrutiny.

1. They found spots of "brilliant green fluorescence" which they interpreted as indicating the presence of double-stranded nucleic acid, either DNA or RNA. As SMITH-SONNEBORN and PLAUT recognized, the fluorescence reaction is complex, and depends on the stacking of the dye. Single-stranded RNA can fluoresce green, as can some proteins and acid mucopolysaccharides (reviewed by KASTEN, 1967).

2. The fluorescent spots were localized over basal bodies, as confirmed using phase contrast microscopy. With the light microscope it is impossible to be sure

that the fluorescence is coming from the basal bodies. This is clearly illustrated by the unexpected localization of the silver stain, which was once thought to stain the basal bodies (see Section II, B).

3. The fluorescence was gone after treatment with DNase, RNase, histone, or protamine. The DNase was presumed to remove the DNA from basal bodies, showing that the green fluorescence was due to DNA rather than RNA, whereas the other proteins were supposed to "interfere with dye binding." Though the conclusion may be correct, it is dangerous to argue that an effect shown by DNase provides evidence for DNA, while the same effect by RNase is artifactual. Spurious results with protein binding are known, and so are artifacts in enzymatic digestion (Bradley, 1966; Kasten, 1967).

4. The green changed to red fluorescence after treatment with 5% acetic acid and then 4% formaldehyde. The treatment was interpreted as making the DNA single-stranded. Such treatment could alter any of the substances likely to cause the fluorescence, so the result has reasonable but not definite interpretations.

Both Randall and Disbrey (1965) and Smith-Sonneborn and Plaut (1967) could detect the green fluorescence only in pellicles isolated during certain portions of the life cycle. Even under the most favorable conditions, only 50—60% of the pellicles from *Paramecium* showed any fluorescence. The inability always to find it suggested that either (a) DNA was only present during part of the life cycle or (b) DNA was always present but masked, either by being too diffuse to be detected or by being covered by something like a basic protein. Both groups favored the latter alternative.

Autoradiography. Several workers have labeled cells with ^{3}H-thymidine and then used autoradiography to seek DNA in basal bodies. E. R. Dirksen, the one person who used this approach with centrioles, was unable to obtain any evidence for centriolar DNA in sea urchin embryos (cited in Mazia, 1961) or embryonic rat tracheas (Dirksen and Crocker, 1966).

The ciliates have been studied most extensively. Rampton (1962) labeled *Tetrahymena*, and calculated that after 40 days' exposure to emulsion he would find a single silver grain over each basal body if each contained 10^{-15} g of DNA. He did not find any labeling over basal bodies. Using the same organism, Stone and Miller (1965) could account for 97% of the cytoplasmic labeling of DNA as being within 1μ of mitochondria. Though these authors were not searching for centriolar DNA, they also certainly did not find any. Pyne (1968) did a detailed electron microscope autoradiographic search for thymidine incorporation, but found no basal-body labeling. He concluded that *Tetrahymena* basal bodies have no DNA or at least much less than a mitochondrion, which has about 4×10^{-16} g of DNA (Suyama and Preer, 1965; see also Hufnagel, 1969b).

Randall and Disbrey (1965) also labeled *Tetrahymena* with ^{3}H-thymidine, but they did light microscope autoradiography with isolated pellicles, and found labeling that seemed to be associated with the kineties (rows of basal bodies). No controls were done. Sukhanova and Nilova (1965) reported incorporation of thymidine into basal bodies of *Opalina*, but their autoradiographs are not compelling.

There have been two detailed autoradiographic studies using *Paramecium*. In the first, Smith-Sonneborn and Plaut (1967) found ^{3}H-thymidine incorporated into material which gave silver grains in rows corresponding more or less to kineties. If the paramecia were labeled for 2 fissions, 70–80% of the label was removed by DNase, but after 4–5 fissions only 40–50% was removed by DNase. Various control experiments all supported the idea that the labeling was not due to contamination of the pellicles by DNA from the macronucleus. There are obstacles in the way of interpreting these experiments as unequivocal evidence for DNA in basal bodies:

a) Thymidine does not specifically label DNA in *Paramecium*. It also labels RNA. Flow of ³H from thymidine into RNA was reported previously (BERECH and VAN WAGTENDONK, 1962), and confirmed by SMITH-SONNEBORN and PLAUT (1967). The lack of specificity may extend beyond nucleic acids, because after 24 hours of labeling most of the DNase-insensitive isotope is not removed by RNase (SMITH-SONNEBORN and PLAUT, 1969). KIMBALL and PERDUE (1962) also found ³H from thymidine in material insensitive to both DNase and RNase. Flow of tritium from thymidine into compounds other than DNA is well known (CLEAVER, 1967).

b) There is a peculiar selectivity to labeling, the significance of which is unclear. Both macronuclei and pellicles are labeled when thymidine is supplied through the medium; macronuclei but not pellicles are labeled when the isotope is fed via prelabeled bacteria (SMITH-SONNEBORN and PLAUT, 1969). Previously BERGER and KIMBALL (1964) observed that feeding paramecia with bacteria prelabeled with ³H-thymidine resulted in more specific incorporation of label into the ciliates' DNA than if the thymidine was supplied through the medium. One possible explanation of the difference, which BERGER and KIMBALL (1964) recognized, is that thymidine from bacteria is probably phosphorylated, whether in a precursor pool or in the bacterial DNA. If this is the case, prelabeled bacteria would provide thymidine nucleotide, which might flow into the DNA of paramecia by a direct route, perhaps more specifically than the nucleoside thymidine, which is not a normal intermediate in DNA synthesis (see CLEAVER, 1967). This speculation suggests the possibility that exogenous thymidine may label pellicles because it is metabolized to other things besides thymidylic acid and does not exclusively label DNA.

c) The lack of specific labeling of DNA by ³H-thymidine does not explain the fraction of the pellicular label that is removed by treatment with DNase. However, the evidence that this material is DNA depends entirely on the assumed specificity of the DNase preparation for DNA. The authors did not investigate the specificity of the reaction. Commercial DNase of the kind used is known to contain traces of RNase (ZIMMERMAN and SANDEEN, 1966), and might also contain other enzymes (e.g., lipases) which have not been sought. Though usually DNase sensitivity indicates DNA, in this case the possibility remains that the DNase was removing some other compound from the pellicles. Other tests, in addition to DNase treatment, should be applied to determine whether the pellicular label is in DNA.

d) The localization of silver grains over basal bodies cannot be done with confidence using the light microscope. In addition, the best evidence for an association of label with the kineties (Figs. 4 and 5 of SMITH-SONNEBORN and PLAUT, 1967) was obtained after 4-5 fissions in ³H-thymidine, when less than half the label was DNase sensitive.

e) The possibility of pellicular contamination with nuclear DNA is seemingly ruled out by SMITH-SONNEBORN and PLAUT (1969), but is a real possibility on the basis of the work of Hufnagel (1969 b), though Hufnagel, for reasons she described, used a different procedure to isolate the pellicles.

The second autoradiographic study of DNA in *Paramecium* basal bodies was done by Dippell, Grimes, and Sonneborn (cited in Sonneborn, 1970; also personal communications). Thin sections of paramecia cultured in medium containing ^{3}H-thymidine were studied by electron microscope autoradiography, which permits more precise localization of the grains and reduces the chance of pellicular contamination. In addition, they used a quantitative procedure to determine the distribution of grains (Salpeter, Bachmann, and Salpeter, 1969). They found some grains over basal bodies, but these grains were due to material which was not removable by DNase, and constituted only about 1/5 of the cortical grains. This study has provided no evidence for a specific DNase-removable label in basal bodies, or anywhere in the pellicle. It has also made it seem unlikely that the grains observed by Smith-Sonneborn and Plaut (1967) were confined to DNA in basal bodies (Sonneborn, 1970).

Conclusion. Of 15 reports of searches for DNA in basal bodies, 8 favor the presence of DNA (Randall and Fitton-Jackson, 1958; Randall, 1959; McDonald and Weijer, 1966; Seaman, 1960; Sukhanova and Nilova, 1965; Randall and Disbrey, 1965; Smith-Sonneborn and Plaut, 1967, 1969), while the other 7 papers found no evidence for DNA. Though the ayes are in a slight majority, the one report based on isolated basal bodies and the 3 based on the Feulgen reaction can probably be discounted. Those 4 that used tritiated thymidine and autoradiography are in doubt, at least until the discordant results of the two groups studying *Paramecium* are resolved. The positive results could be due to label either in molecules other than DNA or in structures other than basal bodies. This leaves as the strongest evidence available the acridine-orange fluorescence study of Smith-Sonneborn and Plaut (1967), but even these careful experiments do not unequivocally exclude the possibility that the yellow-green fluorescence is due to double-stranded RNA or some other non-DNA material in basal bodies, or to DNA that does not belong to the basal bodies. It seems fair to view the evidence for DNA in basal bodies with some caution and skepticism.

2. RNA

RNA has been commonly reported to be associated with centrioles and basal bodies in the isolation, cytochemical, and autoradiographic procedures mentioned above. This has not generally been the question of major interest, and no one seems to have pursued the question to the point of deciding whether the observed material is in fact RNA, and if so whether its association with basal bodies is real or contamination.

Stubblefield and Brinkley (1967; see also Brinkley and Stubblefield, 1970) found that the dense "feet" inside the triplets disappeared when centrioles were treated with RNase. Dippell (1968 and personal communication) found that the dense material which largely fills the lumen of the basal bodies of *Paramecium* disappeared if the specimens were treated with RNase. These results are suggestive of two distinct localizations of RNA in centrioles and basal bodies. They are evidence that a systematic search for centriolar RNA might be worthwhile.

The possible role of nucleic acids in centrioles is discussed in Section IV, D.

III. Morphogenesis

The old ideas of centriole duplication are incorrect. Electron microscopists studying the construction of new centrioles have consistently found that centrioles are not formed using the preexisting centriole as a rubber-stamp template, nor do they necessarily form even in proximity to a preexisting centriolar pinwheel. (Cases of apparent *de novo* appearance of pinwheels are described in Section IV, B.) Since GALL's (1961) paper, hundreds of pages of descriptive detail on the morphogenesis of centrioles have appeared. The main findings may be separated into 3 compartments.

1. *In systems where new centrioles develop next to preexisting centrioles, the new centrioles form adjacent to one specific end of the preexisting centriole but at a right angle to it and separated from it by* 50–100 mµ. GALL (1961), who first showed this clearly, reviews earlier suggestive observations (esp. BERNHARD and DE HARVEN, 1960). GALL found that the formation of a new centriole begins with the appearance of a short, annular structure, oriented perpendicular to the proximal end of the mature centriole and separated from it by a space (Fig. 4). He referred to this "precursor

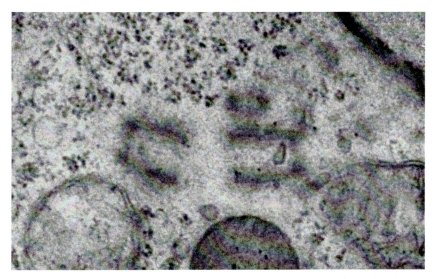

Fig. 4. Centrioles in a chick duodenal cell, with a short procentriole above the right-hand centriole. The procentriole has an electron-dense wall; less dense material connects procentriole and centriole. From SOROKIN (1968). 56,000×

of the daughter centriole" as a procentriole, and found it similar to a mature centriole except for length. GALL suggested, and others have confirmed (e.g., ANDRÉ, 1964; SOROKIN, 1968), that centrioles have a constant polarity, as do basal bodies (e.g., GIBBONS and GRIMSTONE, 1960; GIBBONS, 1961). The cartwheel is always found at the proximal end. The asymmetry of the triplets in transverse section – clockwise vs. counterclockwise pinwheel – also reveals the orientation in the section (e.g., Fig. 2). The distal end of the centriole when it is serving as a basal body is regularly continuous with the flagellum. GALL (1961) considered this end as also probably responsible for the formation of "other centriole products," but cytoplasmic microtubules usually are associated with the proximal end of a centriole. The proximal end is the one associated with a procentriole at the time of duplication; this end is also the one that is itself derived from the procentriole of the preceding duplication. (Notice that according to this mechanism, some centriolar pinwheels could be ancient.)

Gall's account revealed that procentrioles develop adjacent to mature centrioles, but showed no indication that the tubules were arranged by contact with the mature centriole's tubules. Even the perpendicular orientation of mature and developing centrioles argued against a template mechanism. But as Gall noted, "It should be emphasized that procentrioles have never been seen outside the immediate vicinity of a mature centriole." [3]

It is becoming increasingly clear that this pattern of centriole production occurs in many organisms, including those cases where centrioles double in relation to mitosis (André, 1964; Murray, Murray, and Pizzo, 1965; Robbins, Jentzsch, and Micali, 1968). Procentrioles regularly develop adjacent to the proximal end of a mature centriole, perpendicular and at a distance from it, and as time proceeds elongate to approach the length of the mature centriole, at which time the pair is seen as the frequently V-shaped array of the interphase diplosome.

2. *In situations where many centriolar pinwheels are formed simultaneously, the morphogenesis occurs through structural intermediates, one of which often is an amorphous, electron-dense body.* It is possible for more than one centriole to develop around the proximal end of a centriole. For example, Gall (1961) has described stages during the production of multiflagellate sperm in which as many as 20 procentrioles are clustered around a single centriole. That seems to be about the limit of such an arrangement.

A favorable opportunity to study the production of new centrioles has been provided by the embryonic development of ciliated epithelia in vertebrates, where cells with 2 centrioles form about 300 cilia. It was in such a system that Lenhossék (1898) concluded centrioles give rise to basal bodies, presumably by multiplication, and where de Rényi (1924) and others made careful light microscope observations supporting this hypothesis. In such a system, rat tracheal epithelium, Dirksen and Crocker (1966) made the first thorough ultrastructural study of the sequence of events in basal body formation. Their proposed sequence is well supported by electron micrographs. Large masses of loose fibrogranular material appear in association with the preexisting centrioles. From this material develop several electron-dense spherical masses, the "condensation forms." Around each of these condensation forms several procentrioles develop, each oriented with its axis toward the center of the condensation form and separated from it by a space containing fine fibrous material (Fig. 5). As the procentrioles develop, the condensation forms become depleted of material, becoming either smaller or hollow, which suggested to Dirksen and Crocker (1966) that the material is utilized to build the procen-

---

3 Pollister and Pollister (1943) argued, on the basis of an ingenious light microscope study of spermatogenesis in viviparid snails, that kinetochores give rise to centrioles. Gall (1961) studied this system with the electron microscope, and could not support the Pollisters' hypothesis. Although both kinetochores and centrioles are foci around which microtubules assemble, there is no evidence that the two structures are homologous or interconvertible (see Brinkley and Stubblefield's review, 1970).

trioles. After a time the procentrioles separate from the condensation form, elongate, move toward the cell surface, and there participate in cilium formation.

Similar observations, differing in detail, have been made of developing ciliated epithelia of mouse oviduct (DIRKSEN, 1968), mouse nasal passages (FRISCH, 1967), rat lung (SOROKIN, 1968), *Xenopus* trachea and embryonic epidermis (STEINMAN, 1968), and chick trachea (KALNINS and PORTER, 1969). A few of the points of

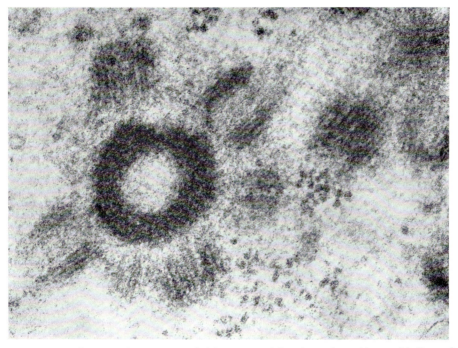

Fig. 5. A condensation form surrounded by procentrioles, from the oviduct epithelium of a 5-day-old mouse. The edge of another condensation form is visible in the upper right. Courtesy of E. R. DIRKSEN. 90,000 ×

difference should be mentioned. STEINMAN called the condensation form a "procentriolar organizer," and though he concurred that it diminished during procentriole formation he emphasized the hypothetical "organizer" function. SOROKIN called it a "deuterosome," and did not believe it decreased in size during procentriole formation. He too postulated a role in organizing procentrioles. KALNINS and PORTER found the electron-dense material had a slightly different morphology, more like a cylindrical core, which disappeared as procentrioles matured.

Observers also differ in the extent to which they implicate the preexisting centrioles in the process. It seems that to the extent they have been followed sufficiently, most of the intermediate forms originate in association with the preexisting centrioles, though in most cases this association does not seem to persist very long. In the chick tracheal epithelium the association with the original diplosomal centrioles is maintained (KALNINS and PORTER, 1969). Little evidence supports the tentative conclusion of Stockinger and Cireli (1965) that the basal bodies form *de novo*.

A developmental sequence similar in kind but quite different in detail has been described for development of flagellated sperm in a fern, *Marsilea* (Mizukami and Gall, 1966). In this and certain other lower plants, no centriolar pinwheel seems to be present (see Section IV, B). In *Marsilea* a "blepharoplast" forms, enlarges to form a hollow sphere, the wall of which comes to be composed of 100 to 150 procentrioles. The wall breaks up to form a cluster of procentrioles, which elongate to become the basal bodies in the sperm. In the cycad *Zamia*, Mizukami and Gall (1966) found a similar blepharoplast but much larger, with a wall containing perhaps 20,000 procentrioles. Though this sequence is quite different, the subunits of the blepharoplast wall are similar to the cylindrical structures observed in the developing chick trachea, as Kalnins and Porter (1969) pointed out.

These studies make it clear that centriolar pinwheels can develop at positions remote from preexisting centrioles, though in most cases the diplosome seems to be associated with initial events. They also show that centriolar pinwheels can develop from dissimilar structures, for which electron-dense masses often serve as foci, or organizing centers, or contribute precursor material — depending on the interpretation of the observer. Finally, they indicate that when cells are going to make many centrioles from a few, they do so through intermediate structures instead of through several rounds of sequential production of procentrioles from centrioles.

3. *The construction of a centriolar pinwheel is a stepwise process which includes the sequential addition of microtubules to form a complete ring of 9 singlets, then doublets, and then triplets.* Microscopists gradually learned that a procentriole develops from amorphous dense material that acquires a 9-fold symmetry and then develops the triplet structure and elongates into a centriole or basal body (e.g., Ehret and de Haller, 1963; Randall and Hopkins, 1963; André, 1964; Mizukami and Gall, 1966). This understanding was greatly extended by Dippell's (1968) precise description of basal body production in *Paramecium*. In this ciliate, the new basal body always develops in a precise position anterior to the preexisting basal body, and oriented at right angles to it, so by reference to adjacent structures it is possible to determine exactly where a developing basal body will be found. The first step in development is formation of a plaque of amorphous electron-dense material roughly 100 mμ away from the proximal end of the mature basal body. In the upper right quadrant of this plaque, a single microtubule appears (Fig. 6A). Additional microtubules are added until a ring of 9 microtubules is formed. These are the A tubules. At this stage the microtubules appear to be connected by spacers (Fig. 6B). Additional microtubular material is added to this ring of singlets to form doublets, again in an orderly fashion (Fig. 6C), and once again to make the triplets (Fig. 6D). When the triplet structure is complete the A—C linkers are still absent and the triplets have a much steeper pitch than they will have in the mature basal body (cf. Figs. 6D and 6E). As the singlets are converted into triplets, the hub-and-spokes cartwheel that will characterize the mature basal body is added.

This remarkably orderly developmental sequence occurs at least to doublets while the procentriole is a disc no more than 70 mμ thick, which can be restricted to a single serial section (Fig. 12 of Dippell, 1968). The newly formed basal bodies, even after they elongate and have microtubules associated with them (Fig. 6D), are still devoid of the internal fibrous and granular material that characterizes the mature basal body.

DIPPELL's conclusions have been extended to the formation of basal bodies from basal bodies in *Tetrahymena* (ALLEN, 1969), and the formation of centrioles

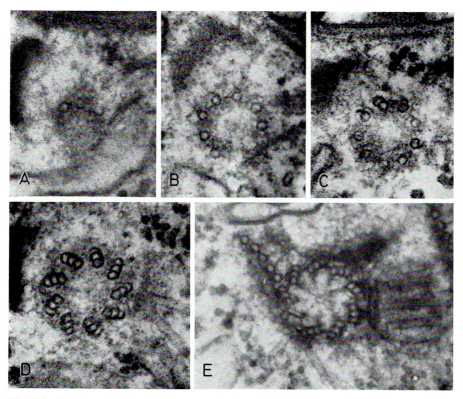

Fig. 6 A—E. Steps in the development of a basal body in *Paramecium*. All except E are printed in similar orientation; the kinetodesmal fibers can be seen as dense landmarks in the mid-to-upper left of A-D. The formation of the first tubule in the dense plaque is shown in A. Most of the ring of singlets has formed in B, and some are doublets in C. The new basal body, D, has a steeper pitch to its tubules than the adult basal body, E. The adult basal body has a cartwheel and basal-body-associated microtubules. In addition, on the right-hand side there is a longitudinal section of a new basal body, forming at right angles and anterior (in the ciliate) to the old one. From DIPPELL (1968). 140,000 ×

from dense masses in chick tracheal epithelium (KALNINS and PORTER, 1969) and in *Xenopus* (STEINMAN, 1968). Others have found examples of basal bodies with singlets and doublets which support the generality of this sequence (JOHNSON and PORTER, 1968; RANDALL et al., 1967; SOROKIN, 1968).

One basic question is how the array of singlets is laid down. As mentioned, DIPPELL has observed a series of spacers around the ring (Fig. 6B). These might align and space the tubules as they condense out of the matrix material, perhaps like the spacer links described by Tilney and Byers (1969) which arrange the micro-tubules in *Echinosphaerium* axonemes. Other observers tend to think that the hub and spokes of the cartwheel are responsible for the initial alignment of the micro-tubules (Stubblefield and Brinkley, 1967; Steinman, 1968; Kalnins and Porter, 1969).

Conclusions. The morphogenesis of procentrioles seems to be quite similar whether they develop next to the proximal part of centrioles or basal bodies or develop through intermediate electron-dense masses (cf. Figs. 4 and 5). In both cases, the procentrioles appear to be separated from the associated structure by a distance of 50 to 100 mµ and to form their long axis perpendicular to the associated structure[4]. In both cases, the procentrioles seem to condense out of granular material, and seem to be associated with the progenitor structure through filamentous material. All of this occurs in a finely fibrous region from which ribosomes and other structures seem to be excluded.

It would seem that the dense masses serve as substitutes for centrioles, whatever the role of these structures might be. It is noteworthy that no centrioles have yet been observed to develop at a distance of more than 0.1 µ from either a preexisting centriole or a dense mass, though this could be due simply to the difficulty of recognizing an isolated procentriole (see Section IV, B). It is tempting to think that centrioles and the intermediate dense masses both serve in some way as an organizing focus for the formation of the structure, rather than as a source of precursor material for the microtubules (see Section V).

Though a preexisting centriole is often involved in morphogenesis, no one has ever observed a developmental sequence suggestive of a template function for or binary fission of the preexisting centriole. The nearest thing to reproduction that could be construed from the morphological observations would be a kind of budding. Although budding is a permissible hypothesis it is not a necessary one, and the hypothesis is weakened by the ability of centriole production to occur through dissimilar intermediates, remote from the preexisting centriole.

The consequences of this mode of centriole morphogenesis for ideas about centriole replication are discussed in Section IV, D.

IV. Inheritance

Self-reproduction, one of the first attributes centrioles were endowed with, was supported by numerous light microscope studies (Wilson, 1925; Lwoff, 1950; Went, 1966a). So great was the impact of these studies that even the idea of "division" has persisted — beyond the discovery of morphological evidence that contradicts it. Many contemporary biologists have expressed the conclusion that centrioles are "self-replicating organelles" (e.g., Fawcett, 1966, p. 49; Loewy and Siekevitz, 1969, p. 59). Yet we are forced to discuss the inheritance of centrioles largely in nineteenth-century terms, since no one has applied the powerful tools of genetic analysis to the problem.

---

4 Stubblefield and Brinkley (1967) have suggested that microtubules coming perpendicularly from the triplets of the parent centriole form the microtubules of the developing procentriole. Their micrographs to support this are not compelling, and their conclusion does not fit the observations of Dippell (1968) and others.

A. Continuity

The first question is whether anything continues from generation to generation. So far the only unequivocal criterion is morphological persistence: continuity of the light microscopist's black dot (which may be too small to see during some parts of the life cycle) or of the electron microscopist's centriolar pinwheel (which may be very hard to find in thin sections during parts of the life cycle). Morphological continuity is not the only, or even the most interesting, kind of continuity, but of necessity it has received almost all the attention. Light microscopists have collected numerous examples of continuity of centrioles from division to division (WILSON, 1925); one of the finest I read was HUETTNER's (1933) study of continuity in *Drosophila*. Electron microscopists have not explored this question in as many organisms, but in many vertebrate somatic cell lines, at least, centriolar pinwheels can be found whenever they are looked for, during interphase or during mitosis, or even in some cells that never will divide again. The studies of centriole construction (Section III) provide evidence for the permanence of centriolar pinwheels throughout the cell cycle.

In ciliates, that classic example of continuity of basal bodies, the light microscope studies require new interpretation because they depend on silver impregnation, which does not actually stain the basal bodies (Section II, B). The electron microscope studies support continuity — though not by division.

B. Exceptions to Morphological Continuity

There is little question that centrioles *can* persist. But there are also several examples of apparent structural discontinuity. The most studied of these are in sea urchin eggs and in amebo-flagellates.

1. *Sea Urchin Eggs*

The attempt to decide about continuity of centrioles in sea urchin eggs has been extensive and frustrating, as a tiny selection of the papers will illustrate. The difficulties of working at or below the limits of resolution of the light microscope are everywhere apparent. Though WILSON and MATHEWS (1895) were able to see centrioles during the maturation divisions of the eggs, they could not find them thereafter. Gradually BOVERI, WILSON, and MORGAN saw centrioles in the sperm aster and in the first cleavage asters. Some drawings show what might be centrioles as we know them today (such as the pairs of dots BOVERI drew in some first cleavages of *Echinus* in his 1901 paper, e.g., Plate 5, Figs. 56—57); but other drawings show finely granular centers that today make no sense in terms of centrioles. The observations at first fit the idea of continuity, the egg having no centrioles but receiving one from the sperm. Then MORGAN discovered that by treating eggs with salt solutions he could induce the formation of one or many cytoplasmic asters, called cytasters. On the basis of a careful study of these cytasters he concluded (1899) that although the question of centrioles is "the most difficult question to decide," "it is fair to conclude that the central bodies form *de novo*."

The discovery of parthenogenesis demonstrated that the egg does not depend on the sperm for normal development. WILSON (1901) studied events during parthenogenesis, and also during the formation of cytasters in nucleate and enucleate eggs; he probably came as close to answering the question as one could hope to do with the light microscope. WILSON concluded that cytasters form centrioles, even in enucleate eggs, though the cytasters do not have centrioles immediately. WILSON's conclusion that "the central bodies of the cytasters are true centrosomes that are formed *de novo* in the cytoplasm" made "untenable" the concept of the centriole as a permanent organelle. WILSON (1925), after reviewing many subsequent observations, retained this conclusion. On the other hand FRY, who made an extensive study of this problem, concluded that centrioles in sea urchins are "coagulation artifacts." In one paper FRY (1929) illustrated some of the various artifacts people had reported. From a contemporary viewpoint one might argue that biologists should have dropped the problem as insoluble, since sea urchin centrioles were simply too small to see, but the question was of sufficient interest that people persisted in spite of the "difficulties" (cf. the question of DNA in centrioles).

The electron microscope left no question that sea urchin sperm (AFZELIUS, 1959) and the poles of the first and subsequent mitotic figures of fertilized eggs (HARRIS, 1961, 1962) have centrioles, which have the typical pinwheel structure (Fig. 3). DIRKSEN (1961) found centrioles even in the asters of artificially activated eggs, so the sperm centriole is not the only possible source of a centriole. However, electron microscopy has not answered all the questions raised by previous work:

a) Do mature eggs contain centrioles? The centriolar pinwheel is too small $(0.2 \times 0.5 \mu)$ to find or see in the cytoplasm with the light microscope, and sea urchin eggs are too large (100μ) to search readily with the electron microscope. Centriolar pinwheels are present in sperm and during cleavage, but whether they are present in unfertilized eggs remains unknown.

b) Is the sperm centriole(s) used in the first cleavage mitotic apparatus? This is an important question, since BOVERI's old idea suggests that centrioles would show paternal inheritance, at least in sea urchins. The sperm centrioles enter the egg, and appear to participate in the formation of the sperm aster (LONGO and ANDERSON, 1968). On the basis of light microscope studies, the sperm aster forms the first mitotic amphiaster, so it has generally and reasonably been concluded that the sperm centrioles end up in the mitotic apparatus (WILSON, 1925). LONGO and ANDERSON's (1968) ultrastructural observations leave unanswered the question whether these amphiastral centrioles come exclusively from the sperm, the egg, or both. It thus remains possible that the cleavage centrioles do not come from, or entirely from, the sperm. In any event, the sperm centriole is quite unnecessary even for complete parthenogenetic development, so the question is whether the sperm centriole normally ends up in the mitotic figure, not whether it must end up there.

c) Do cytasters have centrioles? If so, do the centrioles form *de novo*? DIRKSEN (1961) found that asters formed without sperm have centrioles, but she saw only a few and there is no way to know whether all asters contain a centriole. In fact, she noted subsequently that "centrioles could not be found in the asters shortly after activation but could be found several hours later" (DIRKSEN and CROCKER, 1966; also DIRKSEN, 1964). A great deal hinges on the "could not be found," especially since centrioles would be hard to find in structures this large and no quantitative data are given. If centrioles really form only *after* cytasters, as MORGAN (1899), WILSON (1901), and others also suggested, this would be firm evidence for *de novo* structuring.

Until these important questions are answered the morphological continuity of centrioles in sea urchin eggs remains unsettled, though at present it seems likely that *de novo* formation does occur under certain conditions.

Similar uncertainties and problems exist for the eggs of other species where maturation, fertilization, and early cleavage have been studied (WILSON, 1925; BRIGGS and KING, 1959). For example CONKLIN (1905), in his masterful description of development of ascidian eggs, concluded that "no trace of centrosomes are ever found in connection with the egg nucleus," and that "the cleavage centrosomes may be traced without a break back to the sperm amphiaster, to the sperm aster and finally to the middle piece of the spermatozoan." As CONKLIN said there "could not possibly be a clearer case," except that probably he was not always looking at centrioles as we know them today. For example, at the beginning of cleavage the centrosome "is a small, deeply staining body" which might be a centriole, whereas "in the later stages it becomes much larger and differentiates into the outer granular zone and the central clear area," which is not a description of a centriole (though an electron microscopist would undoubtedly find a centriolar pinwheel in the "central clear area"). This continual uncertainty about whether centrioles are really being described makes even the finest light microscope descriptions of little use in settling the question of centriole continuity.

The electron microscope has also not helped very much because eggs are large and hard to search. It is generally thought that unfertilized, unactivated eggs do not have centriolar pinwheels, but LONGO and ANDERSON (1969) found one in mature eggs of the bivalve mollusc *Mytilus* and emphasized their probable presence in some other eggs. As MAZIA (1961) noticed, it is odd that light microscopists regularly found centrioles early in the maturation divisions, but no centrioles in mature eggs. If they are truly absent, where do they go? Why are they eliminated? In general these questions have received no answer, though RAVEN and his colleagues (1958, 1959) have reported a suggestive light microscope observation in several gastropods. During the final maturation division, the centriole remains undivided and moves to the outer pole of the spindle and into the polar body. The mature egg is thus left without a centriole, which is restored by the sperm. It would be interesting to study this asymmetric division — one pole having an aster and the other anastral — with the electron microscope.

2. *Amebo-flagellates*

The morphological continuity of centrioles has been challenged in several protists. The best known and strongest challenge comes from certain amebae that can transform into flagellates (reviewed by FULTON, 1970). Two somewhat different amebo-flagellates have contributed to the centriole continuity controversy: *Naegleria* and several species of the true slime molds, Myxomycetes.

The question with *Naegleria* is straightforward. *Naegleria* amebae (ca. 15 μ in diameter) do not have obvious centrioles. They can transform into flagellates with about 2 flagella, each of which has an unequivocal basal body which the

13*

electron microscope shows to have a typical centriolar pinwheel and the usual distal-to-proximal structural differentiation (Dingle and Fulton, 1966). What is the source of these basal bodies?

Light microscopists have been unable to agree whether amebae have centrioles, even during mitosis (for references see Fulton, 1970), so we must rely on electron microscope studies. Schuster (1963) and Dingle and Fulton (1966) found no centrioles in amebae. Schuster also reported that "neither a spindle nor centrioles are apparent in the mitotic stages," but based this on two sections of dividing amebae. Outka and Kluss (1967) did not find centrioles in amebae of the related amebo-flagellate *Tetramitus*. Many have accepted this failure to find obvious centrioles in a few sections as conclusive that the centriolar pinwheel of the basal bodies arises *de novo*. We have made an extensive search in *Naegleria*, but found no centrioles or centriole-like structures in hundreds of sections of interphase or transforming amebae (Fulton and Dingle, 1970). In transformation, the appearance of centriole-like structures preceded the outgrowth of flagella by about 10 minutes at 25° C. Study of over 500 sections of amebae (from synchronized cells, where 60—70% of the amebae were in mitosis) showed typical spindle microtubules inside the nucleus but no centrioles or centriole-like structures anywhere in the cell.

We are satisfied that there are no centriolar pinwheels in *Naegleria* amebae, though we are also aware of the problems in defending such a negative conclusion (Fulton and Dingle, 1970). The biggest problem is that if structures with the visibility and depth of procentrioles were present, they might be recognized only if sectioned just right and might appear only in a single section (see examples in Gall, 1961; Dippell, 1968). The problem is compounded by the absence of any known place to look, and by the fact that normally only two basal bodies develop. In spite of the difficulties, the apparent *de novo* formation of centriolar pinwheels during *Naegleria* transformation constitutes the strongest evidence available for *de novo* assembly of this organelle. Another approach will be required to give this negative conclusion more meaning.

In the true slime molds, Myxomycetes, there are two types of division: intranuclear in plasmodia and conventional, with breakdown of the nuclear envelope, in myxamoebae. Centrioles have been found readily in the myxamoebae, but seem to be absent in plasmodia. Thus in these organisms two different types of mitosis can alternate in the life cycle, one with centrioles and one without. The literature on which these conclusions are based has been reviewed recently (Gray and Alexopoulos, 1968; Fulton, 1970); special mention should be made of the ultrastructural studies of Schuster (1965), Aldrich (1967, 1969), and McManus and Roth (1968). No one seems really to have searched for centriolar pinwheels in myxomycete plasmodia, but enough people have looked that if the pinwheels were there and obvious it seems probable they would have been found.

In a paper that appeared while this review was in press, Perkins (1970) has reported the *de novo* construction of centrioles during sporulation in the marine protist *Labyrinthula*. A stage in the construction involves protocentrioles, dense granular aggregates in which cartwheels seem to be assembled spoke by spoke prior to the addition of centriolar microtubules. Intermediate stages in the assembly of centriolar pinwheels have not yet been observed.

Many other protists and lower plants seem not to have centrioles at one stage of their life cycle but to form them at another, though in most organisms this conclusion is based entirely on observations made with the light microscope. SHARP (1921) and LEPPER (1956) have reviewed many of these studies, especially those in plants. In many of the organisms, including *Marsilea* as described by SHARP (1914) and ultrastructurally by MIZUKAMI and GALL (1966), it is clear that basal bodies form through remarkable intermediates but it is less certain that no centriolar pinwheels are present in the cells when these intermediates form.

C. Counting Mechanisms

Though morphological continuity of centrioles occurs in many kinds of cells, some cells do not seem to need to inherit a centriolar pinwheel in order to be able to form one. It is possible, of course, that even cells that form centrioles *de novo* may depend on some self-reproducing cytoplasmic entity for their production. If there is an autonomous self-reproducing particle, whether a centriole or a precursor entity, its reproduction would need to be controlled. Such controlled "reproduction" of centrioles occurs both in division and in flagellum formation. The mechanism by which this counting occurs could tell us much about the inheritance of centrioles, but although the behavior of centrioles tells us that a counting mechanism exists, the mechanism itself remains obscure.

1. Mitosis

Centrioles undergo an extraordinarily precise ritual in relation to mitosis. The interphase cell has 2 centrioles (a diplosome); these separate to the poles at the beginning of division. During the ensuing karyokinesis each centriole duplicates, so that at the end of mitosis each daughter cell receives 2 centrioles[5]. What controls the production of new centrioles, which go from 2 to 4 to 2 during each division cycle? We do not know the answer, but in any event it is clear that the cell exerts control — positive or negative — over the time of production of new centrioles, and over the quantity produced. Whatever the control, it must be very precise, because centriole duplication ordinarily does not get out of phase with nuclear division — neither too few nor too many centrioles are produced.

MAZIA, HARRIS, and BIBRING (1960) did a series of elegant experiments to measure the time of duplication of "mitotic centers" in sea urchin embryogenesis. They defined a center operationally as that which determines a mitotic pole, independently of the visualization of centrioles or other structures. Their basic observation was that if 2-mercaptoethanol, which blocks the division of sea urchin eggs, is added just before the first cleavage metaphase, and eggs are left in it for a while and then removed, the treated eggs divide directly from 1 to 4 cells. Similarly it was possible to manipulate the second division so eggs would divide directly

---

5 Light microscopists have catalogued the time of centriole duplication; some results are tabulated in POLLISTER and POLLISTER (1943). Most observations placed duplication between metaphase and telophase. However, electron microscope studies show that duplication occurs earlier than the light microscope studies suggested — in early prophase (e.g., MURRAY, MURRAY, and PIZZO, 1965) or even prior to prophase (ROBBINS, JENTZSCH, and MICALI, 1968).

from 2 to 8 cells. Mazia and his colleagues argued that what was being manipulated was the duplication of centers, and used this hypothesis to determine when the centers double (prior to the onset of each division). No one has yet attempted the difficult task of determining whether the mitotic centers correspond to centriolar pinwheels. R. E. Kane (personal communication) has seen only a single centriole at each pole of first cleavage mitotic apparatuses isolated in metaphase-anaphase, although the experiments of Mazia, Harris, and Bibring indicate that two mitotic centers are present at each pole by this time.

Other experiments concerned with mitotic centers are discussed by Briggs and King (1959), Mazia (1961), Dalcq (1964), and Went (1966a, b). Went (1966a) reviews the studies of Henshaw, Rustad, and others of the production in sea urchins of multipolar spindles by X-irradiation and of mitotic delay by ultraviolet, both of which have been interpreted as effects on centriole duplication. One clear result of all these studies is that the number of centers, and of mitotic poles, and therefore presumably also of centrioles, can increase in the absence of nuclear division – and thus that these events can be uncoupled[6].

2. *Flagellum Number*

Control of the number of flagella per cell is directly related to control of the number of basal bodies per cell, essentially on a 1:1 basis. Some flagellates virtually always have the same number of flagella, such as the 2 that are characteristic of *Chlamydomonas*. Other flagellates regularly show some physiological variation in flagellum number. For example, flagellates of *Naegleria gruberi* usually have 2 flagella, but flagellates with 1, 3, or 4 flagella are common. It is possible to vary the flagellum number of *Naegleria* in many ways, including clonal strain, growth conditions, stage of division cycle, environment during transformation, ploidy, number of nuclei, and mutation (Fulton, 1970). The most striking variations are produced by exposing *Naegleria* cells to a sublethal temperature shock early in the transformation process (Dingle, 1970). Under these conditions, the cells form more than twice the usual number of flagella and basal bodies. It should be possible to use these variations in flagellum number to study the control of number of basal bodies formed in *Naegleria*.

---

6 Dozens of suggestive experiments pertaining to centers and centrioles have been done using sea urchins. For example, eggs fertilized by more than one sperm often develop multipolar spindles, which has generally been interpreted to indicate that each sperm produces an amphiaster. These amphiasters can divide repeatedly, even in the absence of nuclei and chromosomes (Wilson, 1925, p. 176). Yet the sperm are unnecessary for the formation of amphiasters including, under appropriate conditions, the formation of multipolar spindles (see Wilson, 1925, esp. p. 684). Lorch (1952) found that if the asters were sucked out of one blastomere of a 2-cell embryo, they eventually reformed and division occurred, whereas if both the asters and the nucleus were removed no new asters ever formed. This suggests a role for the nucleus in aster formation. Yet enucleated eggs treated with chemical agents can form numerous cytasters which can increase in number (Wilson, 1901, 1925); even fractions of eggs obtained by centrifugation can form cytasters (Harvey, 1936). These observations implicate various structures in the formation of asters, and presumably of centrioles: sperm, nucleus, cytoplasm, even fragments of cytoplasm. They suggest that almost any part of the sea urchin cytoplasm can form an aster — at least — in response to a suitable stimulus. They remain provocative but inconclusive for our purposes until it is known whether all the asters contain centriolar pinwheels.

Essentially the problem, as in mitosis, is a counting problem. This suggests one hypothesis, that can be developed with variations: there is a self-reproducing precursor entity, G, that duplicates with each cell division so it keeps pace with the cell as centrioles do in cells that have them, going from 2 to 4 before cell division and back to 2 with division. During transformation each G is converted into a basal body, so only as many basal bodies — and thus flagella — can form as there

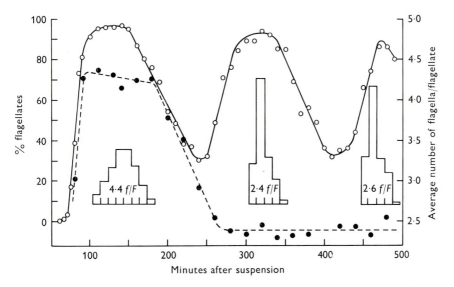

Fig. 7. Change in flagellum number during repeated cycles of transformation in *Naegleria*. The open circles show 3 cycles of the appearance and disappearance of flagellates (cells with flagella) in a population of amebae treated with a sublethal temperature shock for the first 45 minutes after suspension. The average flagellum number, shown in filled circles, is high during the first cycle but at a normal level in succeeding cycles. Frequency histograms indicate the distribution of flagellum number (1, 2, 3, 4, etc.) seen in samples taken at 120, 330, and 480 minutes, with the average of flagella/flagellate shown inside each.
From DINGLE (1970)

are Gs present. The concept of G suggests many experiments, some of which have been tried, but it has not been possible to develop clear evidence for or against this hypothesis. It has not been possible to "cure" *Naegleria* of the capacity to transform (though rare mutants can be obtained; FULTON, 1970), so if a self-reproducing entity G exists it is not easily removed from cells or its loss is lethal.

One experiment, reported by DINGLE (1970), is especially instructive. The temperature-shock treatment was used to produce flagellates with excess flagella. According to the G hypothesis, the temperature shock would result in a transient breakdown in the control of G replication, so more Gs would form and thus more basal bodies and flagella. The flagellates that formed were allowed to revert to amebae and then immediately to retransform into flagellates again (which *Naegleria* does without any treatment in a nonnutrient environment where no growth or division occurs). The first-cycle flagellates had excess flagella per flagellate (Fig. 7). If the cells either (a) retained their first-cycle basal bodies and reused them, or (b) retained and reused the hypothetical excess Gs, then we would expect the second-cycle flagellates would still produce an excessive number of flagella. However Dingle found that only the normal number of flagella is formed in the second round. The cells return to the normal number without any intervening growth. It is clear that unless additional, awkward assumptions are made, this result indicates that the number of flagella formed is not limited simply by the number of entities present.

There have been two studies of the effects of X-rays on basal body formation. In grasshoppers, X-irradiation during spermatogenesis results in the formation of spermatids with excess basal bodies, and thus flagella (Tahmisian and Devine, 1961). So much is happening simultaneously during spermatogenesis that interpretation is even more problematical than in *Naegleria*. On the other hand, Sorokin and Adelstein (1967) reported a clear result. Embryonic rat lungs were X-irradiated with a dose that stopped cell proliferation. This dose caused no delay in the formation of the ciliated border or reduction in the number of cilia, and so was quite without effect on the production of hundreds of basal bodies.

D. Are Centrioles Self-reproducing?

As emphasized in Genesis (Section I), the focus on centrioles as self-reproducing organelles has been a mistake, a historical accident. On the basis of light microscope observations, often of structures now known not to have been centrioles, it was concluded that centrioles are self-reproducing. Because of this, the pioneers who described centriole morphogenesis, even after finding budding rather than fission, continued to write of "Centriole Replication"[7]. Because of this, a devoted search has been made for centriolar DNA. Because of this, the evidence that centriolar pinwheels can form *de novo* is contested. Let it be clearly stated, therefore, that the idea that centrioles are self-reproducing entities was based on inadequate evidence. *There is no overwhelming evidence that centrioles, or basal bodies, are in any way self-reproducing, or even that they contain any self-reproducing entities.*

What are self-reproducing entities? Lwoff (1950, p. 27) described them, with reference to basal bodies:

> One kinetosome is always generated by division of another. We see kinetosomes dividing and have no evidence whatsoever for their formation *de novo*. They are endowed with genetic continuity. It is the custom to refer to such particles as "self-reproducing" or "autocatalytic" units. These terms should not be understood as implying that the particles are independent. A self-reproducing granule is by no means a self-sufficient granule. When we refer to the kinetosomes as self-reproducing units, this means only that the kinetosome never arises *de novo*, and that some specific structure or template, present in the kinetosome, is necessary to organize other molecules into a new kinetosome: like genes, they can only be generated by, or in the presence of, an equivalent structure.

For basal bodies and centrioles, reproduction by division, or any kind of fission, has been ruled out by observation, as have structural template (rubber stamp)

---

7 It could be argued that the term replication is acceptable, and means simply the making of a copy. However, biologists think of replication in a more specific way. As Sager and Ryan (1961, p. 236) stated, "the term replication is employed in modern genetics to mean the origin of a new unit by copying from a pre-existing unit of the same kind, either directly in one step or indirectly with the intervention of complementary intermediates".

models. We now have evidence for their formation *de novo*. But the idea of genetic, if not morphological, continuity persists. The ultrastructural observations still allow a kind of generative mechanism of reproduction, as opposed to a fission mechanism (see MAZIA, HARRIS, and BIBRING, 1960; MAZIA, 1961). According to the generative model, as MAZIA, HARRIS, and BIBRING (1960) pointed out, "the centriole contains a reproducing 'germ' or 'seed' of molecular dimensions. This gives rise to its like, which in turn directs the growth of a replica of the original body." Many have expressed this view. For example, SOROKIN (1968) interpreted "the centriole as a semi-autonomous organelle whose replicative capacity is separable from the characteristic triplet fibre structure of its wall." Such models are permissible. The replicating "germ" could be transmitted through remote intermediates (as in the development of ciliated epithelia) or even where the centriolar pinwheel developed *de novo* (as in amebo-flagellates). The crucial point is that the idea of self-reproduction, stripped to its essentials, carries with it the idea of some replicating entity, such as a DNA molecule, which is transmitted from parent to offspring. Exactly what this entity does is less clear, though it seems reasonable that it would code for at least some aspect of centriole structure or function. In any event, at present the only evidence we have that such an entity exists in centrioles is the evidence for centriolar DNA.

Centrioles do tend to form in association with mature centrioles. The morphogenesis could be some sort of a true budding process, where the parent centriole contributes some material to its offspring centriole. It could be generative reproduction, where the material transferred includes a replicating entity. Electron microscopy cannot tell us whether any material is transferred. The tendency for new centrioles to form in association with old — where old ones are present — could have a simple explanation. Centrioles tend to be foci for the orderly construction of many microtubular structures (see Section V) — why not also of centrioles? A centriole serving as an orienting center for the assembly of materials made elsewhere into a new centriole is a kind of reproduction, though not the kind usually implied by such a term as self-reproduction. Although self-reproduction through some kind of budding or generative reproduction is possible, it is not necessary. As GRIMSTONE (1961) observed, continuity of centrioles "might involve no more than the development of new ones in close proximity to old." There is no reason to imply that any on-site template structure is used or transmitted in this process, or to suggest that the old need contribute any materials to the new — nor is there any reason to assert that these things do not happen.

It seems reasonable to dream that in the future — perhaps not too distant — we will understand the morphogenesis of centrioles in simple terms. This optimism is encouraged by the emerging understanding of morphogenesis in bacteriophage T4 (WOOD et al., 1968), as well as the ability to study the assembly of microtubules *in vitro* (STEPHENS, 1968). It is also encouraged by DIPPELL's (1968) discovery that the assembly of centriolar pinwheels occurs stepwise in an orderly sequence. Possibly the organelle subunits assemble in a precise sequence dictated by their own structures — as do the subunits of phage T4. A frame for the orderly assembly of the tubules into a 9-fold symmetry could be provided by spacer molecules or by self-assembly of a cartwheel (Section III). The preexisting centriole need not provide any of the subunits — these could be made elsewhere (see Section V); it might provide a suitable microenvironment for assembly to occur. Though at present we know almost nothing about this, and terms like self-assembly and microenvironment cover a lot of ignorance, the problem at least seems approachable. And it seems possible, indeed more straightforward, to contemplate centriole morphogenesis without requiring that centrioles contain self-reproducing entities.

Ultrastructural descriptions of centriole morphogenesis do not support the concept of self-reproduction, and do not require even the transmission of a self-reproducing entity. Other searches for evidence for a self-reproducing entity (SRE) have also been unrewarding. The *de novo* assembly of centriolar pinwheels is not relevant, since evidence of morphological discontinuity says nothing about genetic continuity, and a cell could transmit a centriolar SRE without having centrioles. The regular doubling of centrioles (and of mitotic centers) from 2 to 4 in relation to mitosis is suggestive of reproduction, and certainly each preexisting centriole is the focus for assembly of a new one in a closely timed and controlled way. One way to do this would be to have a centriolar SRE which was allowed to reproduce once at a critical time during each cell cycle, and the daughter SRE could then organize one centriole around itself. Centrioles, and basal bodies also, are reproduced with precise spatial and temporal control, but the presence of control need not imply self-reproduction. This is equally true in cases where centrioles, or at least centers, appear to multiply independently of the nucleus; in these cases normal control seems to have lapsed, but again this control need not be control of self-reproduction. In any case, control must come from outside the centriole. So far, a review of the literature has failed to provide any evidence for SRE other than the evidence for centriolar DNA.

In the embryonic development of sea urchins and other organisms there are indications that the sperm centrioles may be the usual source of centrioles for the mitotic centers, though sperm, and thus their centrioles, are not essential. Does this imply paternal inheritance of centrioles? The nature of this question is much influenced by whether or not centrioles contain a SRE. If there is a centriolar SRE, then there may be paternal inheritance; if not, then centrioles from the sperm would simply provide 1 or 2 of the thousands which would form during embryonic development. There is an intermediate hypothesis. Even though a centriole may have no SRE, and thus a paternal centriole would depend entirely on the egg cytoplasm and the zygote genes for its expression, the pattern or other expression of the paternal centriole might influence the way the new centrioles formed. This kind of inheritance through preformed structure, without transmission of genetic information, is what determines cortical pattern in *Paramecium* and other ciliates (Sonneborn, 1967, 1970). However there is no reason to argue that inheritance by transmission of pattern, rather than of material, is ever involved in centrioles.

We have come full circle. DNA was sought to explain genetic continuity of centrioles. Now the finding of DNA would provide evidence for genetic continuity. Yet the evidence for DNA is not conclusive.

If DNA is present, what function would it have, other than providing a SRE? As Hoffman (1965), Sonneborn (1967), and others have emphasized, it would not explain how centrioles make duplicates of themselves. Centriolar DNA is not needed to specify the form of the centriolar pinwheel, or to code for the molecules of which it is built. DNA in basal bodies would not help explain the orderly pattern of cortical inheritance in ciliates (SONNEBORN, 1967, 1970). It has been suggested that centriolar DNA codes for the production of new centrioles, of microtubules, of some parts of cilia, or of some "morphogenetic function" (e.g., RANDALL and DISBREY, 1965; GRANICK and GIBOR, 1967). It even has been suggested that centriolar DNA codes for an RNA which serves to guide the assembly of tubulin into microtubules (BRINKLEY and STUBBLEFIELD, 1970). Except for this last idea, which is discussed in Section V, could not chromosomal DNA code equally well for whatever centriolar DNA is supposed to code for? Of course we cannot answer this question by intellectual attractiveness; all that can be said is that there is no reason at present to suppose centriolar DNA is necessary. (SONNEBORN, 1967, has pointed out that in ciliates where there are thousands of basal bodies, the redundancy provided by thousands of copies of basal body DNA could be useful.)

We are in need of three pieces of concrete information.

1. Is there in fact any DNA, or RNA, or anything self-reproducing in centrioles and basal bodies? Most contemporary discussions of replication hinge on this question, so it is of greatest importance to give it a firm answer. As HUFNAGEL (1969b) emphasized, a decisive answer to this question may require a new approach.

If centrioles contain DNA, it should be possible to see a DNA strand running through them, as has been seen in chloroplasts (e.g., RIS and PLAUT, 1962) and in mitochondria (NASS, NASS, and AFZELIUS, 1965). Several people have tried to see changes in centrioles and basal bodies after treating them with DNase, but with inconclusive results (R. V. DIPPELL, personal communication; STUBBLEFIELD and BRINKLEY, 1967). STUBBLEFIELD and BRINKLEY (1967) have postulated a DNA strand running helically through the inside of centrioles, but no evidence indicates the structure is made of DNA.

We are confronted with a basic dilemma. If we search for something that is not there, we certainly are not going to find it, but our failure is unlikely to completely convince others — or even ourselves — that the something is not there. This is especially true if others have already reported finding the something. This dilemma means that if there is no DNA in centrioles it will be a long time before biologists are satisfied with that conclusion.

2. Do centrioles form *de novo*? The evidence, especially for amebo-flagellates, indicates *de novo* assembly, but the skeptic could argue that assembly occurs from a procentriole-like entity — complete with 9-fold symmetry and maybe a cartwheel or other features of centriolar pinwheels — that has not been found and could be nearly impossible to find. This question, like the first, can never be definitively answered by the "seek and not find" approach.

There is little evidence left for true morphological continuity. Even when morphogenesis occurs next to a preexisting centriole, there is no evidence that the old serves as a template for the new. There is even less continuity when structural intermediates are involved, so the extreme of no continuity is not surprising. If preexisting centrioles are not used as templates, then why are they needed as precursors at all? Possibly what sea urchin eggs and other such systems are trying to tell us is that parental centrioles are not necessary.

3. Are centrioles inherited independently of the nucleus? No one has isolated a mutant centriole, or any cytoplasmic mutation affecting centriolar behavior. In *Chlamydomonas*, an organism that should be especially favorable for seeking such

mutants, mutants with altered flagellar structure have been isolated, but none with altered centrioles (Warr et al., 1966; Randall et al., 1967). If there is a self-reproducing entity in centrioles, then it should be mutable. Certainly centrioles are inherited somewhere in the cell, so mutants must exist. One can postulate that such mutants might usually be lethal, but, if this is the case, conditional lethal mutants can be sought. One can postulate all sorts of problems, but the questions about the inheritance of centrioles can never be answered satisfactorily until someone finds a way to study their genetics – and this means mutants and crosses.

V. Function

A. Guilt by Association

We do not really know anything definite about the function of centrioles and basal bodies. In the absence of specific information, we can seek possible functions only by examining the cellular activities with which centrioles are associated.

1. *Centrioles serve as foci for the production of new centrioles and basal bodies.* Centrioles regularly develop next to centrioles, but the role of the preexisting structure is not clear. Procentrioles also can form through structural intermediates (Section III), and almost certainly can arise *de novo* (Section IV, B).

2. *Centrioles are foci at the poles of the mitotic apparatus, where they are surrounded by microtubules.* Though centrioles are obvious participants in the division of many kinds of cells, it is less clear what the role of that participation might be. Most biologists have thought of centrioles as a functional part of the cell division machinery, having a role either in the movement of chromosomes or in determining the number and position of mitotic apexes. It is also possible that the behavior of centrioles in division is a mechanism to assure the orderly distribution of centrioles.

The participation of centrioles in cell division was reviewed thoroughly by Mazia in 1961. A major shift in thinking during the ensuing decade has been to change from spindle fibers to microtubules. The two types of spindle fibers – the continuous or pole-to-pole fibers and the chromosomal or kinetochore-to-pole fibers – are made of microtubules. Both centrioles and kinetochores seem to play some role in the formation of these microtubules, perhaps in guiding the assembly of continuous and chromosomal fibers, respectively, as discussed provocatively by McIntosh, Hepler, and van Wie (1969) and by Brinkley and Stubblefield (1970).

At the onset of division, the diplosomal centrioles move apart to establish the poles. There are indications that microtubules may be responsible for this movement.[8] In newt fibroblasts, for example, a spindle forms between the centrioles

---

8 Other movements of centrioles have been reported. Several who have observed centrioles in living cells have noted their tendency to move. In *Drosophila* eggs, for example, Huettner and Rabinowitz (1933) reported that centrioles showed a "slight vibratory motion," and "kept shifting their positions with respect to each other". Pfeiffer (1956) reported that in leukocytes during metaphase and anaphase the centrioles oscillated about once a minute, over distances of 2 to 5 μ at each oscillation. Many other cases of abrupt particle movements in cells have been observed, though there is no evidence centrioles are involved in most of these (Rebhun, 1964).

as they separate (Taylor, 1959; see also Stubblefield and Brinkley, 1967). Perhaps the elongation of the continuous microtubules pushes the centrioles apart, though this argument can be reversed – the separating centrioles spinning out microtubules as a spider spins a web.

The pericentriolar region undergoes characteristic changes during the mitotic cycle, including the formation and disappearance of dense masses, which microscopists have interpreted as indicating periods when centrioles are active in the formation of microtubules. Pericentriolar satellites are evident early in mitosis, when the centrioles are separating and the spindle is forming, while increasing numbers of microtubules converge toward the dense material (Robbins and Gonatas, 1964; Robbins, Jentzsch, and Micali, 1968; Murray, Murray, and Pizzo, 1965). Where two centrioles are present at a pole, the satellites surround only one (same references), suggesting that only one centriole – possibly the oldest – is "active." After the end of prophase, when the centrioles have separated, much less of the dense material is found around the centrioles (above references; also de Harven, 1968). Robbins and Jentzsch (1969) suggested that later in division the microtubules depolymerize in the vicinity of the centrioles, which would allow, perhaps aid, the migration of the chromosomes (see also Harris, 1965).

In a morphological sense, the centrioles seem to establish the poles and lie at their centers. The experiments of Mazia and others, discussed in Section IV, C, indicate the existence of mitotic centers, whether centrioles or not. At the beginning of division all mitotic centers separate from one another, leaving the continuous fibers in between. The number of mitotic centers determines the number of poles in the mitotic figure. If mature centrioles are mitotic centers, this indicates a basic role in mitosis.

The direction in which centrioles separate determines the position of the poles which in turn determines the axis of the division. Costello (1961) made a convincing correlation between the orientation of dividing centrioles and the axis of cell division in dividing *Polychoerus* eggs. Costello depended on light microscopy to determine the orientation of the centrioles during duplication, but the details of his figures – including the size of the centrioles, ca. 0.25 μ in diameter, and the orientation at right angles – suggest that he was looking at cylinders based on centriolar pinwheels.

So far, the possible roles of centrioles during division have been emphasized. It is also possible that centrioles are not needed for division, that their regular behavior is to ensure their orderly distribution to daughter cells. In an interesting essay, Pickett-Heaps (1969) has argued that the centriole is "a passenger ensured of equal partitioning by being attached to the spindle in certain cells." A location at the poles would be suitable for this, and centrioles normally but not always travel at the poles. Virtually no protists use centrioles as division centers, even when they are present in the cell (Section VI, A). In *Chlamydomonas* the centrioles sit near the cleavage furrow rather than at the poles during division (Johnson and Porter, 1968). In *Prymnesium*, another algal flagellate, the spindle fibers apparently begin their formation in association with the basal bodies but the basal bodies do not appear at the poles of the spindle (Manton, 1964b). Higher plants accomplish mitosis without centrioles, with only the difficulty that convergence of microtubules toward the poles is "less well defined" than in cells with centrioles (Ledbetter,

1967). Centriolar pinwheels are not required for all mitoses, but that does not indicate whether or not they perform essential functions for mitosis in some cells.

Dietz (1959, 1966) in particular has argued persuasively that centrioles are dispensable, on the basis of his study of meiosis in crane flies. During spermatocyte meiosis, the centrioles become surrounded by centrospheres and well-developed asters, and move to opposite positions near the nuclear membrane, where they participate in karyokinesis in typical fashion. By flattening the cells, and thus preventing the movement of the asters toward their positions next to the nucleus, Dietz was able to prevent the participation of the asters in spindle formation. But even though the asters remained outside, a spindle formed and normal chromosome separation occurred. Thus the asters and their associated centrioles seemed relatively unimportant, at least for chromosome movement.

It seems likely at present that the major function of centrioles in mitosis, where they are present and active, is to determine the number and position of the poles, probably by guiding the formation of the continuous spindle fibers. There is little evidence, other than suggestive, that centrioles play a crucial role in chromosome separation, or that their behavior in mitosis in most cells is simply to ensure their own distribution.

3. *Centrioles are regularly foci from which cytoplasmic microtubules emanate.* This function of centrioles has been described in Section II, C.

4. *Centrioles serve as basal bodies for flagella, cilia, and various modified cilia.* Several of the centriole-associated activities are regularly accomplished by cells without centrioles. Such cells can divide, forming an abundance of mitotic microtubules; they can form cytoplasmic microtubules; some probably can even form centriolar pinwheels.

There is one function that invariably requires a centriole: the formation of flagella and cilia. This activity provides a potential role for the aesthetic form of the centriolar pinwheel: to serve as a "crystallization center" for the outer fibers of cilia and flagella. One can support this speculation, which others also have made (e.g., Grimstone, 1967), with several arguments. The outer fiber doublets of diverse cilia, flagella, and modified cilia are continuous with the inner doublet of the centriole triplet. Other microtubules associated with centrioles and basal bodies usually terminate before they reach the organelle. It is possible to reassociate the microtubule protein from outer fibers *in vitro*, but nucleation with fragments of tubules is necessary if structures resembling tubules are to result (Stephens, 1968, 1970b). Thus it is reasonable to think that outer fiber doublets might begin to assemble on the distal end of a centriole, using the triplets as nucleating centers. Preliminary reports suggest that elongation of flagella occurs mainly by addition to the distal tip (Rosenbaum and Child, 1967; Rosenbaum, Moulder, and Ringo, 1969). A similar mode of assembly occurs in bacterial flagella (Emerson, Tokuyaso, and Simon, 1970). Distal addition suggests that once assembly is underway the distal ends of the flagella become the organizing center. According to this view only initiation of outgrowth would depend on the basal body.

Another observation suggesting an organizing role for basal bodies is that virtually all flagella and variations on flagella have an outer ring of 9 doublets, regardless of whatever variations may occur inside or outside that ring (see Fawcett, 1961; Phillips, 1970; Baccetti, Dallai, and Rosati, 1970).

Flagella are assembled in two distinct ways: internally, as in *Allomyces* (Renaud and Swift, 1964), or while flagellar outgrowth occurs, as in *Naegleria* (Dingle and Fulton, 1966). Some of the recent studies of flagellum development are reviewed by Sorokin (1968) and Phillips (1970). In both modes, the distal end of the centriolar pinwheel serves as the starting point for assembly of the outer fibers.

In many vertebrate cells one of the two centrioles moves to the cell surface and forms a single flagellum (Pollister, 1933; Sotelo and Trujillo-Cenóz, 1958; Sorokin, 1962, 1968). In some tissues most cells have a single flagellum (Dahl, 1963; Rash, Shay, and Biesele, 1968; Dingemans, 1969). Treatments of fibroblasts with colcemid or neural cells with pargyline increased the number of flagella formed (Stubblefield and Brinkley, 1966; Milhaud and Pappas, 1968). Several of these observers have suggested an inverse relationship between mitosis and the presence of the flagella, but almost nothing is known about the factors involved. What induces some cells to form a single flagellum, others to form hundreds of cilia? What causes only one of two centrioles regularly to form a flagellum?

Little or nothing is known about what moves centrioles to such places as the cell surface, though the movement — for example, from nuclear envelope to cell membrane — has been described in many organisms. This movement can be fairly extensive. In the sensory cells of mammalian olfactory epithelium, dendrites $40-80\ \mu$ long but less than $2\ \mu$ in diameter run between a nucleated cell body and a terminal swelling that bears cilia. Clusters of centrioles can be found along this dendrite, and it has been argued that centrioles form next to the nucleus, migrate down the dendrite, and form basal bodies in the terminal swelling (Heist and Mulvaney, 1968; also Frisch, 1967).

In some protozoa the centriole can simultaneously serve as a centriole for division and a basal body for flagella (Section VI, A). The ability to occupy spindle poles while bearing a flagellum is a display of virtuosity, but does not add anything to our understanding of function.

B. Centrioles and Tubulin

DuPraw (1968) was in company with many others when he concluded that "the essential activity of the centriole may be to regulate synthesis and aggregation of protein monomers required for the formation of ... tubular structures." That centrioles are involved in organizing the assembly of microtubules is supported by much circumstantial evidence — centrioles serving as foci for many kinds of microtubular structures. A role in synthesis of the microtubule structural proteins, or tubulins (Mohri, 1968; Stephens, 1970a), seems much less likely. This hypothesis was suggested by de Harven and Bernhard (1956), rejected by Grimstone (1961), and so forth. Although Seaman (1962) pointed to a role for basal bodies in protein synthesis, his conclusion was based on contaminated basal bodies. In general, the centrosphere area is devoid of or depleted of ribosomes, so it looks like a poor place for protein synthesis. This seems to be true in most cells at most times, though Robbins, Jentzsch, and Micali (1968) note that for a brief period in mitosis the centriole mingles with the cytoplasm. There is at present no reason to suppose that centrioles or basal bodies are ever involved in translation.

Centrioles might make messenger RNA for tubulin synthesis, and export it into the cytoplasm for translation into tubulin. In the absence of definitive data on nucleic acids in centrioles, or of any reason to argue that tubulin messages come from centrioles, this argument has little substance.

Brinkley and Stubblefield (1970) have suggested that the centriole is "a *rotary engine* and a classical machine of molecular dimensions," containing a DNA helix that spins out RNA which in turn initiates the winding of protein subunits into rotating microtubules. This imaginative hypothesis lacks only experimental support. One consequence of the hypothesis is that microtubules should contain RNA, but there is no indication that they do (Stephens, 1970b). The work of Stephens (1968, 1970b) also demonstrates that although some tubulins may need a nucleating center to guide polymerization into microtubules, no RNA is required.

It seems most probable that if centrioles do anything with tubulin molecules, centrioles are not a factory that makes them but rather an assembly plant that puts them together into microtubules. The idea that centrioles serve as "orienting centers" has been explored by Inoué (1964; Inoué and Sato, 1967), Grimstone (1961, 1967), and others. Though it is not obvious how centrioles might act as an orienting center, it is clear that to influence their microenvironment they must give out or take up something. Such an interaction could be reversible -- a lending rather than a production of material.

An important question that arises is whether all the centriole-associated microtubules are made of the same tubulin monomer. The possible interconvertibility of microtubules, through a common pool of tubulins, has often been considered, based largely on Inoué's (1964) ideas about a dynamic equilibrium of spindle proteins. Several have suggested, for example, that spindle and cytoplasmic microtubules are interconvertible, both in cells with centrioles (Robbins and Gonatas, 1964) and in cells without (Ledbetter and Porter, 1963). Such ideas depend, in part, on determination of whether the various microtubules are built of a common tubulin monomer.

On the basis of staining characteristics and especially of the conditions necessary for preservation or for isolation, several different classes of microtubules can be distinguished (see esp. Behnke and Forer, 1967; Stephens, 1970b). On the basis of morphology and conditions for preservation, there could be as many as 9 different kinds of tubules associated with the flagellar apparatus of *Naegleria*: the A, B, and C tubules of the basal body, the A and B of outer fibers, the 2 central pair tubules, the spur and the subsurface cytoplasmic tubules. Similar counts could be made in other flagellated and ciliated cells. Stephens (1970a) has shown that even the A and B subfibers from outer fibers differ in amino acid composition and in peptide maps, which indicates a difference in the primary structure of the A and B tubulins. Fulton, Kane, and Stephens (1970) have shown that the mitotic, flagellar, and ciliary microtubules of sea urchins share similar antigenic determinants, though some differences have been detected, even between the outer fiber doublets of sperm flagella and blastular cilia. It remains to be determined how many different tubulins are made, as well as what their cellular loci of transcription and translation are.

A major but probably not sole function of centrioles appears to be to guide the assembly of tubulin molecules made elsewhere into microtubules. At present the way in which centrioles accomplish this is mysterious. They do it in one way for the outer fibers of flagella, where the basal bodies appear to act as a nucleating or crystallization center, in another way for mitotic and cytoplasmic microtubules, where Inoué's term "orienting center" is more applicable, and in still a third way in making new centrioles. The basal-body role would seem to require a pinwheel, but perhaps the mitotic role would not – indeed, that might explain why cells that do not use basal bodies can get along without centriolar pinwheels (Section VI, A). For mitosis, the capacity to mobilize the right tubulin molecules would determine the axis of division, separate the poles, and position the continuous spindle fibers which some argue serve as a framework for chromosome movement.

VI. Evolution

A. Distribution of Centrioles

The centriolar pinwheel is distributed throughout the eucaryotes, with some specific exceptions. The universality of its morphology among phylogenetically diverse organisms is striking. On the other hand, it seems that different groups have placed different emphasis on the activities in which centrioles participate. For these reasons it is instructive to examine the distribution of centrioles and centriolar behavior among different organisms.

Procaryotes. None of the procaryotes have anything resembling a centriolar pinwheel. They have a completely different kind of cell division, and though many bacteria have flagella, these are unlike the 9 + 2 flagella of eucaryotes (Lowy and Spencer, 1968; Iino, 1969). At the base of bacterial flagella are distinctive hook structures, which seem to end on or in the cell membrane. Reports of basal bodies (reviewed in Doetsch and Hageage, 1968) can be explained as artifacts of isolation (Vaituzis and Doetsch, 1969). There is no need to postulate an internal basal body.

Eucaryotic Protists. Numerous protists have centriolar pinwheels, including algae (e.g., Manton, 1959, 1964a; Ringo, 1967a; Turner, 1968), fungi (e.g., Berlin and Bowen, 1964; Renaud and Swift, 1964), and protozoa (ciliates and flagellates). The distribution includes phylogenetically scattered protists, and is not related in any obvious way to likely patterns of evolution.

Eucaryotic protists seem to use centrioles mainly as basal bodies. Grimstone and Gibbons (1966) pointed out that no protozoan is known which has a centriole which does not also serve, during at least part of the life cycle, as a basal body. This generalization can probably be extended to all the protists. Those with centriolar pinwheels form flagella at some stage of their life cycle, and those that never have flagella also do not have centriolar pinwheels.

Protists have diverse forms of mitosis (reviewed by Wilson, 1925; Grell, 1964; Robinow and Bakerspiegel, 1965). Many have intranuclear mitosis, and commonly these do not have centrioles. However, among the fungi there are some water molds in which centrioles persist, and during division are transferred outside the nuclear envelope at the flattened poles, apparently not participating in mitosis (e.g., Ichida and Fuller, 1968; Lessie and Lovett, 1968). The centriolar pinwheels in some algae also do not serve as conventional cell centers (Johnson and Porter, 1968; Turner, 1968). Both *Naegleria* and species of the Myxomycetes have a phase of intranuclear mitosis without centrioles alternating with a phase with centrioles which serve as basal bodies for flagellates (Section IV, B).

Yeast are an example of protists that never have centrioles or form flagella. In their intranuclear mitosis, spindle microtubules converge toward the poles where they approach electron-dense condensations in the nuclear membrane. Robinow and Marak (1966), who described these dense plaques as looking like "large pores of the nuclear envelope," called them "centriolar plaques." Similar plaques have been observed in *Aspergillus* (Robinow and Caten, 1969). These dense plaques look like the dense masses often found at positions from which microtubules emanate (e.g., pericentriolar satellites). They are completely unlike ordinary centrioles and I question, as did de Harven (1968), whether it is informative to call them centriolar plaques.

A number of structures that may act as cell centers in protists definitely do not have the centriolar pinwheel morphology. Drum and Pankratz (1963) found in a diatom a dense sphere with microtubules around it that they considered to be a centrosome. The photograph is of low resolution, but the sphere looks like it might resemble the blepharoplasts of *Marsilea* (Mizukami and Gall, 1966). In another diatom, Manton, Kowallik, and von Stosch (1969) described a quite different, rectangular structure which they called a "spindle precursor"; the same species of diatom can form male gametes with basal bodies (Manton and von Stosch, 1966). There are also the structures in termite flagellates that Cleveland termed centrioles, though their structure is now known not to resemble the centriolar pinwheel (Section II, B). In one of these termite flagellates, *Trichonympha*, there is a true centriole, but it does not participate in mitosis (Grimstone and Gibbons, 1966).

In some flagellates basal bodies sometimes serve as mitotic centrioles while the flagella are still attached (Jahn, 1904); in others the basal bodies produce new basal bodies quite separately from the nuclear division, while the division occurs without centrioles (Outka and Kluss, 1967). Further examples of the behavior of basal bodies in division are described in Wilson (1925, p. 696). Ciliates, though they have thousands of centriole-like basal bodies, have no centriole which acts as a cell center (e.g., Carasso and Favard, 1965; Jenkins, 1967; Tucker, 1967). Amebae that never form flagella do not have centriolar pinwheels (Roth and Daniels, 1962; Bowers and Korn, 1968).

Metaphytes. Many of the mosses and liverworts (bryophytes), ferns, and the cycads and *Ginkgo* (primitive gymnosperms) have flagellated male gametes, and centriolar pinwheels. Some of these lower plants handle their centrioles in unusual ways, constructing them through blepharoplasts (Sharp, 1921; Lepper, 1956; Mizukami and Gall, 1966), arranging them end-to-end (Moser and Kreitner, 1970), and building them into elaborate flagellar apparatuses (Carothers and Kreitner, 1968).

No higher plants (conifers and angiosperms) are known which have a centriolar pinwheel, and none have flagella. These land plants use pollen tubes instead of flagellated male gametes. The mitotic figures have abundant microtubules, however, which converge toward the poles (e.g., Ledbetter and Porter, 1963; Ledbetter, 1967; Pickett-Heaps, 1969).

Metazoa. Virtually all metazoa have centriolar pinwheels, both as division centers and to make cilia and flagella, including ciliated epithelia and flagellated male gametes. Some of the reports of centrioles in metazoa are reviewed by WENT (1966a, p. 11). Centrioles are present in most if not all metazoan cells capable of division, including the cells of nematodes, which as a group are devoid of cilia and flagella (ROGGEN, RASKI, and JONES, 1966; LEE and ANYA, 1967). Most metazoan centrioles are quite typical; only those of *Sciara* are known to be atypical (PHILLIPS, 1967, 1970). During gametogenesis in *Sciara*, normal somatic-line centrioles produce giant centrioles, which are bundles of 60—90 microtubules in an oval array. These produce additional giant centrioles which act as basal bodies in the male gametes.

B. Origin and Phylogeny

We can make few generalizations about the evolution and adaptation of centrioles. One remarkable feature, as many have noted, is the nearly absolute conservation of the basic "9-triplet pinwheel" morphology of centrioles and basal bodies throughout the evolution of the eucaryotic protists, the lower plants, and the multicellular animals. Though there are minor variations, the constancy includes the size, pattern, and spacing of the triplets, and probably even the universal occurrence of a hub-and-spoke cartwheel at the proximal end. The major exceptions are those organisms that never have centriolar pinwheels. However it seems likely that all eucaryotes, except possibly the red algae (MANTON, 1965; KLEIN and CRONQUIST, 1967), are derived from ancestors that had centriolar pinwheels. The odd structures that may serve as cell centers — the dense plaques in yeast and diatoms, for example, or the giant "centrioles" of termite flagellates — are so unlike the pinwheel that they probably represent independent evolution. Attempts to find a reason why the centriolar pinwheel has not evolved more have been uniformly unsuccessful. Perhaps the most satisfying explanation for evolutionary stability is that "successful mutations in the centrosome-basal body are very rare" (BRADFIELD, 1955). SATIR and SATIR (1964) presented an ingenious speculative model based on repeats in a helical protein which, if correct, could explain the ninefold symmetry and its evolutionary stability.

All basically normal flagella, of the "9 + 2" type or one of its variations, have conventional basal bodies at their bases, and it seems that the formation of a flagellum requires a basal body. Since basal bodies have not evolved, it is not surprising that flagellum structure, at least the ring of doublets, has remained constant too.

The origin of the centriolar pinwheel is obscure. All known centrioles have the definitive morphology; there are no evolutionary intermediates on which to build even a hypothetical sequence. Many view one or another of the photosynthetic flagellated algae as representatives of a stem group from which most eucaryotes evolved. These algae have centriolar pinwheels — in the form of basal bodies for their flagella. Several have suggested that the pinwheel structure evolved from a bundle of bacterial flagella (BRADFIELD, 1955; KLEIN and CRONQUIST, 1967), but it is hard to defend this since neither the chemical nor the physical structure of bacterial flagella resembles microtubules (see MOHRI, 1968; IINO, 1969; STEPHENS, 1970b), and since there is no structure in bacteria resembling the basal body. Some

14*

Bradfield, J. R. G.: Fibre patterns in animal flagella and cilia. Symp. Soc. exp. Biol. **9**, 306—334 (1955).

Bradley, D. E.: The fluorescent staining of bacteriophage nucleic acids. J. gen. Microbiol. **44**, 383—391 (1966).

Briggs, R., King, T. J.: Nucleocytoplasmic interactions in eggs and embryos. In: The Cell (Ed.: Brachet, J., Mirsky, E.), Vol. 1, pp. 537—617. New York: Academic Press 1959.

Brinkley, B. R., Stubblefield, E.: Ultrastructure and interaction of the kinetochore and centriole in mitosis and meiosis. In: Advances in Cell Biology (Ed.: D. M. Prescott et al.), Vol. 1, pp. 119—185. New York: Appleton-Century-Crofts 1970.

Burgos, M. H., Fawcett, D. W.: Studies on the fine structure of the mammalian testis, I. J. biophys. biochem. Cytol. **1**, 287—300 (1955).

Carasso, N., Favard, D.: Microtubules fusoriaux dans les micro et macronucleus de ciliés péritriches en division. J. Microscopie **4**, 395—402 (1965).

Carothers, Z. B., Kreitner, G. L.: Studies on spermatogenesis in the hepaticae, II. J. Cell Biol. **36**, 603—616 (1968).

Chatton, E.: Sur les connexions flagellaires des éléments flagellés. C. R. Soc. Biol. (Paris) **91**, 577—581 (1924).

Child, F. M., Mazia, D.: A method for the isolation of parts of ciliates. Experientia (Basel) **12**, 161—162 (1956).

Cleaver, J. E.: Thymidine Metabolism and Cell Kinetics. Amsterdam: North-Holland Publishing Co. 1967.

Cleveland, L. R.: Types and life cycles of centrioles of flagellates. J. Protozool. **4**, 230—241 (1957).

— Function of flagellate and other centrioles in cell reproduction. In: The Cell in Mitosis (Ed.: L. Levine), pp. 3—31. New York: Academic Press 1963.

Conklin, E. G.: The organization and cell-lineage of the ascidian egg. J. Acad. Sci. Philadelphia **13**, 1—119 (1905).

Costello, D. P.: On the orientation of centrioles in dividing cells, and its significance. Biol. Bull. **120**, 285—312 (1961).

Dahl, H. A.: Fine structure of cilia in rat cerebral cortex. Z. Zellforsch. Mikroskop. Anat. **60**, 369—386 (1963).

Dalcq, A. M.: Le centrosome. Bull. Acad. roy. Belgique (Class Sci.) **50**, 1408—1449 (1964).

David-Ferreira, J. F.: Observations préliminaires sur la structure et la cytochimie du centriole. Proc. 5th Intern. Congr. Electron Microscopy (Ed.: S. S. Breese, Jr.), Vol. 2, p. XX—4. New York: Academic Press 1962.

de Harven, E.: The centriole and the mitotic spindle. In: The Nucleus (Ed.: A. J. Dalton and F. Haguenau), pp. 197—227. New York: Academic Press 1968.

— Bernhard, W.: Étude au microscope electronique de l'ultrastructure du centriole chez les vertébrés. Z. Zellforsch. **45**, 378—398 (1956).

de-Thé, G.: Cytoplasmic microtubules in different animal cells. J. Cell Biol. **23**, 265—275 (1964).

Dietz, R.: Centrosomenfreie Spindelpole in Tipuliden-Spermatocyten. Z. Naturforsch **14**b, 749—752 (1959).

— The dispensability of the centrioles in the spermatocyte divisions of *Pales*. Proc. 1st Oxford Conf. on Chromosomes Today (Ed.: C. D. Darlington and K. R. Lewis), Vol. 1, pp. 161—166. New York: Plenum Press (1966).

Dingemans, K. P.: The relation between cilia and mitosis in the mouse adenohypophysis. J. Cell Biol. **43**, 361—367 (1969).

Dingle, A. D.: Control of flagellum number in *Naegleria*. Temperature-shock induction of multiflagellate cells. J. Cell Sci., 463—481 (1970).

— Fulton, C.: Development of the flagellar apparatus of *Naegleria*. J. Cell Biol. **31**, 43—54 (1966).

Dippell, R. V.: The site of silver impregnation in *Paramecium*. J. Protozool. **9**, Suppl., 24 (1962).

— The development of basal bodies in *Paramecium*. Proc. nat. Acad. Sci. (Wash.) **61**, 461—468 (1968).

DIRKSEN, E. R.: The presence of centrioles in artificially activated sea urchin eggs. J. biophys. biochem. Cytol. **11**, 244—247 (1961).
— The isolation and characterization of asters from artificially activated sea urchin eggs. Exp. Cell Res. **36**, 256—269 (1964).
— Observations on centriole formation in the development of ciliated epithelium of mouse oviduct. J. Cell Biol. **39**, 34a (1968).
— CROCKER, T. T.: Centriole replication in differentiating ciliated cells of mammalian respiratory epithelium. J. Microscopie **5**, 629—644 (1966).
DOETSCH, R. N., HAGEAGE, G. J.: Motility in procaryotic organisms. Biol. Rev. **43**, 317—362 (1968).
DRUM, R. W., PANKRATZ, H. S.: Fine structure of a diatom centrosome. Science **142**, 61—63 (1963).
DU PRAW, E. J.: Cell and Molecular Biology. New York: Academic Press (1968).
EHRET, C. F., DE HALLER, G.: Origin, development, and maturation of organelles in *Paramecium*. J. Ultrastruct. Res. Suppl. **6**, 1—42 (1963).
EMERSON, S. U., TOKUYASU, K., SIMON, M. I.: Bacterial flagella: polarity of elongation. Science **169**, 190—192.
FAWCETT, D. W.: Cilia and flagella. In: The Cell (Ed.: J. BRACHET and A. E. MIRSKY), Vol. 2, pp. 217—297. New York: Academic Press 1961.
— An Atlas of Fine Structure. The Cell. Philadelphia: W. B. Saunders Co. 1966.
FISCHER, H., HUG, O., LIPPERT, W.: Elektronenmikroskopische Studien an Forellen-spermatozoen und ihren Zellkernen. Chromosoma **5**, 69—80 (1952).
FRIEDLÄNDER, M., WAHRMAN, J.: Giant centrioles in neuropteran meiosis. J. Cell Sci. **1**, 129—144 (1966).
FRISCH, D.: Fine structure of the early differentiation of ciliary basal bodies. Anat. Rec. **157**, 245 (1967).
FRY, H. J.: The so-called central bodies in fertilized *Echinarachnius* eggs, I—III. Biol. Bull. **56**, 101—158 (1929).
FULLER, M. S., CALHOUN, S. A.: Microtubule-kinetosome relationships in the motile cells of the Blastocladiales. Z. Zellforsch. **87**, 526—533 (1968).
FULTON, C.: Amebo-flagellates as research partners. The laboratory biology of *Naegleria* and *Tetramitus*. In: Methods in Cell Physiology (Ed.: D. M. PRESCOTT), Vol. 4, pp. 341—476. New York: Academic Press 1970.
— DINGLE, A. D.: Basal bodies, but not centrioles, in *Naegleria*. In preparation (1970).
— KANE, R. E., STEPHENS, R. E.: Serological similarity of flagellar and mitotic micro-tubules. In preparation (1970).
FÜRST, E.: Über Centrosomen bei *Ascaris*. Arch. Mikr. Anat. **52**, 97—134 (1898).
GACHET, J., THIÉRY, J.-P.: Application de la méthode de tirage photographique avec rotations ou translations a l'étude de macromolécules et de structures biologiques. J. Microscopie **3**, 253—268 (1964).
GALL, J. G.: Centriole replication. A study of spermatogenesis in the snail *Viviparus*. J. biophys. biochem. Cytol. **10**, 163—193 (1961).
GIBBINS, J. R., TILNEY, L. G., PORTER, K. R.: Microtubules in the formation and develop-ment of the primary mesenchyme in *Arbacia*, I. J. Cell Biol. **41**, 201—226 (1969).
GIBBONS, I. R.: The relationship between the fine structure and direction of beat in the gill cilia of a lamellibranch mollusc. J. biophys. biochem. Cytol. **11**, 179—205 (1961).
— Chemical dissection of cilia. Arch. Biol. (Liège) **76**, 317—352 (1965).
— GRIMSTONE, A. V.: On flagellar structure in certain flagellates. J. biophys. biochem. Cytol. **7**, 697—716 (1960).
GONATAS, N. K., ROBBINS, E.: The homology of spindle tubules and neurotubules in the chick embryo retina. Protoplasma **59**, 377—391 (1964).
GRANICK, S., GIBOR, A.: The DNA of chloroplasts, mitochondria, and centrioles. Progr. Nucleic Acid. Res. Mol. Biol. **6**, 143—186 (1967).
GRAY, W. D., ALEXOPOULOS, C. J.: Biology of the Myxomycetes. New York: Ronald Press 1968.
GRELL, K. G.: The protozoan nucleus. In: The Cell (Ed.: J. BRACHET and A. E. MIRSKY), Vol. 6, pp. 1—79. New York: Academic Press 1964.

Grim, J. N.: Isolated ciliary structures of *Euplotes*. Exp. Cell Res. **41**, 206—210 (1966).
Grimstone, A. V.: Fine structure and morphogenesis in Protozoa. Biol. Rev. **36**, 97—150 (1961).
— Structure and formation of some fibrillar organelles in Protozoa. In: Formation and Fate of Cell Organelles (Ed.: K. B. Warren), Symp. Intern. Soc. Cell Biol. **6**, 219—232. New York: Academic Press (1967).
— Cleveland, L. R.: The fine structure and function of the contractile axostyles of certain flagellates. J. Cell Biol. **24**, 387—400 (1965).
— Gibbons, I. R.: The fine structure of the centriolar apparatus in the complex flagellates *Trichonympha* and *Pseudotrichonympha*. Phil. Trans. Roy. Soc. London, B **250**, 215—242 (1966).
Harris, P.: Electron microscope study of mitosis in sea urchin blastomeres. J. biophys. biochem. Cytol. **11**, 419—431 (1961).
— Some structural and functional aspects of the mitotic apparatus in sea urchin embryos. J. Cell Biol. **14**, 475—487 (1962).
— Some observations concerning metakinesis in sea urchin eggs. J. Cell Biol. **25** (no. 1, part 2), 73—77 (1965).
Hartmann, J. F.: Cytochemical localization of adenosine triphosphatase in the mitotic apparatus. J. Cell Biol. **23**, 363—370 (1964).
Harvey, E. B.: Parthenogenetic merogony or cleavage without nuclei in *Arbacia*. Biol. Bull. **71**, 101—121 (1936).
Heist, H. E., Mulvaney, B. D.: Centriole migration. J. Ultrastruct. Res. **24**, 86—101 (1968).
Henneguy, L.-F.: Sur les rapports des cils vibratiles avec les centrosomes. Arch. Anat. Microscop. Morphol. Exp. **1**, 481—496 (1898).
Hoffman, E. J.: The nucleic acids of basal bodies isolated from *Tetrahymena*. J. Cell Biol. **25**, 217—228 (1965).
Huettner, A. F.: Continuity of the centriole in *Drosophila melanogaster*. Z. Zellforsch. **19**, 119—134 (1933).
— Rabinowitz, M.: Demonstration of the central body in the living cell. Science **78**, 367—368 (1933).
Hufnagel, L.: Cortical ultrastructure of *Paramecium*. J. Cell Biol. **40**, 779—801 (1969a).
— Properties of DNA associated with raffinose-isolated pellicles of *Paramecium*. J. Cell Sci. **5**, 561—573 (1969b).
Ichida, A. A., Fuller, M. S.: Ultrastructure of mitosis in the aquatic fungus *Catenaria*. Mycologia **60**, 141—155 (1968).
Iino, T.: Genetics and chemistry of bacterial flagella. Bact. Rev. **33**, 454—475 (1969).
Inoué, S.: Organization and function of the mitotic spindle. In: Primitive Motile Systems in Cell Biology (Ed. R. D. Allen and N. Kamiya), pp. 549—594. New York: Academic Press 1964.
— Sato, H.: Cell motility by labile association of molecules. J. gen. Physiol. **50**, Suppl. 259—288 (1967).
Jahn, E.: Myxomycetenstudien, 3. Kernteilung und Geisselbildung von *Stemonitis*. Dtsch. bot. Ges. **22**, 84—92 (1904).
Jenkins, R. A.: Fine structure of division in ciliate protozoa, I. J. Cell Biol. **34**, 463—481 (1967).
Johnson, U. G., Porter, K. R.: Fine structure of cell division in *Chlamydomonas*. Basal bodies and microtubules. J. Cell Biol. **38**, 403—425 (1968).
Kalnins, V. I., Porter, K. R.: Centriole replication during ciliogenesis in the chick tracheal epithelium. Z. Zellforsch. **100**, 1—30 (1969).
Kane, R. E.: The mitotic apparatus. Fine structure of the isolated unit. J. Cell Biol. **15**, 279—287 (1962).
Kasten, F. H.: Cytochemical studies with acridine orange and the influence of dye contaminants in the staining of nucleic acids. Intern. Rev. Cytol. **21**, 141—202 (1967).
Kater, J. M.: Morphology and division of *Chlamydomonas*. Univ. Calif. Publ. Zool. **33**, 125—168 (1929).

KIMBALL, R. F., PERDUE, S. W.: Quantitative cytochemical studies on *Paramecium*, V. Exp. Cell Res. **27**, 405—415 (1962).

KLEIN, R. M., CRONQUIST, A.: A consideration of the evolutionary and taxonomic significance of some biochemical, micromorphological, and physiological characters in the Thallophytes. Quart. Rev. Biol. **42**, 105—296 (1967).

KRISHAN, A., BUCK, R. C.: Structure of the mitotic spindle in L strain fibroblasts. J. Cell Biol. **24**, 433—444 (1965).

LANG, N. J.: An additional ultrastructural component of flagella. J. Cell Biol. **19**, 631—634 (1963).

LEDBETTER, M. C.: The disposition of microtubules in plant cells during interphase and mitosis. In: Formation and Fate of Cell Organelles (Ed.: B. WARREN), Symp. Intern. Soc. Cell Biol. **6**, 55—70. New York: Academic Press 1967.

— PORTER, K. R.: A "microtubule" in plant cell fine structure. J. Cell Biol. **19**, 239—250 (1963).

LEE, D. L., ANYA, A. O.: The structure and development of the spermatozoon of *Aspiculuris tetraptera* (Nemesade). J. Cell Sci. **2**, 537—544 (1967).

LENHOSSÉK, M. VON: Ueber Flimmerzellen. Verh. Anat. Ges., Kiel **12**, 106—128 (1898).

LEPPER, R., JR.: The plant centrosome and the centrosome-blepharoplast homology. Botan. Rev. **22**, 375—417 (1956).

LESSIE, P. E., LOVETT, J. S.: Ultrastructural changes during sporangium formation and zoospore differentiation in *Blastocladiella*. Amer. J. Botany **55**, 220—236 (1968).

LOEWY, A. G., SIEKEVITZ, P.: Cell Structure and Function, 2nd ed. New York: Holt, Rinehart and Winston 1969.

LONGO, F. J., ANDERSON, E.: The fine structure of pronuclear development and fusion in the sea urchin. J. Cell Biol. **39**, 339—368 (1968).

— — Cytological aspects of fertilization in the lamellibranch, *Mytilus*, I. J. exp. Zool. **172**, 69—96 (1969).

LORCH, I. J.: Enucleation of sea-urchin blastomeres with or without removal of asters. Quart. J. Microscop. Sci. **93**, 475—486 (1952).

LOWY, J., SPENCER, M.: Structure and function of bacterial flagella. Symp. Soc. exp. Biol. **22**, 215—236 (1968).

LWOFF, A.: Problems of Morphogenesis in Ciliates. The Kinetosomes in Development, Reproduction and Evolution. New York: Wiley and Sons 1950.

MCDONALD, B. R., WEIJER, J.: The DNA content of centrioles of *Neurospora*. Canad. J. Genet. Cytol. **8**, 42—50 (1966).

MCINTOSH, J. R., HEPLER, P. K., WIE, D. G. VAN: Model for mitosis. Nature (Lond.) **224**, 659—663 (1969).

MCMANUS, M. A., ROTH, L. E.: Ultrastructure of the somatic nuclear division in the plasmodium of *Clastoderma*. Mycologia **60**, 426—436 (1968).

MANTON, I.: Electron microscopical observations on a very small flagellate. J. Marine Biol. Ass. U. K. **38**, 319—333 (1959).

— The possible significance of some details of flagellar bases in plants. J. roy. Microscop. Soc. **82**, 279—285 (1964a).

— Observations with the electron microscope on the division cycle in *Prymnesium*. J. roy. Microscop. Soc. **83**, 317—325 (1964b).

— Some phyletic implications of flagellar structure in plants. Advan. Botan. Res. **2**, 1—34 (1965).

— KOWALLIK, K., STOSCH, H. A. VON: Observations on the fine structure of a marine centric diatom, I. J. Microscopy **89**, 295—320 (1969).

— STOSCH, H. A. VON: Observations on the fine structure of the male gamete of *Lithodesimum*. J. Roy. Microscop. Soc. **85**, 119—139 (1966).

MAZIA, D.: Mitosis and the physiology of cell division. In: The Cell (Ed.: J. BRACHET and A. E. MIRSKY), Vol. **3**, pp. 77—412. New York: Academic Press 1961.

— HARRIS, P. J., BIBRING, T.: The multiplicity of mitotic centers and the time-course of their duplication and separation. J. biophys. biochem. Cytol. **7**, 1—20 (1960).

MILHAUD, M., PAPPAS, G. D.: Cilia formation in the adult cat brain after pargyline treatment. J. Cell Biol. **37**, 599—609 (1968).

Mizukami, I., Gall, J.: Centriole replication, II. Sperm formation in the fern, *Marsilea*, and the cycad, *Zamia*. J. Cell Biol. **29**, 97—111 (1966).

Mohri, H.: Amino-acid composition of "tubulin" constituting microtubules of sperm flagella. Nature (Lond.) **217**, 1053—1054 (1968).

Morgan, T. H.: The action of salt-solutions on the unfertilized and fertilized eggs of *Arbacia*. Arch. Entwicklungsmech. Organ. **8**, 448—539 (1899).

Moser, J. W.: Kreitner, G. L.: Centrosome structure in *Anthoceros* and *Marchantia*. J. Cell Biol. **44**, 454—458 (1970).

Murray, R. G., Murray, A. S., Pizzo, A.: The fine structure of mitosis in rat thymic lymphocytes. J. Cell Biol. **26**, 601—619 (1965).

Nass, M. M. K., Nass, S., Afzelius, B. A.: The general occurrence of mitochondrial DNA. Exp. Cell Res. **37**, 516—539 (1965).

Outka, D. E., Kluss, B. C.: The ameba-to-flagellate transformation in *Tetramitus*, II. Microtubular morphogenesis. J. Cell Biol. **35**, 323—346 (1967).

Perkins, F. O.: Formation of centriole and centriole-like structures during meiosis and mitosis in *Labyrinthula*. J. Cell Sci. **6**, 629—653 (1970).

Pfeiffer, H. H.: Über die Mitwirkung rhythmischer Oscillationen der Centrosomen. Protoplasma **46**, 585—596 (1956).

Phillips, D. M.: Giant centriole formation in *Sciara*. J. Cell Biol. **33**, 73—92 (1967).

— Insect sperm: their structure and morphogenesis. J. Cell Biol. **44**, 243—277 (1970).

Pickett-Heaps, J. D.: The evolution of the mitotic apparatus: an attempt at comparative ultrastructural cytology in dividing plant cells. Cytobios **3**, 257—280 (1969).

Pitelka, D. R.: Fibrillar systems in protozoa. In: Research in Protozoology (Ed.: T.-T. Chen), Vol. 3, pp. 279—388. Oxford: Pergamon Press 1969.

Pollister, A. W.: Notes on the centrioles of amphibian tissue cells. Biol. Bull. **65**, 529—545 (1933).

— Pollister, P. F.: The relation between centriole and centromere in atypical spermatogenesis in viviparid snails. Ann. N. Y. Acad. Sci. **45**, 1—48 (1943).

Porter, K. R.: The submicroscopic morphology of protoplasm. Harvey Lect. 1955—56, **51**, 175—228 (1957).

— Cytoplasmic microtubules and their functions. In: Principles of Biomolecular Organization (Ed.: G. E. W. Wolstenholme and M. O'Connor), Ciba Foundation Symp., pp. 308—345. Boston: Little-Brown and Co. 1966.

Pyne, C. K.: Sur l'absence d'incorporation de la thymidine tritiée dans les cinétosomes de *Tetrahymena*. C. R. Acad. Sci. (Paris) **267**, 755—757 (1968).

Rampton, V. W.: Kinetosomes of *Tetrahymena*. Nature (Lond.) **195**, 195 (1962).

Randall, J. T.: The nature and significance of kinetosomes. J. Protozool. **6**, Suppl., 30 (1959).

— Cavalier-Smith, T., McVittie, A., Warr, J. R., Hopkins, J. M.: Developmental and control processes in the basal bodies and flagella of *Chlamydomonas*. Develop. Biol., Suppl. no. 1, pp. 43—83 (1967).

— Disbrey, C.: Evidence for the presence of DNA at basal body sites in *Tetrahymena*. Proc. Roy. Soc. London B **162**, 473—491 (1965).

— Fitton-Jackson, S.: Fine structure and function in *Stentor*. J. biophys. biochem. Cytol. **4**, 807—830 (1958).

— Hopkins, J. M.: Studies of cilia, basal bodies and some related organelles, IIa. Problems of genesis. Proc. Linnean Soc. London **174**, 37—39 (1963).

Rash, J. E., Shay, J. W., Biesele, J. J.: Cardiac cilia. J. Cell Biol. **39**, 177a (1968).

Raven, C. P.: The formation of the second maturation spindle in the eggs of *Succinea*. J. Embryol. exp. Morph. **7**, 344—360 (1959).

— Escher, F. C. M., Herrebout, W. M., Leussink, J. A.: The formation of the second maturation spindle in the eggs of *Limnaea*. J. Embryol. exp. Morph. **6**, 28—51 (1958).

Rebhun, L. I.: Saltatory particle movements in cells. In: Primitive Motile Systems in Cell Biology (Ed.: R. D. Allen and N. Kamiya), pp. 503—525. New York: Academic Press 1964.

Reger, J. F.: Nuclear incorporation of centrioles and mitochondria during spermatogenesis in *Leiobunum*. J. Cell Biol. **43**, 115a (1969).

RENAUD, F. L., SWIFT, H.: The development of basal bodies and flagella in *Allomyces*. J. Cell Biol. **23**, 339—354 (1964).

RÉNYI, G. DE: Untersuchungen über Flimmerzellen. Z. Anat. Entwickl.-Gesch. **73**, 338—357 (1924).

RIEGER, R., MICHAELIS, A., GREEN, M. M.: A Glossary of Genetics and Cytogenetics, 3rd edition. Berlin-Heidelberg-New York: Springer 1968.

RINGO, D. L.: Flagellar motion and fine structure of the flagellar apparatus in *Chlamydomonas*. J. Cell Biol. **33**, 543—571 (1967a).

— The arrangement of subunits in flagellar fibers. J. Ultrastruct. Res. **17**, 266—277 (1967b).

RIS, H., PLAUT, W.: Ultrastructure of DNA-containing areas in the chloroplast of *Chlamydomonas*. J. Cell Biol. **13**, 383—391 (1962).

ROBBINS, E., GONATAS, N. K.: The ultrastructure of a mammalian cell during the mitotic cycle. J. Cell Biol. **21**, 429—463 (1964).

— JENTZSCH, G.: Ultrastructural changes in the mitotic apparatus at the metaphase-to-anaphase transition. J. Cell Biol. **40**, 678—691 (1969).

— JENTZSCH, G., MICALI, A.: The centriole cycle in synchronized HeLa cells. J. Cell Biol. **36**, 329—339 (1968).

ROBINOW, C. F., BAKERSPIGEL, A.: Somatic nuclei and forms of mitosis in fungi. In: The Fungi, an Advanced Treatise (Ed.: G. C. AINSWORTH and A. S. SUSSMAN), Vol. 1, pp. 119—142. New York: Academic Press 1965.

— CATEN, C. E.: Mitosis in *Aspergillus*. J. Cell Sci. **5**, 403—431 (1969).

— MARAK, J.: A fiber apparatus in the nucleus of the yeast cell. J. Cell Biol. **29**, 129—151 (1966).

ROGGEN, D. R., RASKI, D. J., JONES, N. O.: Cilia in nematode sensory organs. Science **152**, 515—516 (1966).

ROSENBAUM, J. L., CHILD, F. M.: Flagellar regeneration in protozoan flagellates. J. Cell Biol. **34**, 345—364 (1967).

— MOULDER, J. E., RINGO, D. L.: Flagellar elongation and shortening in *Chlamydomonas*. J. Cell Biol. **41**, 600—619 (1969).

ROTH, L. E., DANIELS, E. W.: Electron microscopic studies of mitosis in amebae, II. J. Cell Biol. **12**, 57—78 (1962).

SAGAN, L.: On the origin of mitosing cells. J. theoret. Biol. **14**, 225—274 (1967).

SAKAGUCHI, H.: Pericentriolar filamentous bodies. J. Ultrastruct. Res. **12**, 13—21 (1965).

SALPETER, M. M., BACHMANN, L., SALPETER, E. E.: Resolution in electron microscope radioautography. J. Cell Biol. **41**, 1—20 (1969).

SATIR, B., ROSENBAUM, J. L.: The isolation and identification of kinetosome-rich fractions from *Tetrahymena*. J. Protozool. **12**, 397—405 (1965).

SATIR, P., SATIR, B.: A model for ninefold symmetry in α keratin and cilia. J. theoret. Biol. **7**, 123—128 (1964).

SCHUSTER, F.: An electron microscope study of the amoebo-flagellate *Naegleria*, I. J. Protozool. **10**, 297—313 (1963).

— Ultrastructure and morphogenesis of solitary stages of true slime molds. Protistologica **1**, 49—62 (1965).

SEAMAN, G. R.: Large-scale isolation of kinetosomes from *Tetrahymena*. Exp. Cell Res. **21**, 292—302 (1960).

— Protein synthesis by kinetosomes isolated from *Tetrahymena*. Biochim. biophys. Acta (Amst.) **55**, 889—899 (1962).

SHARP, L. W.: Spermatogenesis in *Equisetum*. Botan. Gaz. **54**, 89—119 (1912).

— Spermatogenesis in *Marsilea*. Botan. Gaz. **58**, 419—431 (1914).

— An Introduction to Cytology, Ch. 5. New York: McGraw-Hill (1921).

SLAUTTERBACK, D. B.: Cytoplasmic microtubules, I. J. Cell Biol. **18**, 367—388 (1963).

SMITH-SONNEBORN, J., PLAUT, W.: Evidence for the presence of DNA in the pellicle of *Paramecium*. J. Cell Sci. **2**, 225—234 (1967).

— — Studies on the autonomy of pellicular DNA in *Paramecium*. J. Cell Sci. **5**, 365—372 (1969).

Sonneborn, T. M.: The evolutionary integration of the genetic material into genetic systems. In: Heritage from Mendel (Ed.: R. A. Brink), pp. 375—401. Madison: University of Wisconsin Press 1967.
— Determination, development, and inheritance of the structure of the cell cortex. Symp. Intern. Soc. Cell Biol. (H. A. Padykyla, ed.), vol. 9, in press. New York: Academic Press 1970.
Sorokin, S.: Centrioles and the formation of rudimentary cilia by fibroblasts and smooth muscle cells. J. Cell Biol. 15, 363—377 (1962).
— Reconstructions of centriole formation and ciliogenesis in mammalian lungs. J. Cell Sci. 3, 207—230 (1968).
— Adelstein, S. J.: Failure of 1100 rads of X-radiation to affect ciliogenesis and centriolar formation in cultured rat lungs. Radiation Res. 31, 748—759 (1967).
Sotelo, J. R., Trujillo-Cenóz, O.: Electron microscope study of the kinetic apparatus in animal sperm cells. Z. Zellforsch. 48, 565—601 (also 49, 1—12) (1958).
Steinman, R. M.: An electron microscopic study of ciliogenesis in developing epidermis and trachea in the embryo of Xenopus. Amer. J. Anat. 122, 19—55 (1968).
Stephens, R. E.: Reassociation of microtubule protein. J. molec. Biol. 33, 517—519 (1968).
— Thermal fractionation of the outer fiber doublet microtubules into A- and B-subfiber components: A- and B-tubulin. J. Molec. Biol. 47, 353—363 (1970a).
— Microtubules. In: Biological Macromolecules (Ed.: S. N. Timasheff and G. D. Fasman), Vol. 4, in press. New York: Marcel Dekker 1970b.
Stich, H.: Stoffe und Strömungen in der Spindel von Cyclops. Chromosoma 6, 199 (1954).
Stockinger, L., Cireli, E.: Eine bisher unbekannte Art der Zentriolenvermehrung. Z. Zellforsch. 68, 733—740 (1965).
Stone, G. E., Miller, O. L., Jr.: A stable mitochondrial DNA in Tetrahymena pyriformis. J. exp. Zool. 159, 33—38 (1965).
Stubblefield, E., Brinkley, B. R.: Cilia formation in Chinese hamster fibroblasts in vitro as a response to colcemid treatment. J. Cell Biol. 30, 645—652 (1966).
— — Architecture and function of the mammalian centriole. In: Formation and Fate of Cell Organelles (Ed.: K. B. Warren), Symp. Intern. Soc. Cell Biol. 6, 175—218. New York: Academic Press 1967.
Sukhanova, K. M., Nilova, V. K.: On the synthesis of nucleic acids in the kinetosomes of Opalina. Tsitologiya 7, 431—436 (1965).
Suyama, Y., Preer, J. R., Jr.: Mitochondrial DNA from protozoa. Genetics 52, 1051—1058 (1965).
Szollosi, D.: The structure of centrioles and their satellites in a jellyfish. J. Cell Biol. 21, 465—479 (1964).
Tahmisian, T. N., Devine, R. L.: The influence of X-rays on organelle induction and differentiation in grasshopper spermatogenesis. J. biophys. biochem. Cytol. 9, 29—46 (1961).
Taylor, E. W.: Dynamics of spindle formation and its inhibition by chemicals. J. biophys. biochem. Cytol. 6, 193—196 (1959).
Tilney, L. G., Byers, B.: Studies on the microtubules in Heliozoa, V. J. Cell Biol. 43, 148—165 (1969).
— Porter, K. R.: Studies on the microtubules of Heliozoa, I. Protoplasma 60, 317—344 (1965).
Trager, W.: The kinetoplast and differentiation in certain parasitic Protozoa. Amer. Naturalist 99, 255—266 (1965).
Tucker, J. B.: Changes in nuclear structure during binary fission in a ciliate. J. Cell Sci. 2, 481—498 (1967).
Turner, F. R.: An ultrastructural study of plant spermatogenesis. J. Cell Biol. 37, 370—393 (1968).
Vaituzis, Z., Doetsch, R. N.: Relationship between cell wall, cytoplasmic membrane, and bacterial motility. J. Bact. 100, 512—521 (1969).
Warr, J. R., McVittie, A., Randall, J., Hopkins, J. M.: Genetic control of flagellar structure in Chlamydomonas. Genet. Res. 7, 335—351 (1966).

WENT, H. A.: The behavior of centrioles and the structure and formation of the achromatic figure. Protoplasmatologia **6**, 1—109 (1966a).

— An indirect method to assay for mitotic centers. J. Cell Biol. **30**, 555—562 (1966b).

WILSON, E. B.: The Cell in Development and Inheritance. New York: Macmillan 1896.

— Experimental studies in cytology, I. A cytological study of artificial parthenogenesis in sea-urchin eggs. Arch. Entwicklungsmech. Organ. **12**, 529—596 (1901).

— The Cell in Development and Heredity, 3rd edition. New York: Macmillan 1925.

— MATHEWS, A. P.: Maturation, fertilization, and polarity in the echinoderm egg. J. Morph. **10**, 319—342 (1895).

WOLFE, J.: Structural analysis of basal bodies of the isolated oral apparatus of *Tetrahymena*. J. Cell Sci. **6**, 679—700 (1970).

WOOD, W. B., EDGAR, R. S., KING, J., LIELAUSIS, I., HENNINGER, M.: Bacteriophage assembly. Fed. Proc. **27**, 1160—1166 (1968).

ZIMMERMAN, S. B., SANDEEN, G.: The ribonuclease activity of crystallized pancreatic deoxyribonuclease. Analyt. Biochem. **14**, 269—277 (1966).

Origin and Continuity of Microtubules

Lewis G. Tilney

*Department of Biology, University of Pennsylvania,
Philadelphia, Pennsylvania*

I. Introduction

One of the first signs of differentiation in many cells is the development of an asymmetric cell form. Examples would include the development of a cilium or flagellum, or the elongation of a cell or part of a cell such as occurs in the development of a muscle cell, nerve cell, or lens cell. Coupled with these changes in morphology are changes of the constituents within the cell; for organelles and inclusions become organized and cytoplasmic streaming is compartmentalized. The changes in form or the development of motile mechanisms are, moreover, a product of intracellular forces rather than a result of differential rates of cell division or localized compression. Since the plasma membrane by itself is not recognized either as a rigid form-producing structure or a potential mechanism of motility, attempts have been made to locate intracellular structures that act in form-development or in motility. One intracellular structure which seems to fit these requirements is the microtubule.

Only within the past few years has the microtubule become recognized as a common constituent of eucaryotic cells. This was due to the introduction of improved methods for the fixation of animal and plant cells (SABATINI, BENSCH, and BARRNETT, 1963). Although descriptions of microtubules were published a decade or so earlier, most microtubules are destroyed by standard osmium or acrolein fixation. Furthermore, these tubules depolymerize when the cells are subjected to low temperature (TILNEY and PORTER, 1967; ROTH, 1967). Since prior to 1963 almost all fixations were carried out at $0°$ C in osmium tetroxide, it is no wonder that the discovery of the microtubule is so recent.

The microtubule was originally defined in terms of its morphology: it is a long, unbranched, apparently hollow tube of approximately $210-250$ Å in diameter. Its wall appears to be composed of globular subunits each about 40 Å in diameter. These subunits align into $11-13$ longitudinal strands so that if a microtubule is dried down on a grid, the tips of the tubule tend to splay out into its respective 40 Å strands (PEASE, 1963; ANDRÉ and THIÉRY, 1963; BARNICOT, 1966; SILVEIRA, 1969).

Protein, believed to be derived from microtubules, has been isolated from the outer doublet fibrils of cilia (the microtubules here consist of the well-known

9 + 2 configuration of fibrils which make up the ciliary axoneme; RENAUD, ROWE and GIBBONS, 1968), flagella (SHELANSKI and TAYLOR, 1967; 1968), the mitotic apparatus (STEPHENS, 1968a), and extracts of nervous tissue (WEISENBERG, BORISY, and TAYLOR, 1968). The basic monomeric units appear to be the same in every case both in amino acid composition and calculated molecular weight of about 60,000. Further, in all cases the protein binds GTP and not other nucleotides (WEISENBERG, BORISY, and TAYLOR, 1968; STEPHENS, RENAUD, and GIBBONS, 1967; SHELANSKI and TAYLOR, 1968; STEPHENS, 1968b; YANAGISAWA, HASEGAWA, and MONII, 1968). From the monomeric units derived from the outer doublet fibrils of cilia STEPHENS (1968b) has been able to reform microtubules *in vitro*.

In this review on the origin and continuity of microtubules we will present evidence relating microtubules to the differentiation of cell form and to motility as a background for an extended discussion on the control of assembly of microtubules. This control appears to involve the nucleation of microtubules, factors influencing the equilibrium between polymer and monomer, and intertubule interactions in the form of connecting bridges. These bridges not only appear to be necessary for the stability of microtubule clusters, but also appear as a common feature of tubule-associated motility mechanisms. To illustrate many of the principles involved we will give examples from our own work, some of which is published here for the first time.

II. Functions of Microtubules

Initially, ideas on the functions of microtubules in plant and animal cells were based on the location of these elements within cells or parts of cells. We will list

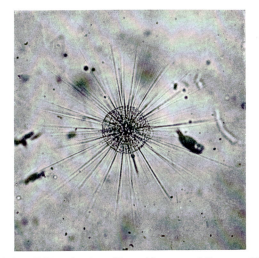

Fig. 1. Living *Echinosphaerium*. (From TILNEY and PORTER, 1965) × 80

six sites where microtubules are commonly encountered and then relate what these cells or portions of these cells have in common: 1. cilia and flagella, 2. nerve processes, 3. the mitotic apparatus, 4. the cortex of meristematic plant cells (LEDBETTER and PORTER, 1963), 5. elongating cells such as during the formation of the lens (BYERS and PORTER, 1964) or during spermatogenesis of certain insects (MOSES, WILSON, and WYIRCK, 1968), 6. selected structures in protozoa such as the axostyle of parasitic flagellates (GRIMSTONE and CLEVELAND, 1965), the axoneme of *Echinosphaerium* (TILNEY and PORTER, 1965), the fiber systems of *Stentor* (BANNISTER and TATCHELL, 1968), and the cytopharyngeal basket of *Nassula* (TUCKER, 1968). In

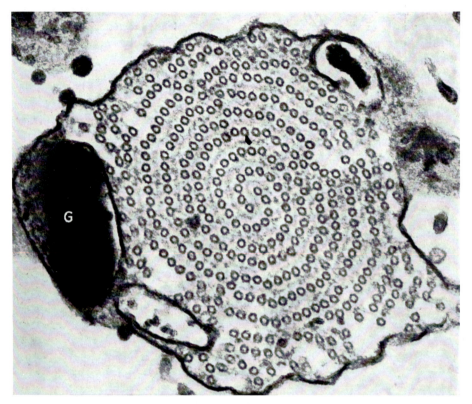

Fig. 2. Transverse section through an axopodium of *Echinosphaerium* midway between its base and its tip. Peripheral to the double coiled array of microtubules which makes up the axoneme, yet within the limiting membrane is a dense granule (G) believed to be involved in prey capture. This particle in living organisms streams rapidly in the narrow space between the limiting membrane and the axoneme. × 110,000

all these examples the microtubules are associated spatially with movement such as the undulation of cilia, flagella, and the axostyle, cytoplasmic streaming as in plant cells, the mitotic apparatus, neurons, and the axopodia of *Echinosphaerium*, and movement of particles such as chromosomes in the mitotic apparatus or food in the gullet of *Nassula*. Moreover, in all cases the microtubules are aligned parallel to the asymmetry of the cell or cell process and thus could be associated with its support. It is recognized, of course, that these two functions, support and movement, cannot always be separated, as for example in the elongation of a cell or process.

More information than just the physical presence of microtubules in relation to these functions now exists. Much of this information is based upon experimentally induced disassembly of the microtubules of the protozoan, *Echinosphaerium nucleofilum* (formerly *Actinosphaerium*). This organism will be described in some detail for it appears to be a model system for a description of the principles of microtubule assembly.

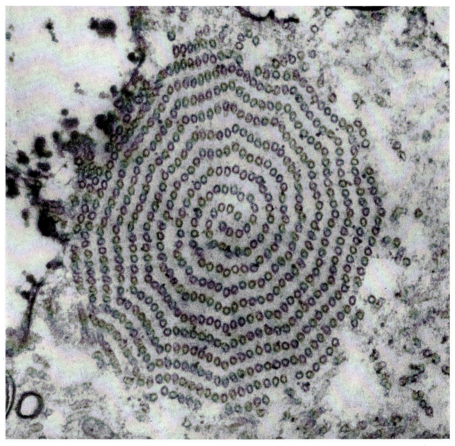

Fig. 3. Transverse section of an axoneme at the base of an axopodium of *Echinosphaerium*. It is possible in this micrograph to resolve the axoneme into twelve sectors. (From TILNEY and PORTER, 1965) × 70,000

Extending in all directions from the surface of *Echinosphaerium* are long, slender pseudopodia (axopodia) which can exceed 400 μ in length, yet which measure only 5–10 μ in diameter (Fig. 1). Food, generally a ciliate or flagellate, is captured with these axopodia. After a prey organism adheres to the axopodia it is conveyed to the surface of the cell body by "melting" and retraction of the involved axopodium or axopodia. Here it is engulfed in a food vacuole. Within each axopodium is a birefringent core, the axoneme, which extends from the tip of the axopodium to the medullary region. Transverse sections of the axoneme (Fig. 2) show it to

15 Cell Differentiation, Vol. 2

be composed of two interlocking coils of microtubules. The number of tubules making up the double coil reaches a maximum of about 500 at the base of the axopodium (Fig. 3) and gradually decreases towards the tip. Within the axoneme

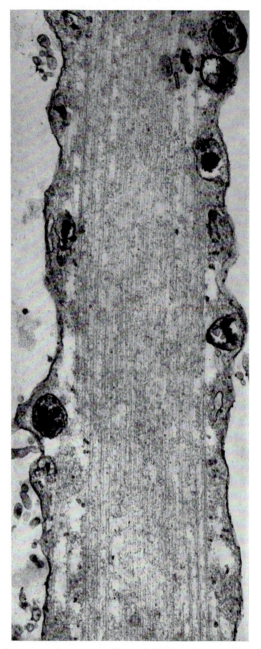

Fig. 4. Longitudinal section of an axopodium of *Echinosphaerium*. Peripheral to the parallel array of microtubules comprising the axoneme are dense granules which undergo saltations. (From TILNEY and PORTER, 1967) × 45,000

the microtubules are precisely spaced, a separation of 70 Å between microtubules in each coil, and a 300 Å separation between adjacent coils. In longitudinal sections the microtubules parallel the long axis of the axoneme (Fig. 4); those in the center

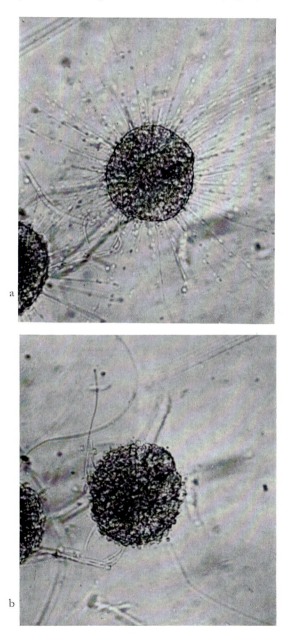

Fig. 5. a) Light micrograph of a living *Echinosphaerium* during treatment with hydrostatic pressure (3 minutes at 6 000 psi). Note the prominent beading along the axopodia. (From TILNEY, HIRAMOTO, and MARSLAND, 1966) × 120, b) Light micrograph of the same organism after 10 minutes treatment with hydrostatic pressure (6 000 psi). × 120

15*

of the axoneme are thought to traverse the full length of the axoneme, those near the outside, because of the taper from the base to the tip, traverse a relatively short distance. Peripheral to the axoneme, but within the axopodium, two species of particles can be found, dense particles believed to be involved in prey capture, and mitochondria. Other particles, even as small as ribosomes, are rarely found in the axopodia. In living *Echinosphaerium*, the dense particles stream in clearly

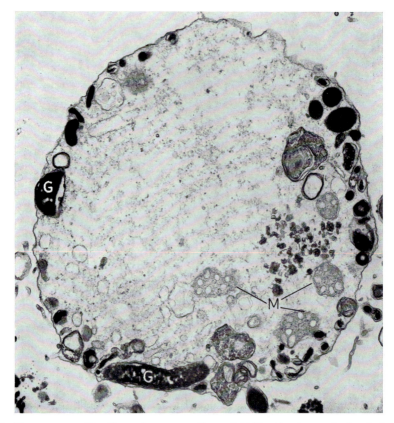

Fig. 6. Transverse section through an axopodium of *Echinosphaerium* at the level of a bead. This organism was fixed under pressure after 3 minutes at 6 000 psi. In the central portion of the bead is a filamentous material, no microtubules. At the periphery are dense granules (G) and mitochondria (M). (From Tilney, Hiramoto, and Marsland, 1966) × 55,000

defined tracks in the narrow space between the limiting membrane and the axoneme. This streaming is a type of saltation (Rebhun, 1964); the particles oscillate for a while, then move rapidly in one direction in a track, stop, fidget, and then either return in the same track, continue on, or jump tracks, now travelling in a new one.

This organism, then, is an ideal cell with which to investigate the functions of microtubules, for there is little else in the slender cell extensions which could reasonably account for their support and the highly patterned cytoplasmic streaming.

In order to test these postulated functions experimentally we applied to *Echinosphae-rium* a variety of agents which are known to cause the disintegration of the mitotic spindle. Since both the axonemes of *Echinosphaerium* and the mitotic apparatus are composed, in large measure, of microtubules, we reasoned that if these elements functioned in support, they would undergo disassembly as fast or faster than axopodial with-drawal, and after the removal of these agents, axopodia would not reform in the

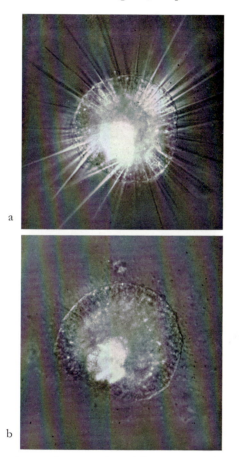

a

b

Fig. 7 a and b. These micrographs were taken of the same organism with a Zeiss polarizing microscope. The large birefringent spot is an ingested rotifer. (From TILNEY and PORTER, 1967). a) *Echinosphaerium* at the beginning of exposure to 4°C. Birefringent axonemes extend from the medulla to the tip of the axopodia. b) After 2 hours at 4°C. Note the loss of birefringent axonemes and axopodia. × 140

absence of oriented microtubules. The results of these experiments were exactly as predicted. Three experimental agents were used: colchicine (TILNEY, 1968b), hydrostatic pressure (TILNEY, HIRAMOTO, and MARSLAND, 1966), and low tempe-rature (TILNEY and PORTER, 1967). Within a minute after the application of hydro-static pressure (6000 psi) the axopodia underwent a prominent beading along the surface (Fig. 5a). The beads were joined by slender connections. If the cells were fixed while the pressure was maintained and sections were cut through the beads, we found a filamentous material in the position formerly occupied by the axoneme (Fig. 6). Peripheral to this central region were the two types of particles, the mitochondria and dense granules. The hydrostatic pressure not only broke down the

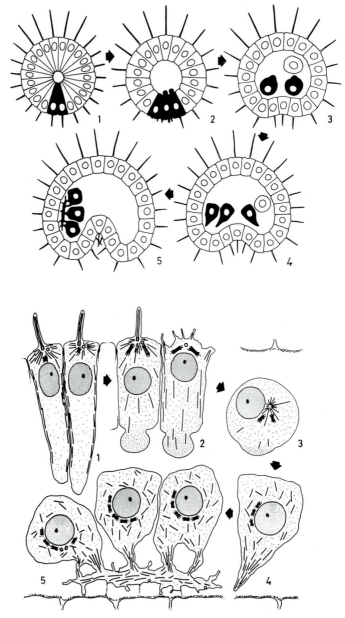

Fig. 8. The upper half of this drawing illustrates the changes in shape in *Arbacia* embryos that primary mesenchyme cells (cells in black) undergo during their formation and differentiation. The lower half of this drawing illustrates the distribution of microtubules (straight rods) at each stage in the sequence. (From Gibbins, Tilney, and Porter, 1969)

microtubules, but also interrupted the patterned movement of the granules, for they were now located exclusively in the beads. Within a few more minutes the beads retracted; most of them were incorporated into the cell body (Fig. 5b). When the pressure was released the microtubules reassembled, axopodia regrew, and in the regrowing axopodia the characteristic streaming resumed.

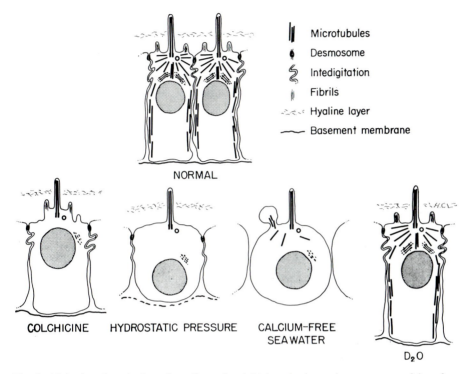

Fig. 9. This drawing depicts the effect of colchicine, hydrostatic pressure, calcium-free sea water, and D₂O on the shape of the ectodermal cells of *Arbacia* and on intracellular filaments, microtubules, lateral interdigitations, the septate desmosomes, the hyaline layer, and the basement membrane. (From TILNEY and GIBBINS, 1969)

Similar results were obtained with low temperature. With this agent it is possible to use polarization microscopy on living organisms to watch the fate of the axoneme during shortening (Fig. 7) and regrowth of the axopodia. Within every reforming axopodium was a birefringent axoneme which, when examined by electron microscopy, was composed of an organized array of microtubules. Furthermore, prior to the initial formation of the axopodia, short axonemes, composed of microtubules (TILNEY and PORTER, 1967; TILNEY and BYERS, 1969), could be found in the cell body; from the cortical tips of these axonemes axopodia began to extend.

From the foregoing observations we concluded that the microtubules are involved in the production and maintenance of cell form, for if the tubules are made to disassemble, the axopodia retract and do not redevelop in the absence of organized arrays of microtubules.

This conclusion applies to multicellular organisms as well. For example, during the development of the primary mesenchyme cells of the sea urchin, *Arbacia punctulata*, microtubules are distributed in patterns which might account for each cell shape in the programmed sequence of shape changes which occurs during the differentiation of this tissue (Fig. 8; Gibbins, Tilney, and Porter, 1969). Calcium-free sea water, D_2O, hydrostatic pressure, or colchicine similarly were applied to these embryos. All agents used blocked development and affected the microtubules as in *Echinosphaerium* (Tilney and Gibbins, 1969). We have concluded that in this tissue too the microtubules are involved in the *production* of cell shape. However, unlike *Echinosphaerium* the microtubules in *Arbacia* appear to play a minor role in the *maintenance* of cell form, this function being taken over by other extra- and intra- cellular specializations. These include attachments to neighboring cells such as desmosomes and lateral interdigitations, extra-cellular coats such as the basement membrane and the hyaline layer, and both extra- and intra-cellular filaments (Fig. 9). Accordingly, if we applied agents (calcium-free seawater) that cause the breakdown of these other factors as well as the microtubules, the cells spherulated. Alternatively, with agents which cause the breakdown of the microtubules and some, but not all of these other factors, partial spherulation occurred.

A correlation can also be made between the absence of microtubules, produced by the application of colchicine, and the failure of a cell to elongate; for example during the development of the lens of the squid (Arnold, 1966) and the formation of skeletal muscle in amphibians (Warren, 1968). In both cases microtubular protein molecules continue to be synthesized, but fail to aggregate into microtubules. Thus we can conclude that in multicellular organisms, as well as in *Echinosphaerium*, microtubules play an important role in the development of cell form, but additional factors may be required in the maintenance of cell form.

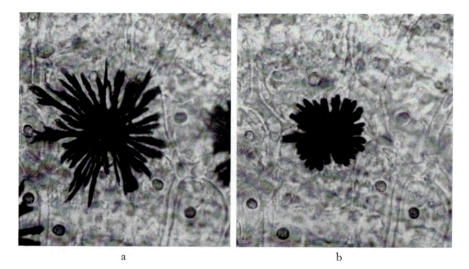

a   b

Fig. 10 a and b. Light micrographs of the same living chromatophore of *Fundulus heteroclitus*. a) The pigment granules are dispersed throughout the processes of the chromatophore as when a fish adapts to a dark background. b) The chromatophore has been induced to concentrate its pigment to the center of the cell. × 500

We have already presented evidence showing that when the microtubules in *Echinosphaerium* are broken down with hydrostatic pressure, cytoplasmic streaming ceases. Before discussing other motile cells or parts of cells and their relation to microtubules, we should make some general statements about the type of streaming that can be associated with microtubules. When microtubules are present in association with the streaming region, there is a saltatory motion (REBHUN, 1964;

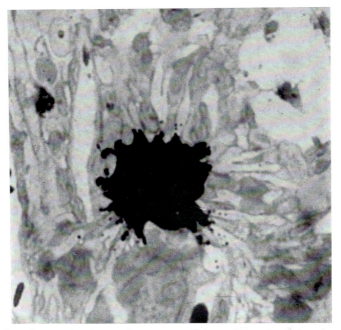

Fig. 11. Light micrograph of a section cut parallel to the surface of a chromatophore of *Fundulus*. Although the pigment granules are concentrated in the center of the cell the processes remain unchanged. (From BIKLE, TILNEY, and PORTER, 1966) × 860

FREED, BHISEY, and LEBOWITZ, 1968) similar to that already described for *Echinosphaerium*. Two features are common to this type of motility: 1. specificity of the particles moved, not indiscriminant movement of the cytoplasm in a stream (cyclosis) such as has been described in *Nitella* (NAGAI and REBHUN, 1966), *Physarum* (WOHL-FARTH-BOTTERMANN, 1964), and the amoeba; 2. movement of the particles in rigidly defined tracks; particles in one track do not seem to interfere with the motion of particles in adjacent tracks even though they are separated by only $0.2\,\mu$. An example of this type of streaming is illustrated by the behavior of pigment granules in the melanophore of the killifish, *Fundulus heteroclitus*. Within this stellate cell, pigment granules, each about $1\,\mu$ in diameter, can migrate into the center of the cell (for example, when the fish adapts to a light background); or the granules can disperse throughout the stellate processes (when the fish adapts to a dark background) (Fig. 10). In both conditions the processes remain unchanged (Fig. 11). Of particular interest to this discussion is the fact that the granules migrate in tracks similar to those already described for *Echinosphaerium* (BIKLE, TILNEY, and Porter, 1966; Green, 1968), and that mitochondria, of comparable size to the pigment granules, do not move from the processes when the granules are induced to migrate to the center of the cell. (Fig. 12 demonstrates the alignment of the granules in rows; microtubules are present between the aligned granules and the

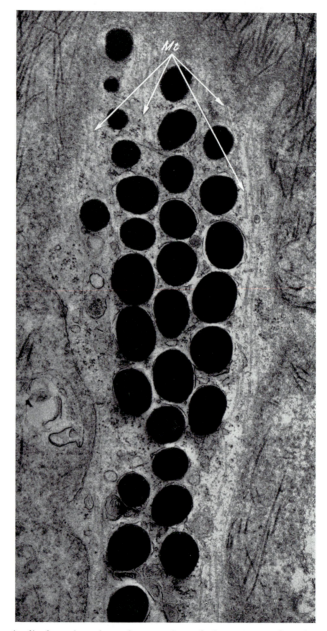

Fig. 12. Longitudinal section through a portion of the process of a chromatophore of *Fundulus*. Note that the pigment granules are aligned in rows or tracks. Microtubules (mt) can be found between the rows. (From Bikle, Tilney, and Porter, 1966) × 20,000

limiting membrane.) Granule movement to the center of the cell can be inhibited with hydrostatic pressure; with this agent the melanin granules are no longer confined to tracks. Furthermore, colchicine inhibits streaming when applied to certain other types of cells. Examples include the cessation of granule movement in the developing lens of the squid (ARNOLD, 1966), and the inhibition of saltatory motion in cultured cells (FREED, BHISEY, and LEBOWITZ, 1968). Both colchicine (INOUÉ, 1952) and hydrostatic pressure (PEASE, 1941) inhibit the movement of chromosomes of the mitotic apparatus. On the other hand, cytoplasmic movements which are not of the saltatory type, such as streaming in *Nitella* or pseudopodia formation in amoeba, are unaffected by colchicine (unpublished observations). In these instances the basis of the streaming appears to reside in filaments rather than in the microtubules (NAGAI and REBHUN, 1966; BHOWMICK, 1967). Thus we have presented evidence that relates microtubules to saltatory movements. In a subsequent section we will suggest a possible mechanism for these movements.

We have neglected to discuss undulatory movements such as are displayed by cilia and flagella, in the axostyle, and in the pharyngeal basket of protozoa as well as in certain types of sperm (SILVEIRA and PORTER, 1964; ROBISON, 1966). We will delay this discussion until later.

III. Characteristics of Microtubule Assembly and Disassembly

A. Differences in the Stability of Microtubules

Although microtubules in different cell types appear to be similar morphologically and chemically, it is now clear that they differ somewhat in their stability. This fact was first recognized about 5 years ago when it was noted that conventional osmium tetroxide fixation preserved ciliary and flagellary microtubules, microtubules of the axostyle of protozoa (GRIMSTONE and CLEVELAND, 1965), and certain clusters of cortical microtubules present in protozoa (PITELKA, 1963), but not the majority of cytoplasmic microtubules. It is now clear that these more stable microtubules are insensitive to colchicine (BEHNKE and FORER, 1967; TILNEY and GIBBINS, 1968), hydrostatic pressure (TILNEY and GIBBINS, 1968), and low temperature (BEHNKE and FORER, 1967; TILNEY, unpublished observations). Furthermore, the central pair of microtubules in the ciliary axoneme, although more stable than cytoplasmic microtubules, is less stable than the nine peripheral pairs (see BEHNKE and FORER, 1967). Moreover, STEPHENS (1968a) has stated that tubule protein isolated from the nine outer doublet fibrils of cilia will reassemble either at low temperature or in the presence of colchicine. How relevant these observations are to the *in vivo* assembly of the doublet microtubules can be questioned, for it is known that regrowth of amputated cilia can be inhibited by colchicine (ROSENBAUM and CARLSON, 1969).

B. Precursor Pool

Since *Echinosphaerium* is constantly changing the length of its axopodia during feeding or under a variety of environmental influences, we might expect that there is a pool of unpolymerized monomer. We are using the word "monomer" loosely here; actually it may be that the disassembled form of the polymer exists in the cell not as a monomer, but as a dimer of the basic 60,000 molecular weight unit (Stephens, 1968b). One way of getting a qualitative view of the size of the pool is by applying deuterated water (D_2O) to *Echinosphaerium*. If *Echinosphaerium* is treated with D_2O, within a minute numerous tubules aligned parallel to existing axonemes can be found in the cytoplasm (Tilney, 1968a). Likewise, the number of tubules in the mitotic apparatus increases 2—3 times with D_2O. The action is extremely rapid, being 80% complete in 40 seconds, so that the increase in tubule number cannot be accounted for by net protein synthesis (Inoué and Sato, 1967).

The most convincing evidence for a precursor pool comes from recent studies of Rosenbaum, Moulder, and Ringo (1969) on the flagellate *Chlamydomonas*. Following flagellar amputation and in the presence of a potent inhibitor of protein synthesis, cycloheximide, the flagella will regenerate to half their normal length. When the cells are released from this drug they elongate to the untreated length.

C. Re-Use of Microtubule Protein

We have already mentioned *in vitro* studies of Stephens (1968a) which demonstrate that tubule protein isolated from ciliary outer doublet fibrils can be reassembled. Re-use of the monomer has also been demonstrated *in vivo* with inhibitors of protein synthesis; for example, cold-treated *Echinosphaerium* (leading to the disassembly of the microtubules) will reform their axonemes and axopodia in the presence of cycloheximide when returned to room temperature. Similar results have been described for the formation of the mitotic apparatus (Inoué and Sato, 1967).

D. Requirement of GTP for Tubule Assembly

In studies on the isolated protein from the outer doublet fibrils of protozoan cilia and sperm flagella it was found that there is 1 mole of bound guanine nucleotide for every mole of 60,000 molecular weight protein. Reassembly of this protein is greatly enhanced in the presence of GTP; ATP is ineffective (Stephens, 1968a). At this stage we cannot conclude that GTP is essential for assembly *in vivo*. We do know, however, that some energy source is required for the assembly and maintenance of the axoneme of *Echinosphaerium*. The following observations support this statement. First, inhibitors of oxidative phosphorylation (cyanide, dinitrophenol, FCCP) block the reformation of axonemes following their disassembly with low temperature. Secondly, anaerobic conditions, induced by the use of the enzymes glucose oxidase and catalase, reversibly prevent the reformation of axonemes and axopodia following low temperature retraction (Tilney, unpublished observations). Thus oxygen is needed for the maintenance of axonemes and axopodia. It is also known that dinitrophenol induces a reversible loss in birefringence of the mitotic apparatus (Sawada and Rebhun, 1969) and an inhibition of mitosis. In both *Echinosphaerium* and the mitotic apparatus, then, blockage of oxidative phosphorylation may ultimately be affecting microtubules by reducing the concentration of GTP, which is formed by transphosphorylation from ATP and GDP. We cannot exclude the possibility, nevertheless, that the ATP required for this step may be wholly, or in part, exerting an effect upon bridges (to be discussed in detail subsequently) which connect microtubules in *Echinosphaerium* and in the mitotic apparatus.

E. Equilibrium between Monomer and Polymer

We have already indicated that microtubules in cilia and flagella appear to be extremely stable and under normal circumstances do not break down and reassemble. There is, however, one exception; this occurs just prior to cytokinesis in flagellated protozoa when duplication of the flagella takes place (TAMM, 1967).

Turnover (disassembly followed by reassembly) of the microtubules or portions of microtubules in *Echinosphaerium* and in the mitotic apparatus, on the other hand, is a common occurrence. For example, during feeding, during movement of the organism (WATTERS, 1968), and when subjected to a variety of external factors the axopodia of *Echinosphaerium* shorten. Likewise, the birefringence of the mitotic apparatus, which is a fairly accurate measure of the presence of microtubules (REBHUN and SANDER, 1967), can be altered with low or high temperature (INOUÉ, 1964), by the concentration of divalent cations (GOLDMAN and REBHUN, 1969), or hydrogen ions (GOLDMAN and REBHUN, 1969; FORER and GOLDMAN, 1969), by micromanipulation (NICKLAS, 1967), and by ultraviolet irradiation (FORER, 1965). Also, as previously indicated, the microtubules of *Echinosphaerium* are temperature, pressure, and D_2O sensitive as is true of the microtubules of the mitotic apparatus (INOUÉ and SATO, 1967; ROTH, 1967; TILNEY, unpublished observations). From the behavior of the microtubules under these diverse influences it appears that in both of these systems there is an equilibrium between the microtubules in the polymerized state and in the monomeric condition, a suggestion made some time ago (WENT, 1960; INOUÉ, 1964). Furthermore, since minor changes such as lowering the temperature a few degrees (TILNEY and PORTER, 1967; TILNEY and BYERS, 1969) result in alterations in the lengths of the microtubules, we are led to predict that this equilibrium can be easily shifted. Quantitation of the thermodynamic parameters associated with this equilibrium has been attempted by INOUÉ and SATO (1967) using birefringence as a measure of the amount of oriented material. Measurements were made on the mitotic spindle when subjected to changes in temperature (INOUÉ, 1964; CAROLAN, SATO, and INOUÉ, 1965, 1966) and, assuming an equilibrium between monomeric and polymeric forms of the mitotic spindle, values for the entropy and enthalpy could be calculated. They found a large increase in entropy when the temperature was lowered and the tubules depolymerized. This result, at first contrary to what one would intuitively predict, appears to arise as a solvent effect, i.e., from differences in the hydration between the polymer and its monomeric units. This conclusion is based on similar changes in these thermodynamic parameters that occur in model systems, namely the polymerization of tobacco mosaic virus protein and the transformation of G to F actin. These simple systems depolymerize on either lowering the temperature or increasing the hydrostatic pressure (IKKAI and OOI, 1966; STEVENS and LAUFFER, 1965). Therefore, it appears that a common feature in these assembly systems is the predominance of a specific kind of non-covalent hydrophobic bonding (see HARRINGTON and JOSEPHS, 1968).

To illustrate the dynamic nature of this equilibrium we shall describe some experiments designed to eliminate not only the pool of monomer, but also any monomer that might become available through the partial or total disassembly of the polymer. In this way it will be possible to visualize any flux of monomer into or out of the polymer, for as soon as any polymer breaks down the resulting monomer will be unable to repolymerize. The first experiment involves the use of colchicine. From the investigations of Rosenbaum, Moulder, and Ringo (1969), Bischoff and Holtzer (1968), and Warren (1968) we know that colchicine, while not affecting protein synthesis (which would include the synthesis of microtubule monomer) prevents the assembly of microtubules. Since colchicine has no effect

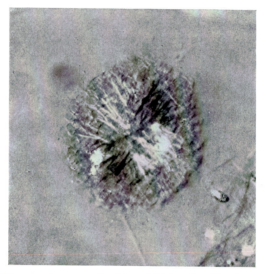

Fig. 13. Living *Echinosphaerium* treated with colchicine for 40 minutes and viewed with a Zeiss polarizing microscope. Note that the bases of the axonemes remain unchanged, while both the axopodia and the axoneme portion within the axopodia have disappeared. The amount of birefringence is comparable to that in the bases of an untreated organism.
× 175

on existing flagella and cilia (Rosenbaum and Carlson, 1969; Tilney and Gibbins, 1968; Behnke and Forer, 1967) or on the basal portion of the axoneme of *Echinosphaerium* (Fig. 13; Tilney, 1968b), it seems reasonable to suggest that colchicine binds strongly to the monomer and thus prevents reassembly. This suggestion is further supported by the fact that D_2O, known to favor the polymerized state of microtubule protein, inhibits the action of colchicine (Marsland and Hecht, 1968). The *in vitro* binding of colchicine to microtubular precursors (Borisy and Taylor, 1967a; 1967b; Shelanski and Taylor, 1967; Taylor, 1965) is also consistent with this interpretation. We presume that the effect of colchicine on microtubules in the axopodia of *Echinosphaerium* is similar. Each time a bit of polymer is transformed transiently to monomer, it is immediately bound by colchicine which prevents its reassembly into the polymer. Thus the equilibrium between the monomer and polymer is shifted toward monomer as the pool diminishes resulting in a net disassembly of microtubules. In time, therefore, all the microtubules in the axopodia disassemble. Since it normally takes 30 minutes for the axopodial microtubules to disassemble, we feel that this is a measure of the flux from polymer to monomer.

Another approach has been used in an attempt to measure turnover rates of microtubule protein. This involves the effect of oxygen depletion on the axonemes and thus on the microtubules of *Echinosphaerium*. Since STEPHENS (1968a) has shown that GTP is necessary for the reassembly of microtubule protein *in vitro*, it seems reasonable to measure turnover rates by depleting the cell of oxygen which ultimately depletes the cell of GTP (provided of course the glycolytic pathway is minor). Oxygen is removed using the enzymes, glucose oxidase and catalase. Under these conditions axopodia withdraw to tiny stubs within 100 minutes. These experiments are not only totally reversible when the cells are put into fresh medium, but the enzymes themselves have no effect without substrate. Since it takes 12 minutes to remove any measureable oxygen from the sealed solution containing *Echinosphaerium* and probably an equal amount of time to deplete the pools of ATP or GTP in the cell, the maximum rate of microtubule depolymerization would be about 75 minutes. Since we have not affected anaerobic glycolysis, nor taken into account the quantity of dissolved oxygen in the cell, this figure must be reduced somewhat. If these factors are taken into consideration, the observed value of 30 minutes for axopodial withdrawal in the presence of colchicine is of the same order of magnitude as this value of less than 75 minutes. We are led to conclude, therefore, that the cytoplasmic microtubules are in dynamic equilibrium, constantly polymerizing and depolymerizing.

Experiments by FORER (1965) using a source of microbeam irradiation on the mitotic apparatus provide analogous data on the constant flux of polymer to monomer. He found that if he irradiated a spindle fiber in metaphase, the result was a local loss in birefringence. Of considerable interest, however, was the observation that this region of low birefringence migrated slowly to the pole, being replaced by birefringent material. He interpreted this result as indicating that there is a flux of spindle precursor into the spindle fiber at the centromere and a withdrawal of it at the pole. It is interesting that in the cases mentioned so far (*Echinosphaerium*, the mitotic apparatus, and developing systems of many kinds) the equilibrium between polymer and monomer is so delicately balanced that rather minor changes in the cytoplasmic environment are capable of displacing the equilibrium in either direction. In this way, then, the cell may have sensitive control over the length and abundance of microtubules.

F. Possible Relationship between Microtubules, Filaments, and 340 Å Tubules

From investigations on the self-assembly of model systems it is now apparent that the same monomeric units can assemble into different forms depending upon the local conditions. These include not only the chemical environment such as the type and concentration of salts, the pH, and the relative concentration of protein, but also such factors as the temperature and pressure. Examples would include the assembly of tobacco mosaic virus protein or certain types of viral structures (see CASPER, 1966; KLUG et al., 1966).

When *Echinosphaerium* is placed at low temperature and thin sections are examined with the electron microscope, we frequently find a randomly arranged population of tubules which are approximately 340 Å in diameter (Fig. 14); the 220 Å microtubules have completely disassembled (TILNEY and PORTER, 1967). Conversely, upon warming the cell, these larger tubules disappear and the 220 Å tubules reappear. Since the wall thicknesses of these two tubules are of similar dimensions, and the 340 Å tubule appears when the 220 Å microtubule has disassembled, it seems reasonable to expect that they may be derived from one another. Based on

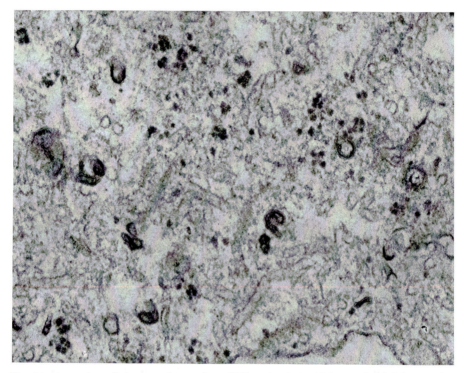

Fig. 14. A portion of the cytoplasm of an *Echinosphaerium* which was fixed 2 hours after the initiation of treatment with low temperature. Of particular interest is the random population of tubules, each measuring approximately 340 Å in diameter. These tubules are believed to be made up of the same monomer found in the 220 Å microtubules.
× 100,000

the substructure of the microtubule, a 40 Å globular unit which probably represents the 60,000 molecular weight monomer, we have suggested how the two types of tubules may be formed — both, as seen (Fig. 15), from the same subunits. Others have described tubular structures in cells which have dimensions similar to the 340 Å tubule just described. These include the dividing cells of *Psilotum* (Allen and Bowen, 1966), the developing chloroplasts of *Chara* and *Volvox* (Pickett-Heaps, 1967), and certain protozoa (Allen, unpublished observations). We would interpret these findings as well as the reports on *Echinosphaerium* as indicating that under favorable conditions, the monomer can assemble into a variety of polymeric forms as is known for the simple viral systems.

The microtubular monomer may assemble into still another form, a filament. This subject is complicated by the different sized filaments in cells. In general there appear to be two populations of filaments within most cells; filaments about 40–50 Å in diameter and filaments approximately 100 Å in diameter. When nervous tissue has been treated with mitotic inhibitors such as colchicine or vinblastine, the microtubules disassemble and in their place appear 100 Å filaments (WISNIEWSKI

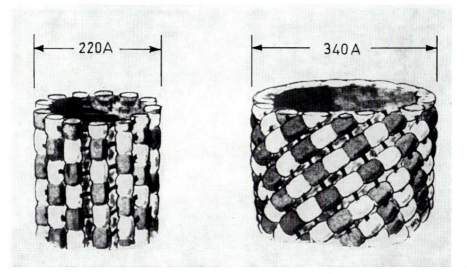

Fig. 15. In this drawing we have illustrated how we believe that the same subunits can make up a 220 Å microtubule and a 340 Å low temperature induced tubule. (From TILNEY and PORTER, 1967)

and TERRY, 1968). When these drugs are removed the filaments gradually disappear and microtubules reform (WISNIEWSKI, SHELANSKI, and TERRY, 1968). It has been suggested from these observations that the filament is made up of the same protein as the microtubule (WISNIEWSKI, SHELANSKI, and TERRY, 1968) and that these are alternate forms of the microtubule monomer. The suggestion has also been made, on even less evidence, that the 50 Å filaments are also formed from microtubules (FAWCETT, 1966; O'BRIEN and THIMANN, 1966; TILNEY, 1968a). Admittedly, there is insufficient evidence either to affirm or contradict these speculations.

IV. Control of Microtubule Pattern in Cells

In the first section we concluded that microtubules appear to function in the development of cell form and in certain types of motility such as chromosomal movement or the translocation of particles. Microtubules may also function by defining streaming channels and thus ordering regions of the sol. If it is true that microtubules are involved in the production of cell shape, in compartmentalizing cytoplasmic movements, and in the distribution of cytoplasmic particles, then those factors which govern the distribution of microtubules in the cell would assume tremendous importance, for they could serve to control cell differentiation.

A. Nucleating Centers

In principle the easiest way to control ordered patterns of microtubules is to control the initiation of microtubule assembly. This point can perhaps be most readily appreciated by analogy to what may be a prototype of the microtubule, the bacterial flagellum. Polymerization of this tubular structure can be carried out under physiological conditions by "seeding" a solution of the monomer with small bits of mature flagella (Asakura, Eguchi, and Iino, 1964). If these seeds, chosen from strains of bacteria which possess curly flagella, are added to monomer from wild type bacteria which possess straight flagella, the resulting flagella are curly. Thus nucleating material not only initiates polymerization, but also exerts some control over the nature of the resultant polymer (Asakura, Eguchi, and Iino, 1966).

An obvious counterpart of this principle in eucaryotic cells is the formation of the ciliary or flagellary axoneme. The axoneme appears not only to be derived from the basal body, but to grow out of it (Renaud and Swift, 1964; Rosenbaum, Moulder, and Ringo, 1969; Outka and Kluss, 1967). Thus the basal body can be considered as a nucleating site. Embedded in the wall of each basal body and making up a prominent part of it are short segments of microtubules arranged to form 9 triplets. Transverse sections of cilia and flagella show a similar organization; in this case 9 doublets surround a central pair. Each doublet in longitudinal section is continuous with two of the microtubules of each triplet. Because tubules in these doublet and triplet configurations share a common wall (Behnke and Forer, 1967; Phillips, 1966; Ringo, 1967), a situation not present in most configurations of tubules, they illustrate the principle first mentioned in regard to bacterial flagella: that is, the type of nucleating site may have considerable control over the morphology of the resultant polymer.

There are other loci in cells on which ends of many microtubules make contact, such as the surfaces of centrioles (Gibbins, Tilney, and Porter, 1969), small densities associated with centrioles (de Thé, 1964; Szollosi, 1964; Robbins, Jentzsch, and Micali, 1968), bodies existing free in the cytoplasm (Fig. 16; Bowers and Korn, 1968), membrane-associated bodies (Satir and Stuart, 1965; Bassot and Martoja, 1965), portions of the nuclear envelope as in dividing yeast cells (Robinow and Marak, 1966), the kinetochore (Brinkley and Nicklas, 1968), the diffuse density seen at the midbody of animal cells (Robbins and Gonatas, 1964; Byers and Abramson, 1968) and in the phragmoplast of plant cells (Hepler and Jackson, 1968). This association between microtubules and discrete loci has led to the suggestion that these sites, like the basal body, may control the distribution of microtubules by acting as nucleating centers (Porter, 1966; Gibbins, Tilney, and Porter, 1969; Tilney, 1968a; Inoué and Sato, 1967). Yet, unlike the basal body or centriole, none of these loci contain anything that morphologically resembles a microtubule. Furthermore, unlike ciliary microtubules, the cytoplasmic microtubules may not be as precisely ordered relative to the focus point (Robbins, Jentzsch, and Micali, 1968).

The concept of initiating sites is an attractive one. By controlling the nucleation of microtubules the cell can have precise control over their location, size (the 220 Å tubule rather than 340 Å), and direction of growth. Also by initiating the polymerization of microtubules biological use can be made of tubule polarity which may account for unidirectional motion of particles such as takes place during chromosome movement. Polarity may be important in controlling the length of tubules. There is evidence indicating that addition or subtraction of monomer

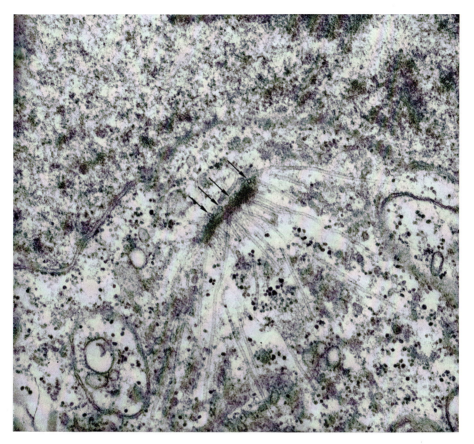

Fig. 16. Section through an elongate aggregation of dense material of *Acanthamoeba castellanii*. On this material numerous microtubules insert. Note the cross sections of microtubules along the upper edge (see arrows). (Courtesy of BOWERS and KORN, 1968)

occurs from only one end of the tubule. The most direct evidence that polymerization is carried out from only one end of a polymer comes from the work of ASAKURA, EGUCHI, and IINO (1968) on the growth of bacterial flagella *in vitro*. They demonstrated that when a solution of flagellin (flagella precursor protein) is seeded with small bits of flagella, growth occurs unidirectionally. This growing tip in an intact cell corresponds to the distal end of the flagellum. In eucaryotic cells, studies by ROSENBAUM and his coworkers (ROSENBAUM and CHILD, 1967; ROSENBAUM, MOULDER, and RINGO, 1969) on the regrowth of a protozoan flagellum indicate a similar pattern of growth from the distal tip. These studies were carried out using autoradiography. It is possible that depolymerization of microtubules

16*

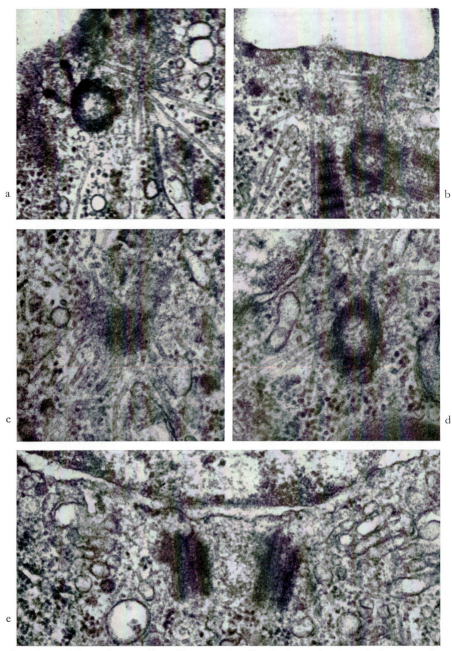

Fig. 17 a—e. These five sections illustrate the distribution of microtubules at various stages in the formation and differentiation of the primary mesenchyme cells of *Arbacia*. (From Gibbins, Tilney, and Porter, 1969). a) and b) Apical portion of the ectodermal cells of an early blastula prior to their migration into the blastocoel to form the primary mesenchyme cells (Stage 1 of Fig. 8). (a) is cut through the basal body; in (b) the section passes parallel to the basal body. The microtubules insert on satellites associated with the basal body. c) and d) These sections are cut through the centrosphere of newly formed

might likewise preferentially take place at the distal tip, although under drastic conditions depolymerization could occur at a number of places along the tubule at once. The evidence for this consists of observations on the shortening of the axoneme of *Echinosphaerium* when subjected to agents such as colchicine (TILNEY, 1968b). Since the microtubules in the basal ends of the axonemes do not depolymerize even though the tubules in the axopodial portion of the axoneme do (Fig. 13), and since the axoneme gradually increases in diameter (thus containing progressively more microtubules) from the tip to the base of the axopodium, the most likely method for the depolymerization of the microtubules, which in the center of the axoneme appear to extend from the basal end of the axoneme to its distal tip, is by the disassembly of the distal tips of the microtubules.

Finally, by activating or repressing nucleating sites the cell could dictate changes in shape or function which could be dependent upon a programmed assembly and disassembly of microtubules. For example, during mitosis the kinetochore and centriole appear to be involved in spindle and astral fiber formation (for evidence see INOUÉ and SATO, 1967; CAMPBELL and INOUÉ, 1965; BRINKLEY and NICKLAS, 1968), but the spindle and astral fibers, which are composed largely of microtubules, do not form in interphase cells. A second example of tubules associated with different particles in the cytoplasm at different times occurs in *Arbacia* during the formation and differentiation of the primary mesenchyme (Fig. 17). Prior to primary mesenchyme formation the microtubules in the presumptive mesenchyme cells (Stage 1, Fig. 8) converge on loci (satellites) associated with the basal body. In the newly formed mesenchyme cells (Stage 3, Fig. 8) the microtubules make direct contact with the centrally situated centriole and from here radiate in all directions. Finally (Stage 5, Fig. 8), when the mesenchyme cells have fused and are beginning to elaborate a skeleton, the microtubules are no longer associated with the centriole. Thus at different stages in the formation and differentiation of the primary mesenchyme the microtubules are attached to different regions of the cell. Our interpretation of these results is that the loci (nucleating centers) involved are sequentially turned on or turned off. This activation and repression then would ultimately control the precise sequence of shape changes that these cells must undergo (see Fig. 8) during their formation and differentiation.

The recent evidence indicating the existence of DNA in basal bodies (RANDALL and DISBREY, 1965; SMITH-SONNEBORN and PLAUT, 1967) complicates this discussion in an exciting way, for we must now consider that initiating sites may be part of a cytoplasmic system which possibly contains DNA. It is interesting in this regard to examine any similarities between the basal bodies and other initiating sites. The only information available is from comparative morphology. Most of the sites upon which microtubules attach have not been precisely defined, but all seem to be made up of an electron dense material which grades off into the cytoplasm. The centriole or basal body appears to be made up of similar material,

---

mesenchyme cells (Stage 3 of Fig. 8). The microtubules make direct contact with the centriole. The centriole in (c) is cut near its basal surface. e) When the mesenchyme cells have fused to form the cable syncytium (Stage 5 of Fig. 8), the microtubules do not appear to converge on focal points in the cells. Thus the centrioles do not have cytoplasmic microtubules associated with them. × 70,000

only in this case it contains 27 small segments of microtubules (the 9 triplets). This contention is substantiated by the studies on the formation of centrioles or basal bodies as occurs in the epithelium of the uterus or trachea, or in protozoa (DIRKSEN and CROCKER, 1966; KALNINS, 1968; DIRKSEN, 1968; SOROKIN, 1968; OUTKA and KLUSS, 1967); this dense material appears first to condense in the form of a cylinder, the embedded bits of microtubules (9 triplets) appearing subsequently. The chemical nature of this material requires further investigation. Nevertheless, the facts that these sites have automony as demonstrated by transplantation experiments on the ciliate cortex (BEISSON and SONNEBORN, 1965), and that they all appear to be able to replicate prior to cytokinesis, may indicate that indeed such a cytoplasmic system of initiating sites exists. By controlling the nucleation of microtubules this system could have considerable control over their spatial and temporal distribution.

Until now only a small proportion of microtubules in cells has been seen to be associated with these loci. The explanation for this may be that not only are these sites poorly characterized possibly reflecting poor preservation for electron microscopy, but also the technical problem of locating the ends of microtubules in thin sections is staggering. Even so it is entirely possible that all microtubules do not insert on initiating sites but instead, provided the conditions are favorable, some may spontaneously polymerize.

B. Bridges

It has only recently become apparent that many microtubules are joined by thin, threadlike bridges. These connections can be seen only with high resolution electron microscopy in thin sections. They appear to be easily disrupted by fixatives. Bridges have been described in cilia and flagella where they connect the peripheral doublets to the central region (ALLEN, 1968; GIBBONS, 1961; SILVEIRA, 1969), in the axostyle of flagellates (GRIMSTONE and CLEVELAND, 1965), in the bundle of microtubules surrounding the fowl spermatid nucleus (MCINTOSCH and PORTER, 1967), between the tubules of the mitotic apparatus (WILSON, 1969; LU, 1967), connecting neurotubules in axons (KOHNO, 1964; PALAY et al., 1968), in the cell plate of dividing plant cells (HEPLER and JACKSON, 1968), in the axonemes of *Echinosphaerium* (MACDONALD and KITCHING, 1967; TILNEY and BYERS, 1969), and in numerous members of the ciliata such as *Stentor* (BANNISTER and TATCHELL, 1968; GRAIN, 1968) and *Nassula* (TUCKER, 1968). The function of these bridges is still largely unknown; nevertheless there is evidence that they give pattern to clusters of microtubules as well as playing a major role in motility. The latter function will be discussed separately.

To illustrate the idea that associations between tubules via bridges might lead to an ordered pattern of microtubules we will present data obtained from recent studies on *Echinosphaerium*. We mentioned earlier that the microtubules in the axoneme are separated from their neighbors by 70 Å and 300 Å. These are the linear dimensions of the bridges which are illustrated in Fig. 18. We asked ourselves whether the pattern of microtubules in the axoneme might be determined in large measure by these two species of bridges and the substructure of the microtubule, that is a structure having a 12 or 13 fold symmetry derived from the subunits which make up the wall of each tubule. To examine this hypothesis we disassembled the microtubules of *Echinosphaerium* with low temperature and then subjected these organisms to conditions which favored polymerization. By examining early stages in reassembly we hoped that many of the patterns of tubules which appeared might reflect inherent bonding properties of the units (bridges and tubule substructure).

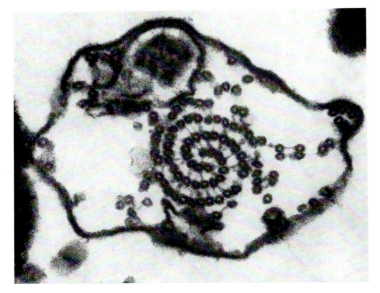

Fig. 18. Transverse section through an axopodium of *Echinosphaerium*. This section was heavily stained with $KMnO_4$ to emphasize the bridges which connect the microtubules of the axoneme. The long bridges connecting the tubules of adjacent rows show up clearly.
$\times$ 110,000

To shift the equilibrium towards the polymerized state we applied to the organism heat (22° C) and D_2O, two agents which favor the polymerized state of the microtubule protein. Since until recently we were unable to observe the bridges routinely, we carefully analyzed the separations between tubules in the clusters that formed shortly after raising the temperature and adding D_2O. If the spacings between tubules approximated 70 Å or 300 Å, we predicted that the bridges would be present although not preserved by our methods. We tried to reconstruct the patterns that formed using scale models built from two sizes of bridges and plastic cylinders to represent tubules. We milled 12 equally spaced grooves in each cylinder to represent the subunit bonding sites for the bridges (see Fig. 19). As presented in Fig. 19 and 20, the clusters of tubules that formed shortly after the addition of D_2O and the application of heat can be interpreted in terms of these two species of bridges and the 12 bonding sites on the tubule even though most of these clusters do not resemble any part of an axoneme. Further stages which could readily be interpreted as early stages in the assembly of an axoneme (Fig. 20) can likewise be considered as arising from interactions between the bridges and the microtubules. The 12-sided symmetry of large axonemes (see Fig. 3) can also be inter-

preted by these models and thus described solely by considering the two types
of bridges and the substructure of a microtubule. This point can best be appreciated
in Fig. 21 which is a half model and half tracing of a micrograph of an axoneme.
For simplicity we have inserted only long bridges. The tubules within each of
the 12 sectors are bound to each other by 2 long bridges, those on the radii by 3.

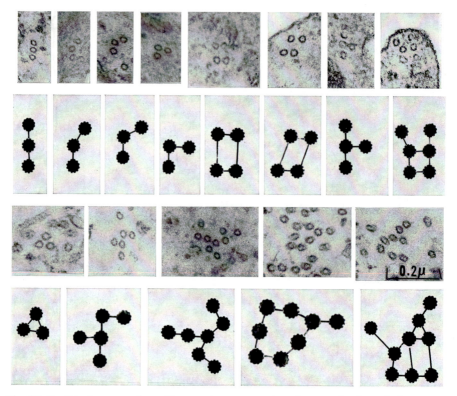

Fig. 19. In this figure we have illustrated representative clusters of microtubules present
in *Echinosphaerium* which has been exposed to 0°C for 3 hours, then fixed at a very early
stage after their removal to room temperature (top row) or to warm 70% D_2O (third row).
In the second and fourth rows we have built models of these clusters. The tubules are
represented as small cylinders in whose surface are cut 12 grooves to represent the subunit
bonding sites. Adjacent tubules are connected by plastic pieces of two sizes to represent
the two sizes of bridges connecting adjacent tubules. (From TILNEY, 1968a.) × 75,000

Because the long bridges within each sector are parallel, and thus form 30 degree
angles with those in adjacent sectors, each row within the sector, instead of lying
on an arc of a circle actually forms the secant. This spider web pattern then is a
product of the parallelism of the long bridges in each sector which in turn is determined by the proportionality between the lengths of the bridges and the diameter
of the tubules.

From the above considerations it appears that two species of bridges and the
substructure of a microtubule may contain sufficient information to account for axone-

mal pattern. Even so there is one clearly unexplained observation. This is related to the frequent occurrence of non-axonemal arrays of microtubules in the early phases of axonemal formation. Since these configurations occur infrequently in cells fully recovered from induced axoneme breakdown, there must be a mechanism for eliminating incorrect assemblies. This could be carried out by a mechanism

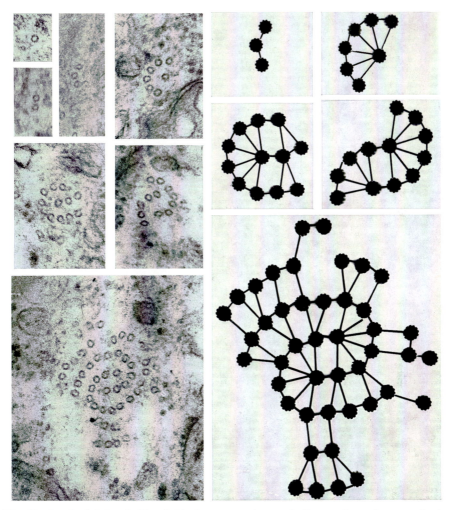

Fig. 20. On the left hand side of this figure are what we believe to be early stages in the formation of an axoneme of *Echinosphaerium* following induced axoneme breakdown with low temperature. To the right are the corresponding models. (From TILNEY, 1968a)

other than the system of bridges and tubules. Alternatively, the existence of an equilibrium between the polymer and monomer may result in the elimination of non-axonemal assemblies by selecting that configuration which is energetically most stable. If the axonemal pattern is thermodynamically most stable then in time this pattern would be selected above all other possible configurations of tubules; the less stable forms would completely depolymerize. In support of this possibility is the observation from studies with models that the normal axoneme would contain more bridges per tubule than any other cluster we have found. As seen

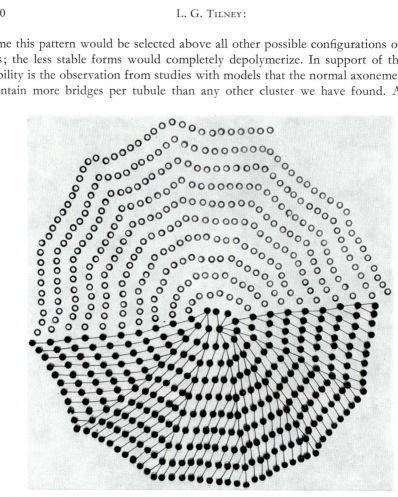

Fig. 21. In the upper part of this figure is a tracing of an electron micrograph of half of an axoneme of *Echinosphaerium*; in the lower half we have built a model of the other half of the axoneme. From this figure it can be readily appreciated that the twelve sided symmetry of large axonemes (Fig. 3) can be interpreted in terms of the distribution (parallelism) of the long bridges. Within each sector the long bridges are parallel to each other; the bridges of adjacent sectors form a 30° angle with each other. We believe that this arrangement of long bridges is related to the substructure of each tubule, that is having 12 equivalent bonding sites on the surface of the tubules. × 75,000

in Fig. 20 the tubules at the inner ends of the two interlocking coils have 7 bridges per tubule, the next 8 have 5 bridges per tubule. In all other clusters depicted here or illustrated in more detail in Tilney and Byers (1969) the number of bridges per tubule is significantly less. It is conceivable, therefore, that the perfect axoneme may be selected above any other cluster because it is the energetically most favorable configuration. Referring to earlier discussion in the observation that the axonemal bases do not break down in the presence of colchicine, we would assume that the stability of these bases is a result of their containing the largest number of cross bridges.

Thus, in *Echinosphaerium* we have described the involvement of the substructure of tubules and of specific types of bridges in the determination and maintenance of the pattern of an axoneme. As we mentioned earlier there are many other systems where bridges are found. These have not been analyzed carefully during their formation, but it seems likely that patterns of microtubules in the axostyle of *Saccinobacculus* and in the gullet of *Nassula* are governed by phenomena similar to those described above. In the formation of the ciliary or flagellary axoneme once growth of the microtubules has been initiated at the basal body then intertubular bridges would exert an important influence in the over-all form of this structure along the length of the axoneme.

C. Control by the Environment

Recent studies on properties of the isolated mitotic apparatus (FORER and GOLDMAN, 1969; GOLDMAN and REBHUN, 1969) reveal that the integrity of the mitotic apparatus, analyzed by comparing the birefringence of the *in vivo* mitotic apparatus to the isolated mitotic apparatus, is highly dependent on the pH of the isolation medium. For example, if the mitotic apparatus is isolated at pH 6.1, it appears rather stable and can be stored for several weeks. If the mitotic apparatus is isolated at pH 6.3, however, it is unstable leading to a gradual loss in birefringence (GOLDMAN and REBHUN, 1969). The presence of divalent cations slows the loss of birefringence from the isolated mitotic apparatus. Thus regulation of the hydrogen ion and cation concentration in the isolation medium can determine the stability of the spindle. Other factors which may affect the isolated spindle have not been investigated. Since the birefringence of the spindle depends in large measure upon the integrity of microtubules (REBHUN and SANDER, 1967), an obvious interpretation of these results is that these simple factors (hydrogen ion and cation concentration) affect the equilibrium between the polymer and the monomer. Thus low pH and high magnesium ion concentration will shift the equilibrium to favor the formation of polymer so that monomer cannot be formed and thus the isolated mitotic spindle would appear stable.

In living cells as well it seems reasonable to expect that by regulation of the internal environment the cell can have sensitive control over the equilibrium which in turn would regulate the length of microtubules and thus cell shape, assuming, of course, that there is an available pool of precursor material (monomer, bridges, and the requisite energy source, presumably GTP). It is further possible that the microenvironment immediately surrounding a cell or cell process can in turn exert enough of an effect upon the internal environment to influence this equilibrium either directly or indirectly. Of course all cells are separated from their external environment by a membrane which efficiently maintains a relatively constant cytoplasmic environment. Nevertheless, the striking changes in shape could result from environmental gradients in developing systems. For example, low O_2 tension stimulates the proliferation of microtubule-rich processes in tracheole formation in insects (see PORTER, 1966), low divalent cation concentration causes rounding of the mesenchymal processes in *Arbacia* (TILNEY and GIBBINS, 1969), and nerve growth factor initiates a rapid proliferation of nerve cell processes (LEVI-MONTAL-CINI, 1965). Even changes in the ratios of sodium and potassium ion concentration can affect pigment migration in chromatophores which, as mentioned earlier, seems to be under the control of microtubules. In addition, alterations in the limiting membrane, caused by a variety of lipid soluble agents such as nembutal, ether, and halothane can affect the integrity of the mitotic apparatus and the axopodia of *Echinosphaerium* (HEILBRUNN, 1956; ALLISON and NUNN, 1968; TILNEY, unpublished observations). These agents are not physiological, but it seems reasonable to expect that biologically occurring molecules may be effective in regulating the internal environment which in turn would influence the equilibrium between the polymer and the monomer. Thus, once initiation of microtubule polymerization has begun, the length of the resulting tubules as well as their direction of growth could be regulated by rather subtle environmental factors and/or stabilizing bridges. These considerations may prove to be of considerable importance in differentiation.

V. Motility Mechanisms, in Particular Their Relation to Bridges

We have included this section in the hope that it may correlate many of the preceeding observations and discussions on motility by showing that bridges may be a common feature of tubule-associated motility. Since the potential significance of bridges as well as the frequency of their occurrence are now just becoming clear, much of this discussion is highly speculative.

Situations where the presence of bridges has been noted have been mentioned earlier. To initiate this discussion let us consider motility mechanisms in which the undulation of clusters of microtubules is recognized. Such situations include the beating of cilia and flagella, the undulations of the axostyle, and the movements of the pharyngeal basket of *Nassula*. As is well known, most flagella and cilia possess as part of their axonemes the $9 + 2$ configuration of tubules. There are, however, motile cilia and flagella that differ in this pattern, for example the $9 + 1$ pattern in flatworms (SILVEIRA, 1969) and in certain insect cells (PHILLIPS, 1969), the $9 + 0$ pattern in certain plant and insect cells (PHILLIPS, 1969), and the bizarre combinations in many insect sperm (PHILLIPS, 1969). The axostyle and the pharyngeal basket, on the other hand, are made up of parallel rows of microtubules (Fig. 22). All these systems have one common feature: *the microtubules are connected by bridges*. In the $9 + 2$, $9 + 1$, and $9 + 0$ combinations the outer doublets are connected to a central ring which surrounds the central pair if present (ALLEN, 1968; SILVEIRA, 1969; PHILLIPS, 1969). Furthermore the outer doublets are interconnected. (This observation is based on the fact that when the axoneme is isolated the 9 peripheral doublets remain equally spaced, although the central tubules frequently disassemble). Likewise the tubules in the axostyle (GRIMSTONE and CLEVELAND, 1965; see Fig. 22) and the tubules in the pharyngeal basket are connected by threadlike bridges. When cilia beat, the microtubules do not appear to shorten, instead they slide past one another (SATIR, 1965; 1968). Since the bridges are the one common feature of all these diverse types of undulatory systems and since motion is carried out by sliding the tubules past one another, it seems plausible that they may act to affect motion in an analagous way to the myosin bridges in skeletal muscle.

There are other instances where the presence of bridges between tubules can be correlated spatially and temporally with motility. Yet in these other cases the

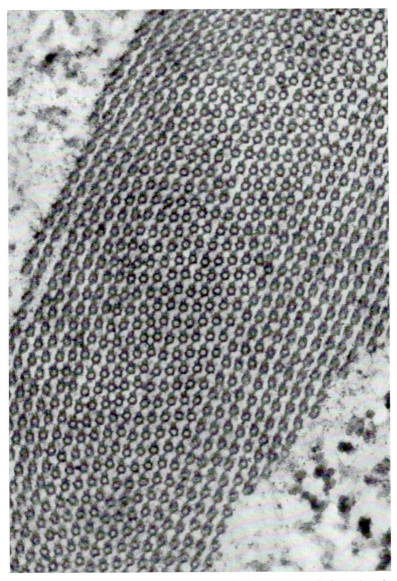

Fig. 22. Transverse section through a portion of the axostyle of *Saccinobacculus*. This micrograph illustrates advantageously the bridges between the tubules. (Courtesy of Dr. KEITH R. PORTER)

types of resultant motility differ from the undulatory motion just described. For example, during the elongation of the nucleus of the chicken spermatid a pair of interlocking microtubules which coil around the nucleus are thought to slide relative to each other and in this way transform the nucleus from a sphere into a cylinder (MCINTOSCH and PORTER, 1967). The bridges between these tubules are thought to provide the requisite force for this sliding. Another example in which bridges between microtubules might provide, as in muscle, the motile force might be chromosomal movement in dividing cells. In this case the bridges (recently described by Wilson, 1969) may affect movement by causing the chromosomal spindle fibers to slide relative to the interzonal fibers, both fibrous systems of course being composed of microtubules. Such an idea of relative movement of these two fibers was suggested years ago by Belar (1929) and recently by Subriana (1968) and Bajer (1968). Subriana points out that such movement would require opposite polarity for the two systems. Such a requirement, as mentioned earlier, could be brought about by the existence of different nucleating sites, in this case the kineto-chore and the centriole.*

A third type of motility, cytoplasmic streaming, which, when associated with microtubules appears to be of the saltatory type, may also be the result of inter-actions between tubules and bridges. This type of streaming is characterized by the specificity of the particles to be translocated and by rigidly defined paths for the streaming particles (see the discussion in part III). At present there is no evidence for bridges between microtubules and particles which stream in living cells. Never-theless, because of the specificity of the streaming particles it seems possible that bridges on the surface of microtubules may attach to sites on the surface of specific particles and thereby move the particles. After one bridge has moved the particle, either another bridge would attach to the particle or it might be released. This type of mechanism not only accounts for the specificity of the particles moved, but also for the jumps or saltations of the particles. Furthermore, this model is consistent with the sensitivity of this type of cytoplasmic streaming to the action of colchicine. Obviously if the tubules were to disassemble, streaming would be interrupted.

Taken all together, it seems to us that interactions between tubules or between tubules and particles via bridges not only provides a common basis for microtubule-associated motility, but also accounts for the diverse types of movements observed.

VI. Summary and Conclusions

Evidence was presented that linked microtubules to two major functions: the production of cell shape and motility. If this relation is true, and considerable evidence exists to substantiate this conclusion, then factors controlling the distri-bution of microtubules must ultimately regulate the differentiation of cell form as well as control a variety of specific motility mechanisms.

Basic to any discussion on the control of microtubule distribution in cells is the dynamic equilibrium (Inoué, 1964) which must exist between the polymer and its monomer. For cytoplasmic microtubules in contrast to the stable micro-tubules of cilia, flagella, the axostyle, and certain other clusters found in protozoa, we envision that under the conditions that exist in the cytoplasm during mitosis or in *Echinosphaerium* a constant flux between the polymer and monomer occurs at the tip of each microtubule. Thus, if one were to remove the monomer as fast

---

* Since this review was written in June, 1969, an interesting model of chromosomal movement involving bridges has been suggested [McIntosch, R., P. K. Hepler and D. G. van Wie. Nature **224**, 659—663 (1969)].

as it formed, within a short period of time little polymer would remain. We believe that colchicine acts in such a way on cells by binding to the monomer and so removing it from the system. It is essential for maintaining this dynamic equilibrium that the monomer can be re-used, that within a cell resides an adequate pool of this material, and that there is an available energy source, presumably GTP.

By keeping the equilibrium delicately balanced biological use can be made of the readily reversible transition between the monomer and the polymer: for example, during feeding in *Echinosphaerium*, during the programmed series of shape changes that occur in primary mesenchyme development, and during chromosomal movements. In these instances and in many others addition or subtraction of monomer to or from existing microtubules can perform a useful function. Changes in the length of the existing microtubule, then, might be controlled by the transient conditions within the cell that could affect the equilibrium between tubule monomer and polymer. This does not mean that the addition or substraction of monomer necessarily provides the entire mechanism for these phenomena, but instead it may supply a vital component, as for example during the movement of the chromosomes to the pole.

The control of the distribution of microtubules in cells must also involve loci which affect the nucleation of microtubules. These loci are thought to include the basal body, the centriole, satellites associated with the basal body and centriole, the kinetochore, and a host of other dense particles, many of which are membrane-associated. By activating and repressing specific loci the cell could control the distribution of microtubules, provided of course that the equilibrium for assembly was favorable and that the requisite molecules were available. Therefore, during differentiation we suggest that there may be a sequential turning on or turning off of these loci which we assume act as sites to initiate tubule assembly. These in turn regulate the sequential changes in cell shape that cells undergo during their formation and differentiation. It is possible that these sites are part of a cytoplasmic system which behaves as an autonomous unit regulating the differentiation of cell form.

Bridges which connect adjacent microtubules also appear to be of considerable importance in the control of the distribution of microtubules in cells as well as being active in motility. This exciting chapter in the development of our understanding of how microtubules function has just begun to become unravelled. Bridges have now been found connecting the microtubules of *Echinosphaerium*, the mitotic apparatus, nerve processes, developing spermatids, cilia and flagella, the axostyle, and a wide variety of structures in protozoa. In *Echinosphaerium* two sizes of bridges, the substructure of the microtubules, and the selection by the equilibrium of the most energetically stable configuration of tubules can account for the double coiled pattern of its axoneme. Bridges are undoubtedly important in controlling the distribution of microtubules in many other systems composed of ordered arrays of microtubules. Furthermore, bridges have now been found associated with almost all microtubule-associated motile mechanisms. We have suggested that these bridges may provide the active force in these diverse motile processes. Thus, motion might be carried out by the relative sliding of tubules past each other (given the requisite polarity, a factor presumably determined by nucleating sites), or in the case of saltatory motion by the sliding relative to the involved tubules of specific particles whose surfaces contain sites which can attach to the bridges.

Acknowledgements

I wish to thank EDWARD SALMON, JAN GODDARD, RICHARD RODEWALD, and BERNARD GERBER who spent a great deal of time criticizing and thus improving this manuscript. Generous support from the National Science Foundation grant GB 7974 is gratefully acknowledged.

References

ALLEN, R. D.: A reinvestigation of cross sections of cilia. J. Cell Biol. **37**, 825—831 (1968).
— BOWEN, C. C.: Fine structure of *Psilotum nudum* cells during division. Caryologia **19**, 99—123 (1966).
ALLISON, A. C., NUNN, J. F.: Effects of general anaesthetics on microtubules. A possible mechanism of anaesthesia. Lancet 1968, 1326—1329.
ANDRÉ, J., THIERY, J. P.: Mise en évidence d'une sous-structure fibrillaire dans les filaments axonematiques des flagelles, J. Microscopie **2**, 71—80 (1963).
ARNOLD, J. M.: Squid lens development in compounds that affect microtubules. Biol. Bull. **131**, 383 (1966).
ASAKURA, S., EGUCHI, G., IINO, T.: Reconstitution of bacterial flagella *in vitro*. J. molec. Biol. **10**, 42—56 (1964).
— — — *Salmonella* flagella: *in vitro* reconstruction and over-all shapes of flagellar filaments. J. molec. Biol. **16**, 302—316 (1966).
— — — Unidirectional growth of *Salmonella* flagella *in vitro*. J. molec. Biol. **35**, 227—236 (1968).
BAJER, A.: Chromosome movement and fine structure of the mitotic spindle. Symp. Soc. exp. Biol. **22**, 285—310 (1968).
BANNISTER, L. H., TATCHELL, E. C.: Contractility and the fibre systems of *Stentor coeruleus*. J. Cell Sci. **3**, 295—308 (1968).
BARNICOT, N. A.: A note on the structure of spindle fibres. J. Cell Sci. **1**, 217—222 (1966).
BASSOT, J. M., MARTOJA, R.: Présence de faisceaux de microtubules cytoplasmiques dans les cellules du canal éjaculateur du criquet migrateur. J. Microscopie **4**, 87—90 (1965).
BEHNKE, O., FORER, A.: Evidence for four classes of microtubules in individual cells. J. Cell Sci. **2**, 169—192 (1967).
BEISSON, J., SONNEBORN, T. M.: Cytoplasmic inheritance of the organization of the cell cortex in *Paramecium aurelia*. Proc. nat. Acad. Sci. (Wash.) **53**, 275—282 (1965).
BELAR, K.: Beitrage zur Kausalanalyse der Mitose III. Untersuchungen an den Staubfadenhaarzellen und Blattmeristemzellen von *Tradescantia virginica*. Z. Zellforsch. **10**, 73—134 (1929).
BHOWMICK, D. K.: Electron microscopy of *Trichamoeba villosa* and amoeboid movement. Exp. Cell Res. **45**, 570—589 (1967).
BIKLE, D., TILNEY, L. G., PORTER, K. R.: Microtubules and pigment migration in the melanophores of *Fundulus heteroclitus* L. Protoplasma **61**, 322—345 (1966).
BISCHOFF, R., HOLTZER, H.: The effect of mitotic inhibitors on myogenesis *in vitro*. J. Cell Biol. **36**, 111—127 (1968).
BORISY, G. G., TAYLOR, E. W.: The mechanism of action of colchicine. Binding of colchicine-^{3}H to cellular protein. J. Cell Biol. **34**, 525—533 (1967a).
— — The mechanism of action of colchicine. Colchicine binding to sea urchin eggs and the mitotic apparatus. J. Cell Biol. **34**, 535—548 (1967b).
BOWERS, B., KORN, E. D.: The fine structure of *Acanthamoeba castellanii*. 1. The trophozoite. J. Cell Biol. **39**, 95—111 (1968).
BRINKLEY, B. R., NICKLAS, R. B.: Ultrastructure of the meiotic spindle of grasshopper spermatocytes after chromosome micromanipulation. J. Cell Biol. **39**, 16 A (1968).
BYERS, B., ABRAMSON, D.: Cytokinesis in HeLa: post-telophase delay and microtubule-associated motility. Protoplasma **66**, 413—435 (1968).
— PORTER, K. R.: Oriented microtubules in elongating cells of the developing lens rudiment after induction. Proc. nat. Acad. Sci. (Wash.) **52**, 1091—1099 (1964).

CAMPBELL, R. D., INOUÉ, S.: Reorganization of spindle components following UV micro-irradiation. Biol. Bull. **129**, 401 (1965).

CAROLAN, R. M., SATO, H., INOUÉ, S.: A thermodynamic analysis in the effect of D_2O and H_2O on the mitotic spindle. Biol. Bull. **129**, 402 (1965).

— — — Further observations on the thermodynamics of the living mitotic spindle. Biol. Bull. **131**, 385 (1966).

CASPER, D. L. D.: Design principles in organized biological structures. In: "Principles of Biomolecular Organization" (Ed.: G. E. W. WOLSTENHOLME and M. O'CONNOR), pp. 7—35. London: Churchill 1966.

DE-THÉ, G.: Cytoplasmic microtubules in different animal cells. J. Cell Biol. **23**, 265—275 (1964).

DIRKSEN, E. R.: Observations on centriole formation in developing ciliated epithelium of the mouse oviduct. J. Cell Biol. **39**, 34 A (1968).

— CROCKER, T. T.: Centriole replication in differentiating ciliated cells of mammalian respiratory epithelium. An electron microscopic study. J. Microscopie **5**, 629—644 (1966).

FAWCETT, D. W.: The Cell. An Atlas of Fine Structure. Philadelphia: Saunders 1966.

FORER, A.: Local reduction of spindle fiber birefringence in living *Nephrotoma suturalis* (Loew) spermatocytes induced by ultraviolet microbeam. J. Cell Biol. **25**, 95—117 (1965).

— GOLDMAN, R. D.: Comparison of isolated and *in vivo* mitotic apparatus. Nature (Lond.) **222**, 689—691 (1969).

FREED, J. J., BHISEY, A. N., LEBOWITZ, M. M.: The relation of microtubules and micro-filaments to motility of cultured cells. J. Cell Biol. **39**, 46 A (1968).

GIBBINS, J. R., TILNEY, L. G., PORTER K. R.: Microtubules in the formation and develop-ment of the primary mesenchyme in *Arbacia punctulata* I. The distribution of microtu-bules. J. Cell Biol. **41**, 201—226 (1969).

GIBBONS, I. R.: The relationship between the fine structure and the direction of beat in gill cilia of the lamellibranch mollusc. J. biophys. biochem. Cytol. **11**, 179—205 (1961).

GOLDMAN, R. D., REBHUN, L. I.: The structure and some properties of the isolated mitotic apparatus. J. Cell Sci. **4**, 179—209 (1969).

GRAIN, J.: Les systémes fibrillaires chez *Stentor igneus* Ehrenberg et *Spirostomum ambiguum* Ehrenburg. Protistologica **4**, 27—36 (1968).

GREEN, L.: Mechanism of movements of granules in melanocytes of *Fundulus heteroclitus*. Proc. nat. Acad. Sci. (Wash.) **59**, 1179—1186 (1968).

GRIMSTONE, A. V., CLEVELAND, L. R.: The fine structure and function of the contractile axostyles of certain flagellates. J. Cell Biol. **24**, 387—400 (1965).

HARRINGTON, W. F., JOSEPHS, R.: Self-association reactions among fibrous proteins: the myosin polymer system. Devel. Biol. Suppl. **2**, 21—62 (1968).

HEILBRUNN, L. V.: The Dynamics of Living Protoplasm. New York: Academic Press 1956.

HEPLER, P. K., JACKSON, W. T.: Microtubules and early stages of cell-plate formation in the endosperm of *Haemanthus katherinae* Baker. J. Cell Biol. **38**, 437—446 (1968).

IKKAI, T., OOI, T.: The effects of pressure on F-G transformation of actin. Biochem. J. **5**, 1551—1560 (1966).

INOUÉ, S.: The effect of colchicine on the microscopic and submicroscopic structure of the mitotic spindle. Exp. Cell Res. (Suppl.) **2**, 305—318 (1952).

— Organization and function of the mitotic spindle. In: Primitive Motile Systems in Cell Biology. (Ed.: R. D. ALLEN and N. KAMIYA), pp. 549—594. New York: Academic Press 1964.

— SATO, H.: Cell motility by labile association of molecules. The nature of mitotic spindle fibers and their role in chromosome movement. J. gen. Physiol. **50**, 259—288 (1967).

KALNINS, V. I.: Centriole replication during ciliogenesis in the chick tracheal epithelium. J. Cell Biol. **39**, 70 A (1968).

KLUG, A., FINCH, J. T., LEBERMAN, R., LONGLEY, W.: Design and structure of regular virus particles. In: Principles of Biomolecular Organization. (Ed.: G. E. W. WOLSTEN-HOLME and M. O'CONNOR), pp. 158—189. London: Churchill 1966.

KOHNO, K.: Neurotubules contained within the dendrite and axon of Purkinje cell of
  frog. Bull. Tokyo Med. Dent. Univ. **11**, 411—442 (1964).
LEDBETTER, M. C., PORTER, K. R.: A "microtubule" in plant cell fine structure. J. Cell
  Biol. **19**, 239—250 (1963).
LEVI-MONTALCINI, R.: The nerve growth factor: its mode of action on sensory and
  sympathetic nerve cells. Harvey Lectures **60**, 217—259 (1965).
LU, B. D.: Meiosis in *Coprinus lagopus:* A comparative study with light and electron micro-
  scopy. J. Cell Sci. **2**, 529—536 (1967).
MACDONALD, A. C., KITCHING, J. A.: Axopodial filaments of Heliozoa. Nature **215**,
  99—100 (1967).
MARSLAND, D., HECHT, R.: Cell division: combined anti-mitotic effects of colchicine and
  heavy water on first cleavage in the eggs of *Arbacia punctulata*. Exp. Cell Res. **51**,
  602—608 (1968).
MCINTOSCH, J. R., PORTER, K. R.: Microtubules in the spermatids of the domestic fowl.
  J. Cell Biol. **35**, 153—173 (1967).
MOSES, M. J., WILSON, M., WYIRCK, D.: Assembly of microtubules in the flagellate sperm
  of a coccid. J. Cell Biol. **39**, 96 A (1968).
NAGAI, R., REBHUN, L. I.: Cytoplasmic microfilaments in streaming *Nitella* cells. J. Ultra-
  struct. Res. **14**, 571—589 (1966).
NICKLAS, R. B.: Chromosome micromanipulation II. Induced reorientation and the expe-
  rimental control of segregation in meiosis. Chromosoma **21**, 17—50 (1967).
O'BRIEN, T. P., THIMANN, K. V.: Intracellular fibers in oat coleoptile cells and their possible significance in cytoplasmic streaming. Proc. nat. Acad. Sci. (Wash.) **56**, 888—894
  (1966).
OUTKA, D. E., KLUSS, B. C.: The amoeba-to-flagellate transformation in *Tetramitus rostratus*
  II. Microtubular morphogenesis. J. Cell Biol. **35**, 323—346 (1967).
PALAY, S. L., SOTELO, C., PETERS, A., ORKAND, P. M.: The axon hillock and the initial
  segment. J. Cell Biol. **38**, 193—201 (1968).
PEASE, D. C.: Hydrostatic pressure effects upon the spindle figure and chromosome move-
  ment 1. Experiments on the first mitotic division of *Urechis* eggs. J. Morph. **69**, 405—441
  (1941).
— The ultrastructure of flagellar fibrils. J. Cell Biol. **18**, 313—326 (1963).
PHILLIPS, D. M.: Substructure of flagellar tubules. J. Cell Biol. **31**, 635—638 (1966).
— Exceptions to the prevailing pattern of tubules $(9+9+2)$ in the sperm flagella of certain
  insect species. J. Cell Biol. **40**, 28—43 (1969).
PICKETT-HEAPS, J. D.: Microtubule-like structures in the growing plastids or chloroplasts
  of two algae. Planta **81**, 193—200 (1968).
PITELKA, D. R.: Electron-Microscopic Structure of Protozoa. New York: Pergamon
  Press Inc., 1963.
PORTER, K. R.: Cytoplasmic microtubules and their function. In "Principles of Biomole-
  cular Organization" (Ed.: G. E. W. WOLSTENHOLME and M. O'CONNOR), pp. 308—345.
  London: Churchill 1966.
RANDALL, J. R., DISBREY, C.: Evidence for the presence of DNA at basal body sites in *Tetra-
  hymena pyriformis*. Proc. roy. Soc. London **162**, 473—491 (1965).
REBHUN, L. I.: Saltatory particle movements in cells. In: "Primitive Motile Systems in
  Cell Biology" (Ed.: R. D. ALLEN and N. KAMIYA), pp. 503—525. New York: Academic
  Press 1964.
— SANDER, G.: Ultrastructure and birefringence of the isolated mitotic apparatus of
  marine eggs. J. Cell Biol. **34**, 859—883 (1967).
RENAUD, F. L., SWIFT, H.: The development of basal bodies and flagella in *Allomyces
  arbusculus*. J. Cell Biol. **23**, 339—354 (1964).
— ROWE, A. J., GIBBONS, I. R.: Some properties of the protein forming the outer fibers
  of cilia. J. Cell Biol. **36**, 79—90 (1968).
RINGO, D. L.: The arrangement of subunits in flagellar fibers. J. Ultrastruct. Res. **17**,
  266—277 (1967).
ROBBINS, E. R., GONATAS, N. K.: The ultrastructure of a mammalian cell during the mitotic
  cycle. J. Cell Biol. **21**, 429—463 (1964).

ROBBINS, E. R., JENTZSCH, G., MICALI, A.: The centriole cycle in synchronized HeLa cells. J. Cell Biol. **36**, 329—339 (1968).

ROBINOW, C. F., MARAK, J.: A fiber apparatus in the nucleus of the yeast cell. J. Cell Biol. **29**, 129—151 (1966).

ROBISON, W. G., JR.: Microtubules in relation to the motility of a sperm syncytium in an armored scale insect. J. Cell Biol. **29**, 251—266 (1966).

ROSENBAUM, J. L., CARLSON, K.: Cilia regeneration in *Tetrahymena* and its inhibition by colchicine. J. Cell Biol. **40**, 415—425 (1969).

— CHILD, F. M.: Flagellar regeneration in protozoan flagellates. J. Cell Biol. **34**, 345—364 (1967).

— MOULDER, J. E., RINGO, D. L.: Flagellar elongation and shortening in *Chlamydomonas*. The use of cycloheximide and colchicine to study the synthesis and assembly of flagellar proteins. J. Cell Biol. **41**, 600—619 (1969).

ROTH, L. E.: Electron microscopy of mitosis in ameba. III. Cold and urea treatments: a basis for test of direct effects of mitotic inhibitors on microtubule formation. J. Cell Biol. **34**, 47—59 (1967).

SABATINI, D. D., BENSCH, K., BARRNETT, R. J.: Cytochemistry and electron microscopy. The preservation of cellular ultrastructure and enzymatic activity by aldehyde fixation. J. Cell Biol. **17**, 19—58 (1963).

SATIR, P.: Studies on cilia. II. Examination of the distal region of the ciliary shaft and the role of filaments in motility. J. Cell Biol. **26**, 805—834 (1965).

— Studies on cilia, III. Further studies on the cilium tip and a "sliding filament" model of ciliary motility. J. Cell Biol. **39**, 77—94 (1968).

— STUART, A. M.: A new apical microtubule-associated organelle in the sternal gland of *Zootermopsis nevadensis* (Hagen), Isoptera. J. Cell Biol. **24**, 277—283 (1965).

SAWADA, N., REBHUN, L. I.: The effect of dinitrophenol and other phosphorylation uncouplers on the birefringence of the mitotic apparatus of marine eggs. Expt. Cell Res. **55**, 33—38 (1969).

SHELANSKI, M. L., TAYLOR, E. W.: Isolation of a protein subunit from microtubules. J. Cell Biol. **34**, 549—554 (1967).

— — Properties of the protein subunit of central-pair and outer-doublet microtubules of sea urchin flagella. J. Cell Biol. **38**, 304—315 (1968).

SILVEIRA, M.: Ultrastructural studies on a "nine plus one" flagellum. J. Ultrastruct. Res. **26**, 274—288 (1969).

— PORTER, K. R.: The spermatozoids of flatworms and their microtubular systems. Protoplasma **59**, 240—265 (1964).

SMITH-SONNEBORN, J., PLAUT, W.: Evidence for the presence of DNA in the pellicle of *Paramecium*. J. Cell Sci. **2**, 225—234 (1967).

SOROKIN, S. P.: Reconstruction of centriole formation and ciliogenesis in mammalian lungs. J. Cell Sci. **3**, 207—230 (1968).

STEPHENS, R. E.: On the structural protein of flagella outer fibres. J. molec. Biol. **32**, 277—283 (1968a).

— Reassociation of microtubule protein. J. molec. Biol. **33**, 517—519 (1968b).

— RENAUD, F. L., GIBBONS, I. R.: Guanine nucleotide associated with the protein of the outer fibers of flagella and cilia. Science **156**, 1606—1608 (1967).

STEVENS, C. L., LAUFFER, M. A.: Polymerization-depolymerization of tobacco mosaic virus protein. IV. The role of water. Biochemistry **4**, 31—37 (1965).

SUBRIANA, J. A.: Role of spindle microtubules in mitosis. J. theoret. Biol. **20**, 117—123 (1968).

SZOLLOSI, D.: The structure and function of centrioles and their satellites in the jellyfish, *Phialidium gregarium*. J. Cell Biol. **21**, 465—479 (1964).

TAMM, S. L.: Flagellar development in the protozoan, *Peramena trichophorum*. J. exp. Zool. **164**, 163—186 (1967).

TAYLOR, E. W.: The mechanism of colchicine inhibition of mitosis. I. Kinetics of inhibition and the binding of H³-colchicine. J. Cell Biol. **25**, 145—160 (1965).

TILNEY, L. G.: The assembly of microtubules and their role in the development of cell form. Devel. Biol. Suppl. **2**, 63—102 (1968a).

Tilney, L. G.: Studies on the microtubules in Heliozoa. IV. The effect of colchicine on the formation and maintenance of the axopodia of *Actinosphaerium nucleofilum* (Barrett). J. Cell Sci. **3**, 549—562 (1968b).
— Byers, B.: Studies on the microtubules in Heliozoa V. Factors controlling the organization of microtubules in the axonemal pattern in *Echinosphaerium* (*Actinosphaerium*) *nucleofilum*. J. Cell Biol. **43**, 148—165 (1969).
— Gibbins, J. R.: Differential effects of antimitotic agents on the stability and behavior of cytoplasmic and ciliary microtubules. Protoplasma **65**, 167—179 (1968).
— — Microtubules in the formation and development of the primary mesenchyme in *Arbacia punctulata* II. An experimental analysis of the role of these elements in the development and maintenance of cell shape. J. Cell Biol. **41**, 227—250 (1969).
— Hiramoto, Y., Marsland, D.: Studies on the microtubules in Heliozoa. III. A pressure analysis of the role of these structures in the formation and maintenance of the axopodia of *Actinosphaerium nucleofilum* (Barrett). J. Cell Biol. **29**, 77—95 (1966).
— Porter, K. R.: Studies on the microtubules in Heliozoa. I. Fine structure of *Actinosphaerium* with particular reference to axial rod structure. Protoplasma **60**, 317—344 (1965).
— — Studies on the microtubules in Heliozoa. II. The effect of low temperature on these structures in the formation and maintenance of the axopodia. J. Cell Biol. **34**, 327—343 (1967).
Tucker, J. B.: Fine structure and function of the cytopharyngeal basket in the ciliate, *Nassula*. J. Cell Sci. **3**, 493—514 (1968).
Watters, C.: Studies on the motility of the Heliozoa I. The locomotion of *Actinosphaerium eichorni* and *Actinophrys* sp. J. Cell Sci. **3**, 231—244 (1968).
Warren, R. H.: The effect of colchicine on myogenesis *in vivo* in *Rana pipens* and *Rhodnius prolixus* (Hemiptera). J. Cell Biol. **39**, 544—555 (1968).
Weisenberg, R. C., Borisy, G. G., Taylor, E. W.: The colchicine binding protein of mammalian brain and its relation to microtubules. Biochemistry **7**, 4466—4479 (1968).
Went, H. A.: Dynamic aspect of mitotic apparatus protein. Ann. N. Y. Acad. Sci. **90**, 422—429 (1960).
Wilson, H. J.: Arms and bridges on microtubules in the mitotic apparatus. J. Cell Biol. **40**, 854—859 (1969).
Wisniewski, H., Shelanski, M. L., Terry, R. D.: Effects of mitotic spindle inhibitors on neurotubules and neurofilaments in anterior horn cells. J. Cell Biol. **38**, 224—230 (1968).
— Terry, R. D.: Experimental colchicine encephalopathy I. Induction of neurofibrillary degeneration. Lab. Invest. **17**, 577—587 (1968).
Wohlfarth-Bottermann, K. E.: Differentiations of the ground cytoplasm and their significance for the generation of the motive force of ameboid movement. In: "Primitive Motile Systems in Cell Biology" (Ed.: R. D. Allen and N. Kamiya), pp. 79—108. New York: Academic Press 1964.
Yanagisawa, T., Hasegawa, S., Monii, H.: Bound nucleotides of the isolated microtubules of sea urchin sperm flagella and their possible role in flagellar movement. Exp. Cell Res. **52**, 86—100 (1968).

Origin and Continuity of Desmosomes*

Richard D. Campbell and John H. Campbell

*Department of Developmental and Cell Biology,
University of California, Irvine, California*

and

*Department of Anatomy, School of Medicine,
University of California, Los Angeles, California*

I. Introduction

The problems of desmosome formation and maintenance are different from those of the other cell components discussed in this volume. Desmosomes, and other cell attachment devices, are not organelles but rather are modifications of the cell periphery. No specific chemical or ultrastructural components have been identified in desmosomes which are not common to the rest of the peripheral cytoplasm and membrane of the cell. The problem of desmosome origin, therefore, seems to be one of localization and organization of ubiquitous components, a process less well understood than the reproduction of true organelles.

In order to give a wide scope to this article we will consider all types of specializations of the cell periphery related to adhesion and attachment and will call these cell junctions. Cell junctions present a variety of morphological forms. The relationships between them are not well understood and the interested reader is referred to Farquhar and Palade (1963), Brightman and Palay (1963), Fawcett (1966) and Kelly and Luft (1966). Kelly and Luft (1966) have proposed that the attachment specializations represent a spectrum of forms (Fig. 1). Toward one end are the elaborate junctions, which contain accumulations of cytoplasmic materials subjacent to the membrane. These include the desmosome (node of Bizzozero or macula desmosome; Farquhar and Palade, 1963), the zonula adherens (Farquhar and Palade, 1963), hemidesmosomes (Kelly, 1966), and intercalated discs (Sjöstrand, Andersson-Cedergren and Dewey, 1958). Their principal cytoplasmic components may include filaments and a plaque, which is a densely staining lamella a few millimicrons thick and very close to the membrane. The filaments frequently are arranged in bundles running either perpendicular or parallel to the membrane. The extracellular components of cell junctions appear in electron micrographs as a fairly amorphously stained matrix

* Supported by NIH grants GM-14443, GM-17391 and AI-07847 and Research Career 1-K04-GM42595.

between the cells. In highly elaborate junctions they form a distinct "middle lamella" bisecting the intercellular gap. In general, the more elaborate a junction is in terms of the accumulated cytoplasmic materials, the more densely staining is the intercellular material and also the wider is the intercellular gap of that junction.

Toward the other end of the spectrum are the morphologically simpler junctions, generally with little or no associated increased cytoplasmic density. These include

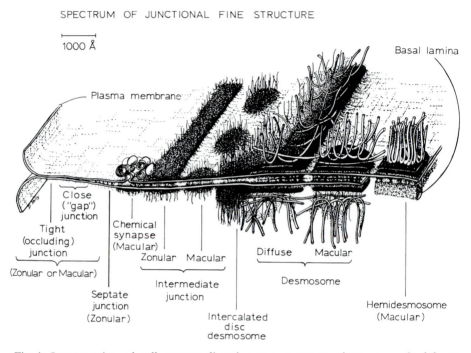

Fig. 1. Interpretation of cell contact diversity as a spectrum of structure. At left are portrayed the structurally simple contact types; at right are the elaborate contact types. Courtesy of Dr. Douglas E. Kelly

tight and close junctions (Muir and Peters, 1962; Rosenbluth, 1965; Revel and Karnovsky, 1967), gap junctions (Brightman and Reese, 1969), synapses, and the septate desmosome (Wood, 1959). The intercellular gap is generally less than 150 Å in width. Tight junctions have no visible intercellular gap at all, the outer leaflets of the two unit membranes fusing together to form a quintuple-layered structure. Close junctions have an intercellular gap of 20–30 Å, and septate desmosomes have a gap of 150–200 Å traversed by regularly spaced particles or bars to give a ladder-like profile in cross section.

In the spectrum of morphological forms of junctions, unmodified apposed cell membranes would occupy a middle portion. This indicates that if cell junctions arise from unmodified cell membranes, the two halves of the spectrum may represent different processes of membrane modification; elaborate junctions (e.g. desmosomes) may form by accretion of materials while simple junctions may develop by removal or thinning of these components. Accordingly, it is useful to recognize a dichotomy of forms and to speak of "elaborate" and "simple" junctions as generic terms.

All types of junctions may exist in various geometrical forms; as foci which may be as small as a few millimicrons in extent, as maculae large enough to be seen by light microscopy and as belts or zonulae. These in turn may associate into larger units. Along the lateral surfaces of cells focal junctions may occur in long rows resembling zonulae. The apical border of epithelial cells is generally encircled by a characteristic series of simple and elaborate zonulae, visualized as a terminal bar in the light microscope and a junctional complex (FARQUHAR and PALADE, 1963) in the electron microscope (Fig. 2).

II. Time of Development

Cell junctions arise at characteristic times during ontogeny. Septate desmosomes between sea urchin blastomeres appear between the eighth and eleventh hours of cleavage (DAN, 1960). In the rat embryo they begin to appear at the two cell stage, and definite junctional complexes become visible at the morula stage (POTTS, 1966). In the chick blastoderm desmosomes begin to appear near the embryonic axis at stages 5–6, and then form in progressively more lateral regions (OVERTON, 1962). During histogenesis junction formation is temporally related to the initial acquisition of tissue structure.

There is evidence that different junctions of a cell develop at different characteristic times. This is the case, for example, in the complicated ameloblast which has several distinct junctional regions (KALLENBACH, CLERMONT, and LEBLOND, 1965). For epithelia with a typical junctional complex encircling the apical cell margin, it has been reported that the most apical zone, the tight junction (zonula occludens), forms first, and that the adherens junctions form later (FARQUHAR and PALADE, 1964; TRELSTAD, HAY, and REVEL, 1967). DEANE and WURZELMANN (1965a) reported that in seminal vesicle epithelium the lateral desmosomes scattered between adjacent cells form after the junctional complex proper, a situation apparently occurring also in the hepatocyte (WOOD, 1965). This general apico-basal time course of desmosome development is consistent with the presence in some epithelia of only the most apical portions of the junctional complex without lateral desmosomes ever appearing. However, the generality of an apico-basal time course of development remains to be critically verified.

III. Morphology of Developing Junctions

Studies on developing junctions have been made in a variety of tissues. These studies indicate that maturation is more a process of quantitative alteration than of qualitative metamorphosis. Development of the more elaborate junctions involves a gradual increase in membrane-associated electron-dense material, finally reaching the degree characteristic of the particular type. Simple junctions form by the two plasma membranes coming into intimate contact. We will first report observations which indicate the morphologies of developing junctions, and then examine the mechanisms proposed to explain the observations.

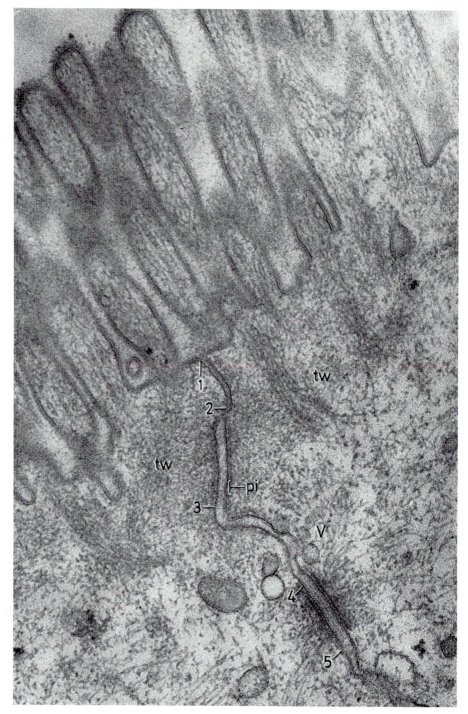

Fig. 2

A. Chick Blastoderm and Other Simple Epithelia

The initial recognizable stage of desmosome formation in electron micrographs of various embryonic tissues and reaggregating cells is a region of membrane apposition with diffuse but heightened staining of the subjacent cytoplasm (Overton, 1962; Overton and Shoup, 1964; Patrizi, 1967a; Wartiovaara, 1966; Wood, 1965; Paweletz, 1969) (Fig. 3). Overton's (1962) account of subsequent development in chick blastoderm is the most detailed. The extended patches of dense, close membrane become restricted in area. The first cytoplasmic structures to become recognizable are the plaques and small fibrous tufts which extend deeper into the cytoplasm. Subsequently, the main bundles of cytoplasmic fibrils arise. The longitudinal fibrils oriented parallel to the membrane appear before the radial ones directed into the cytoplasm. Thus, organization of the cytoplasm seems to occur from the membrane inwards. On the extracellular side the space between cells enlarges, perhaps to its final width before it becomes electron dense. As the cytoplasmic components of the incipient desmosome organize, the extracellular matrix becomes progressively more dense and structured.

It is unfortunate that the fine structure of these early stages of condensation has not been examined at resolutions attained in studies of mature desmosomes because knowledge of the fibrillar packing arrangements is probably critical for understanding the mode of formation. Even Overton's observations do not adequately reveal the initial configurations or sequence in which components arise.

B. Keratinizing Epithelium

An orderly adult tissue in which a time sequence of desmosome structures has been described is keratinizing epithelium (Oland and Reed, 1967). Thick skin shows three definitive layers or strata of living cells. These are capped by a sharply demarcated stratum corneum of dead anucleate keratinized cells. New cells are produced in the most basal layers of skin and passively migrate up to

---

Fig. 2. Junctional complex between two cells in the epithelium of the intestinal mucosa (rat). The tight junction *(zonula occludens)*, located nearest the lumen, extends from arrow 1 to arrow 2. The narrowing of the apparent intercellular "gap" (~ 90 Å) is clearly visible, but the fusion line of the two apposed membranes cannot be clearly distinguished at this magnification. — The intermediate junction *(zonula adhaerens)* extends from arrow 2 to arrow 3. A relatively wide intercellular space (~ 200 Å) is maintained throughout the junction. Extensive condensation of cytoplasmic fibrils occurs as a fine feltwork along either side of the junction. This condensation is continuous with the terminal web *(tw)* into which the filamentous rootlets of the microvilli penetrate. Plate-like densifications within the cytoplasmic feltwork can be seen along part of the junction, especially along the right side *(pi)*. — The limits of a desmosome are marked by arrows 4 and 5. This element is characterized by a wide intercellular space (~ 240 Å) bisected by an intermediate line. Bundles of cytoplasmic fibrils, coarser (diameter ~ 80 Å) and more distinct than those of the terminal web, converge into dense plates on each side of the desmosome. These plates are separated from the inner leaflets of the cell membrane by a zone of low density. Similar fibrils appear throughout the remainder of the field below the terminal web. — Between the intermediate junction and the desmosome, the two apposed cell membranes are separated by an irregular space of varying width and show membrane invaginations and associated vesicles (v). $\times$ 87,000. Courtesy of Dr. Marilyn G. Farquhar; from Farquhar and Palade (1963)

the surface as other cells are formed below them. Therefore, the age of a cell can be related to its position in the epithelium.

Desmosomes are abundant throughout keratinizing epithelium. Between adjacent living cells they take the form of ovoid elaborated areas of opposing cell surfaces

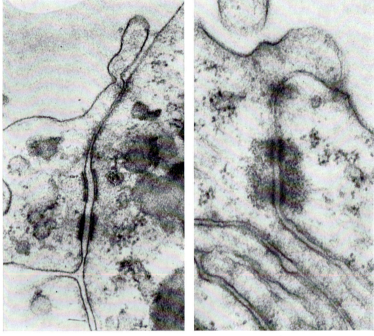

a   b

Fig. 3 a and b. Desmosome development in chick extraembryonic tissue. a) early stage. b) mature desmosomes with associated fiber bundles. × 50,000. Courtesy of Dr. JANE OVERTON; from OVERTON (1962)

and cytoplasm. Between the lowest cells and the basement lamella the attachments involve only one cell membrane and, as their name hemidesmosome implies, they appear similar to one-half of a desmosome between two cells (KELLY, 1966).

In the lowest layers of cells, desmosomes must be continuously forming or increasing in number at a rate commensurate with the rate of cell division. Cells of the stratum spinosum (an intermediate layer of cells) are reported to have a larger number of desmosomes than cells of the underlying stratum basale, indicating that formation of new desmosomes may occur to some extent throughout the epithelium. However, no initial stages of formation have been described in these layers. Possibly desmosomes form very rapidly (perhaps just after cell division, on the newly formed cell surface, BUCK and KRISHAN, 1965), and have simply not been noticed in the relatively few electron microscope studies on epithelial cells in mitosis. Alternatively, they may arise by fragmenting or splitting of pre-existent desmosomes (MERCER, 1961). This origin would be compatible with the rather broad range of sizes of desmosomes present in these cells. Other possibilities are that immature desmosomes are so similar or so different in their basic appearance from mature desmosomes that they simply have not been recognized.

Throughout the living layers of skin, desmosomes present a fairly uniform appearance. Their most conspicuous features are bundles of intracellular tonofibrils which are anchored to the desmosomes and which may extend from one desmosome to another (WILGRAM, CAULFIELD, and MADGIC, 1965). The region of apposed membranes characteristically appears as seven electron dense layers, the outer ones representing the cytoplasmic plaques (KARRER, 1960). The central line is the middle lamella between the two cells, and the remaining two pairs of lines represent the cytoplasmic membranes.

The only striking changes displayed by these desmosomes accompany keratinization in the upper layers of skin. These events probably represent degeneration to allow the cell eventually to slough. Initially, the changes occur so rapidly that the desmosomes between the highest living cells and the lowest keratinized cells are asymmetrical (ODLAND and REED, 1967), an unusual quality for an intercellular attachment apparatus. HORSTMANN and KNOOP (1958) and LISTGARTEN (1964) indicate that desmosomes progressively lose contrast and definition of structure in electron micrographs. First the middle lamellae disappear. Subsequently, the tonofilaments become less apparent, either disappearing or losing their distinctive staining characteristics. At the upper layer of the stratum corneum the boundaries of the cell membranes become indistinct. The cytoplasmic plaque, the most persistent desmosomal structure, finally loses contrast at the level where sloughing occurs.

SNELL (1965, 1966) has described a slightly different sequence subsequent to the disappearance of the middle lamella. The extracellular matrix over the entire cell membrane increases in density and this increase extends through the desmosomes. Concomitantly, the tonofibrils lose their attachment on the desmosome plaque. Later the extracellular material in the desmosome separates from that over the adjacent regions. It tends to round up as a "fusiform body," with a trace of a membrane or other discrete lining surrounding it. When this plug-like connection between the desmosome halves loses contrast and appears to rupture, the cells slough.

C. Intercalated Discs

SJÖSTRAND, ANDERSSON-CEDERGREN, and DEWEY (1958) have shown that intercalated discs represent elaborated desmosomes. The developmental sequence leading from ordinary desmosomes to the discs has been described by a number of investigators for a variety of animals (FAWCETT and SELBY, 1958; CHALLICE and EDWARDS, 1960; MUIR, 1957; GRIMLEY and EDWARDS, 1960; MANASEK, 1968a). Intercalated discs begin their development as small patches of apposed lateral membranes under which the cytoplasm is slightly condensed. These regions, called desmosomes or "protodiscs," occur predominantly at the level of the Z bands (see Fig. 4). Myofilaments become inserted in these desmosomes and aggregate into definitive myofibrils. The desmosomes then shift from lateral to end positions and are called intercalated discs. Further development involves a thickening and widening of the structure together with a widening of the sarcomere by the acquisition of additional myofibrils.

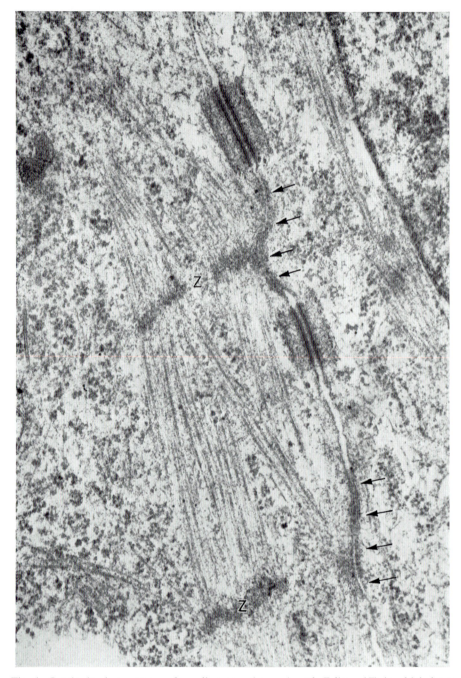

Fig. 4. Continuity between zonulae adherentes (arrows) and Z lines (Z) in chick heart myoblast. $\times$ 54,000. Courtesy of Martin Hagopian; from Symp. Soc. Cell Biol. 6, 71 (1968)

It is widely felt that in some way the membrane aids in organizing the Z bands of cardiac and skeletal muscle or in aligning and attaching the nascent myofilaments to the Z band (see SPIRO and HAGOPIAN, 1967; KELLY, 1969; RASH, SHAY and BIESELE, 1969). Since an important component of many desmosomes is the bundles of attached filaments, it is possible that myofibrillar organization is related to the organization of fibrils in desmosomes of other cell types. The comparative fine structure of desmosomes and intercalated discs is presently too poorly known to permit one to go beyond this conjecture. KELLY (1969) emphasizes the similarity of fibrillar configurations in the two structures. MANASEK (1968a) even finds that the first intercalated discs arise from the preexisting junctional complex in the early chick myocardial epithelium. However, he distinguishes between desmosomes and initial stages of intercalated discs. Although he does not clearly describe the differences which lead him to suppose their non-identity, many reports which illustrate desmosomes and intercalated discs in the same cell suggest that the desmosomes tend to have two distinct fibrillar or granular zones while the intercalated discs are more homogeneous. Thus, while intercalated disc formation appears, in general, to involve a transformation of desmosomes, the fine structural aspects of this process remain controversial. Within a myofiber normal desmosomes and intercalated discs probably represent branching lines of development from a similar initial point (CHALLICE and EDWARDS, 1960).

D. Tight Junctions and Septate Desmosomes

Simple junctions have not been studied from a developmental point of view as extensively as the elaborate desmosomes. Contacts between embryonic or aggregating cells are often in the form of "focal tight junctions" only a few millimicrons in extent (TRELSTAD, HAY, and REVEL, 1967; OVERTON, 1962; LESSEPS, 1963). There is some reason to think that focal tight junctions enlarge by either simple expansion or fusion of large numbers of adjacent foci (TRELSTAD, REVEL, and HAY, 1966; TRELSTAD, HAY, and REVEL, 1967) (Fig. 5). For example, tight junctions in mature cells frequently have focal separations in line with this interpretation.

Embryonic sea urchin blastomeres are connected by septate desmosomes. WOLPERT and MERCER (1963) report that these junctions appear sometime between the eighth and eleventh cleavage division. While no intermediate stages have been described, the junctions apparently form initially in their typical location near the cell margins, as WOLPERT and MERCER (1963) also present evidence that membranes are fairly stationary in these embryos.

E. Morphology of Developing Junctions: Summary

To summarize, the observations on the morphology of forming contacts support the idea that developing junctions do not differ greatly from mature ones. The main difference between young and old elaborate junctions lies in the quantities and degree of compaction of intracellular and extracellular membrane-associated materials. Dense and highly oriented fiber bundles of mature desmosomes appear to be derived from more loosely organized, less dense arrangements. In some cell types junctions increase in area as they develop (e.g., FAWCETT and SELBY,

1958). In other cells, young desmosomes are believed to be broad in extent, and to undergo diminution in area during formation (e.g., Overton, 1962). In both cases, maturation involves both an increase in the total amount of materials present and an increase in density of the elements.

Almost without exception, junctions of all types and at all stages are symmetrical in the two cells involved. Only when the two cells are dissimilar are asymmetrical

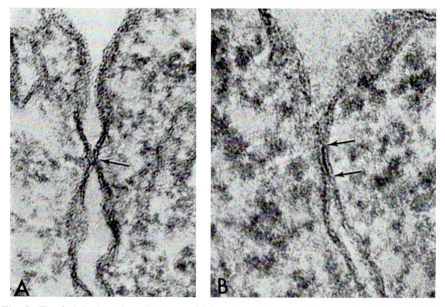

Fig. 5. Focal tight junctions between chick embryonic mesenchyme cells. (*A*) Focal tight junction (arrow) of embryo stage 4. (*B*) More extensive tight junction at stage 18. These resemble stages in a possible sequence of focal tight junction expansion. × 150,000.
Courtesy of Dr. Robert L. Trelstad; from Trelstad, Hay, and Revel (1967)

desmosomes found. Under normal circumstances the two halves are in perfect register in the two cells. Thus, by some means the developmental activities of adjacent cells must be exactly synchronized to produce desmosomes symmetrically. Furthermore, cells may form junctional complexes with only particular neighboring cells; in reaggregating kidney cells they develop between laterally adjacent but not apically adjacent cells (Wartiovaara, 1966).

IV. Interpretation of Ultrastructure and Geometry

Mercer (1965) has pointed out that the origin of desmosomes poses two problems: the source of the materials comprising the desmosomes and the source of the organization of these materials into desmosomes at particular sites on the cell surface. The latter issue is the more interesting because it is a problem of cellular organization instead of mere biochemical origin. Several hypotheses concerning this aspect of desmosome formation have been proposed. Each rests on indirect evidence and can be neither proven nor refuted at this time. However, the various ideas are based upon different assumptions about the principal sources of the precursor materials. Thus, it is pertinent to discuss the individual morphological aspects of desmosomes with respect to the probable source of their constituent molecules as a prelude to the broader question of the mechanism of desmosome genesis.

A. Intercellular Matrix and Lamellae

The extracellular components of elaborate junctions are almost certainly composed of the materials similar to those which coat the rest of the cell ("glycocalyx", "extraneous coat", "fuzzy coat") but which have become compacted. Special stains for carbohydrates (PEASE, 1966; KELLY, 1966; RAMBOURG and LEBLOND, 1967) show the continuity between the extracellular components of desmosomes and the extraneous coat of the rest of the cell. When viewed at very high magnifications, the cell coat and the extracellular components of desmosomes appear to have the same texture (KELLY, 1966) and to merge at their junction by a gradual

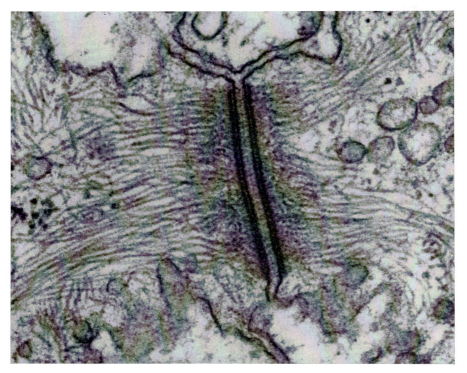

Fig. 6. Fine structure of a desmosome in newt epidermis. The intercellular gap of the desmosome is occupied by moderately dense material which displays a discontinuous midplane density. The cell membranes appear as single dense lines. Subjacent desmosomal plaques are represented as dense lines, separated from the cell membranes by thin lucent bands. In several places the plaque has a beaded appearance. Tonofilaments approach plaque from cytoplasm and loop back away at varying distances from the membrane. Courtesy of Dr. DOUGLAS E. KELLY; from KELLY (1966). × 93,000

transition in degree of compaction. Moreover, there is a continuous series of desmosome types ranging from those with an extracellular portion indistinguishable from the adjacent cell coat, to forms in which the extracellular materials are so compacted that they exhibit a lamellar organization (Fig. 6). It is believed that the intercellular lamella represents a region where fibers or compacted particles from the two membranes overlap, rather than a distinct layer of intercellular material (Kelly, 1966). Easty and Mercer (1962) have even developed an experimental model for the compaction of membrane-bound molecules into intercellular lamellae using erythrocytes and anti-erythrocyte agglutinating antibodies. When erythrocytes coated with a heavy layer of antibody are compacted, a distinct lamellar structure appears midway between the two membranes of adjacent cells.

The extracellular component of desmosomes increases progressively in amount during formation. This is indicated both by developmental studies and by a correlation among the apparent degree of development of a desmosome, the width of the intercellular gap and the intensity with which the material in the gap stains. Adherent zones which are only weakly elaborated have gaps of 150–250 Å, similar to the distance between the membranes in adjacent unspecialized regions. The extracellular material within the gap accepts little or no stain by conventional treatments. Highly developed junctions may have a gap as wide as 500 Å which is very densely staining.

If the extracellular material of desmosomes does represent condensed cell glycocalyx, then desmosome formation must involve either *in situ* deposition or a recruitment of this material from adjacent regions. Glycocalyx is thought to arise together with new membrane from Golgi vesicles (Rambourg, Hernandez, and Leblond, 1969; Hicks, 1966). Golgi activity simultaneous with developing junctional complexes has been reported (Wartiovaara, 1966; Wood, 1965) but such observations are not striking or numerous.

B. Cytoplasmic Plaques or Lamellae

The cytoplasmic portion of highly elaborated desmosomes characteristically contains one or more "plaques"[1] (Jurand, 1965; Karrer, 1960). Their visible organization is lamellar and parallel to the membrane surface. Several authors consider this lamellar organization to be secondary, with the principal substructures being vesicles, bent and twisted fibers, or fibers perpendicular to the plane of the lamella (Kelly, 1966; Sedar and Forte, 1964; Challice and Edwards, 1960; Chapman and Dawson, 1961) (see Fig. 7). The lamellar appearance predominates because fibers are thicker at points which are all in register or because a matrix (perhaps similar to keratohyalin) is deposited along the fibers in this region. Milhaud and Pappas (1966) have shown a plaque-like structure of a neural synapse to be composed of individual particles or vesicles which are densely packed. Thus, the lamellar structures of desmosomes can be interpreted as resulting from compaction and condensation of elements ubiquitously present in the cortex of the cell, with perhaps some *in situ* deposition of a homogeneous matrix.

---

1 The earlier use of the term "plaque" included both the cell membrane and adjacent cytoplasmic densities. In this paper, as in the more recent literature, the term "plaque" refers to a specific very dense cytoplasmic lamella adjacent to the membrane (Kelly, 1966).

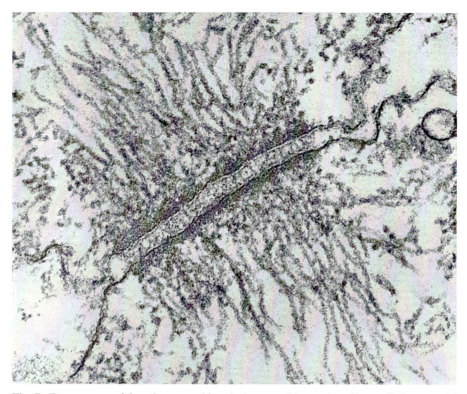

Fig. 7. Desmosome of larval newt epidermis in very thin section. Extracellular material is irregular and filamentous. The dense, desmosomal plaques subjacent to the membrane appear granular or fibrous. × 126,000. Courtesy of Douglas E. Kelly; from Kelly (1966)

C. Intracellular Fibrils

Fibrils are a major constituent of the most highly developed elaborate desmosomes and hence may be crucial elements for deducing the mechanism of desmosome formation. Currently, the concepts on the construction of macromolecular complexes from subunits are more highly developed for fibrils than for other subcellular structures such as membrane, glycocalyx and plaques. Characteristically, fibrils course through desmosomes rather than ending on them (Figs. 1, 6). The fibrils may come from deeper regions of cytoplasm and loop outwards at the desmosome (Pflugfelder and Schubert, 1965; Kelly, 1966) or run through the desmosome parallel to the surface membrane (Fawcett, 1958; Rogers, 1965; Kelly, 1966). No fibril has yet been shown actually to terminate at a desmosome — a feature expected if fibrils grew from a catalytic site on the desmosomal plaque (Mercer, 1965).

The attachment of fibers at internal sites along their length and the lack of precise order among the tangle of fibers in desmosomes are compatible with the idea that the accumulation of fibers depends upon their chance occurrence in the local region of cytoplasm recruited and condensed to form the attachment zone. Supporting a more specific mechanism are very high resolution electron micrographs (Kelly, 1966) which show extremely fine filaments connecting the fibrils to the desmosomes. Conceivably, these fine filaments could have pulled the loops of larger fibrils to the desmosomes from interior regions of cytoplasm.

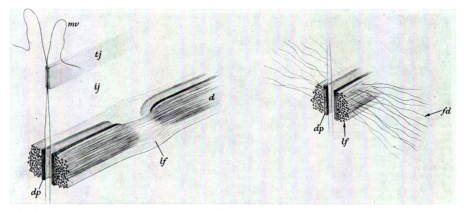

Fig. 8 a. Diagrammatic reconstruction of the terminal junction in the chick wing bud epithelium in transverse section. The drawing on the left shows all three parts, i.e. the tight junction (*tj*), the intermediate junction (*ij*) and a part of the desmosome chain (*d*) represented by two units. The dense plaques (*dp*) are separate for each unit, but the units are connected by the bundles of longitudinal fibres (*lf*) continuous between the units. — The drawing on the right shows the reconstruction of a desmosome unit together with the associated cytoplasmic fibers (*fd*) running roughly at right angles to the bundles of the longitudinal fibres (*lf*). (Approx. mag. × 100,000)

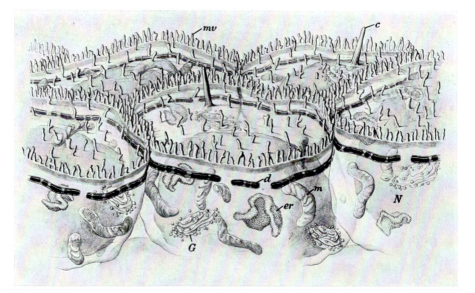

Fig. 8 b. Diagrammatic reconstruction of a few epithelial cells of an early chick limb bud showing the characteristic organelles, i.e. microvilli (*mv*), cilia (*c*), terminal junctions with chains of desmosomes (*d*), Golgi groups (*G*), endoplasmic reticulum (*er*), mitochondria (*m*) and nuclei (*N*). (Approx. mag. × 5,000) Fig. 8a and 8b courtesy of Dr. A. Jurand; from Jurand (1965)

Bundles of fibers running parallel to the plasma membrane may join many maculae into a band-like structure (Fig. 8). Since only favorably oriented sections (those parallel to the skeins) will reveal these "trains," this relationship between maculae is usually not obvious. However, tracts of longitudinal fibers are characteristic of adherent desmosomes and trains of maculae have been observed in many cell types (for example, JURAND, 1965; ROGERS, 1965; GORGAS, 1968). This indicates that desmosome formation is concerted in widely spaced areas of a cell even though desmosomes may appear discrete in electron micrographs.

D. Other Cytoplasmic Materials

A number of other cytoplasmic structures are occasionally related to cell junctions. The terminal web is perhaps the cytoplasmic structure best known for its

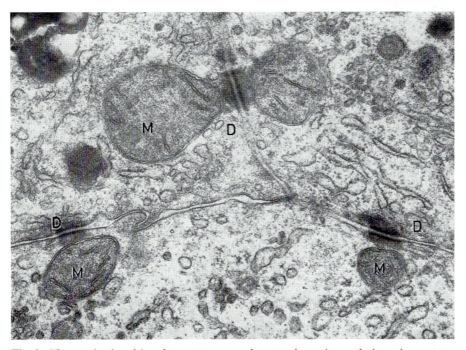

Fig. 9. Three mitochondrion-desmosome complexes and portions of three hepatocytes from a needle biopsy specimen obtained from a 51-year-old woman without known liver disease. D, desmosome; M, mitochondrion. × 35,000. Courtesy of Dr. IRMIN STERNLIEB; from STERNLIEB (1968)

association with desmosomes (SAUER, 1935; LEBLOND, PUCHTLER and CLERMONT, 1960). Less regularly, mitochondria have been reported to be intimately associated with desmosomes in a variety of embryonic tissues (DEANE and WURZELMANN, 1965a; DEANE, WURZELMANN and KOSTELLOW, 1966; and see STERNLIEB, 1968, for further references) (Fig. 9). Since mitochondria are thought to arise only from other mitochondria, one must view their occurrence as an example of recruitment. In

18*

tunicate endostyles, ciliary rootlets terminate at the lateral plasma membrane at or near desmosomes (OLSSON, 1962; LEVI and PORTE, 1964) (Fig. 10). This attachment frequently involves considerable bending of the rootlet which, in most cell types, is notably straight. Whether the desmosomal connection arises before or after rootlet

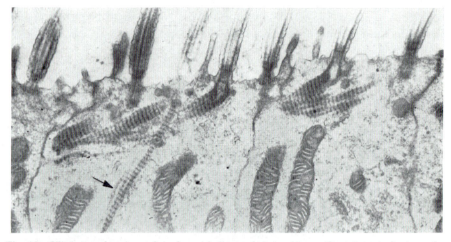

Fig. 10. Ciliary rootlets in endostyle epithelium of the tunicate *Ciona intestinalis*, inserting on plasma membrane just under junctional region. Occasional rootlets (arrow) are not attached to membrane and extend deep into cell. × 8,400

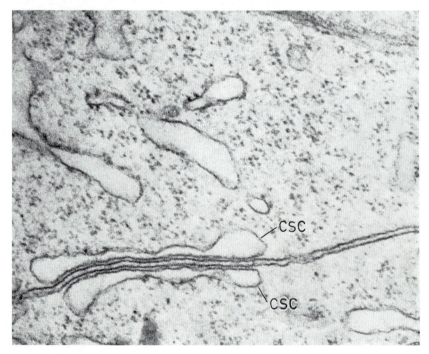

Fig. 11. Confronting subsurface cisternae (CSC) in apposed chick embryo neuroblasts. × 21,800. Courtesy of PEDDRICK WEIS; from WEIS (1968)

formation, there must be forces in the cells of the endostyle which systematically move cytoplasmic or membrane structures. Such forces are a prerequisite for generation of desmosomes by recruitment or by specific transport of precursors from distant sites of synthesis.

Cytoplasmic structures called "subsurface cisternae" are flattened vesicles, sometimes continuous with the endoplasmic reticulum, immediately subjacent to the plasma membrane (Fig. 11). They occur in a wide variety of cell types and are relevant to desmosome formation in two respects. First, they are sometimes associated with cell contact zones which do not have other elaborated cytoplasmic elements (COPELAND, 1966; SJÖSTRAND, ANDERSSON-CEDERGREN, and DEWEY, 1958). Thus, they may represent a distinct class of cell junction (BERGER, 1967). Second, they mimic several aspects of desmosome organization. The apposed membranes (representing opposite sides of the cisternae) may be fused in a tight junction, and they may show a close or a tight junctional relation with the plasma membrane (ROSENBLUTH, 1962). Subsurface cisternae also may be "confronting" (WEIS, 1968) in adjacent cells, that is paired and in register beneath the plasma membranes of adjacent cells (Fig. 11). In this configuration they show the existence of forces or influences which embrace the periphery of both cell surfaces, just as does symmetrical desmosome formation.

V. Models of Development

Several ideas have been devised to explain how junctions develop. A complete theory would rationalize the specific locations of junctions and the morphology of these specializations. However, most speculations on junction formation are restricted to only one aspect or the other.

A. In situ Synthesis

One straightforward explanation for the local occurrence of a structure is that it is made from locally synthesized components. In the case of desmosomes, this possible mechanism has some but not extensive support. Indirect evidence for local formation involves the observation that mitochondria may be found very close to desmosomes in developing, but not adult, epithelia (DEANE and WURZEL-MANN, 1965a; DEANE, WURZELMANN, and KOSTELLOW, 1966; STERNLIEB, 1968). These investigators have suggested that "young" desmosomes have an energy requirement perhaps associated with their construction. However, there are no concentrations of ribosomes, polysomes or rough endoplasmic reticulum in regions forming cell junctions (WARTIOVAARA, 1966). Autoradiographic experiments have yielded no evidence of noteworthy incorporation of any isotopically labeled compounds at these sites (RAMBOURG and LEBLOND, 1967, for example).

In line with the general absence of synthetic machinery from developing junctions, the ultrastructural organization of desmosomes does not suggest an obvious way for orderly *in situ* synthesis. For example, the extracellular matrix increases during desmosome elaboration yet in highly developed stages there is such an accumulation of intracellular materials it is hard to imagine locally synthesized extracellular materials passing through the membrane. Also during desmosomal development fibrils seem to accumulate gradually and their configurations pose serious problems with regard to *in situ* synthesis. Where radial and longitudinal fibers of a desmosome are interwoven, development involves an increase in the density of the fiber tangle. In properly oriented sections, the radial fibers approach the desmosome plaque and then loop away. It is difficult to account for such patterns by local synthesis.

The possibility that all organization of desmosomes arises by *in situ* synthesis is virtually untenable because no apparent mechanism exists by which such processes could be exactly synchronized spatially and temporally in adjacent cells. At least the first event must have its origin in the interaction of membranes of adjacent cells. If nothing more, such interaction would provide a time and spatial point at which synthesis could begin. Considering the complexity of desmosome organization, *in situ* formation is unlikely as the primary mechanism for development. No investigator has strongly defended such a mechanism and existing morphological data do not recommend it.

B. In situ Transformation

Local synthesis need not occur if desmosomes arise through the local transformation of some other cell structure. One of the earliest attempts to explain the origin of desmosomes stemmed from light microscopic observations on the relationship between mitotic division and terminal bar appearance (Schneider, 1902, 1908) (see Fig. 12). When an epithelial cell divides, the furrow begins at the basal surface and cuts upward towards the tissue surface. The midbody (daughter cell connection containing the mitotic spindle remnant near the end of the cell cleavage) is thus in the position within the tissue typical for apical desmosomes. Schneider (1902) expressed the belief that the midbody actually represented the young terminal bar connecting the two daughter cells. Sauer (1937) confirmed the plausibility of this proposal and explored it in greater detail. He indicated that after forming from

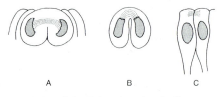

A   B   C

Fig. 12. Model of desmosome and junction-associated filaments (terminal web) arising through compaction and transformation of mitotic spindle. A—C represent successive stages in epithelial cell division. Redrawn after Sauer (1937)

the midbody, the terminal bar became more densely staining and more uniformly spread over the apical junction due to some continuing process. In a further study (Sauer, 1937) the terminal web also appeared to originate from the mitotic spindle.

The limit of resolution of the light microscope is insufficient to eliminate the possibility that the proteinaceous mitotic spindle and the terminal bar (which stains as though it were proteinaceous, Puchtler and Leblond, 1958) were simply in close proximity. Recently Allenspach and Roth (1967) have paralleled the studies of Sauer but at the electron microscopic level. Their observations leave no doubt as to the intimacy between the midbody and the desmosomes (Fig. 13). They suggest that during lateral furrowing, terminal bars between the dividing and adjacent cells are pulled inwards toward the midbody. However, the midbody and terminal bars clearly are adjacent rather than continuous with one another and there

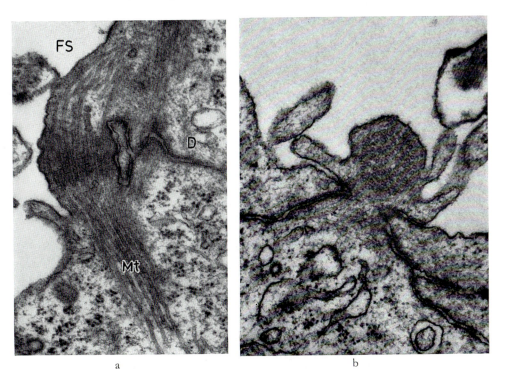

a   b

Fig. 13 a and b. Relation between desmosomes and midbody at end of mitosis in chick embryonic epithelia. (a) Longitudinal section through midbody with microtubules (Mt) compacted within it. Desmosome (D) is near basal edge of midbody. FS, free surface of epithelium. × 36,500. (b) Cross section through similar midbody. × 46,500. Courtesy of Dr. ALLAN L. ALLENSPACH; from ALLENSPACH (1967)

is no evidence that the midbody transforms into a desmosome. Even so, the midbody region is complicated geometrically, and the exact origin of the telophase junction there is not clear.

The idea that terminal bars form during mitosis is insufficient to account for the embryonic formation of secondary epithelia from mesodermal mesenchyme. SAUER (1937) discussed this problem and concluded that there must be some mechanism for terminal bar formation not related to mitosis or alternatively that "mesenchymous" tissues retain well-developed terminal bars during mitosis. Cell reaggregation experiments definitely show that desmosomes can form in the absence of mitosis while TRELSTAD, HAY, and REVEL (1967) have verified the persistence of desmosomes in mesenchyme.

Scattered reports (e.g., PATRIZI, 1967a) indicate the occasional presence of microtubules near cell junctions. Recently, KAVANAU (1965) has suggested that locally available microtubules are transformed into some of the materials of desmosomes, an idea also presented by Sandborn et al. (1965). From studies of regeneration of corneal epithelium Blümcke, Rode, and Niedorf (1969) have described how very fine (30 Å) fibrils assemble at sites and penetrate the membrane. Some remain in "tufts" from which hemidesmosomes are presumed to develop.

The observation that mitochondria are sometimes found associated with developing desmosomes raises the possibility that the mitochondria provide local concentrations of calcium ions which in turn precipitate cytoplasmic materials into desmosome constituents (Deane and Wurzelmann, 1965; Sternlieb, 1968).

C. Local Membrane Inhomogeneities

It has been proposed that certain discrete regions of membrane have unique physical or chemical properties which cause local transformation of cytoplasmic components into desmosomes. Although the morphological evidence indicates a continuity between the extracellular matrix of desmosomes and the adjacent extraneous coat, careful analysis shows that the extraneous coat and the cell membrane are by no means uniform over an entire cell. Numerous histochemical studies have shown enzyme activity associated with plasma membrane in certain parts of a cell and not others. For example, phosphatase activities in membranes may be restricted to the apex (Essner, Novikoff, and Masek, 1958) of a cell, or to the lateral surfaces (Kaye et al., 1966; Farquhar and Palade, 1966; Kaye and Pappas, 1965), i. e. in fields limited by the junctional complex. There are even reports of specific enzyme activity occurring along the entire surface of a cell with the exception of the junctional regions (Schmidt, 1968, p. 218; Bartoszewicz and Barrnett, 1964). Thus, it is possible that desmosomes represent unique sites in terms of specific membrane components even though their general composition is similar to that of adjacent areas of cell surface.

Weiss (1950, 1957) has interpreted many aspects of cell shape and behavior in terms of specific arrangements of membrane components. He has suggested that contact junctions between cells may involve microregions with increased permeability (as has been subsequently verified in many cases; see Loewenstein, 1966). The diffusion of extracellular constituents, perhaps ions, into the subjacent layers of cytoplasm could cause condensation of cytoplasmic materials. Reciprocal leakage from two cells might result in a symmetrical deposition of extracellular matrix. Mercer (1961) has added the suggestion that desmosomes may form at membrane areas "highly pierced" by pores or porous sites where adhesive mucopolysaccharides diffuse to the intercellular gap.

The need for *in situ* synthesis of other elements is circumvented by the assumption that the initially condensed cytoplasm acts autocatalytically for further condensation. Unfortunately, little information is available concerning the applicability of local catalysis to desmosome origin. Mercer (1965) has pointed out that intracellular fibrils could represent polymerization from catalytic sites on the membrane surface, but this particular suggestion is contradicted by the finding that the fibrils probably do not contact the membrane (Kelly, 1966) and do not terminate at desmosomes.

The origin of the postulated inhomogeneities is pivotal in considering the catalytic hypothesis. One possibility is put forth by MERCER (1964, p. 177) who states, "It seems necessary to assume that a pattern of sites ("holes", adsorbed enzymes, or porous "spots") exist in biological membranes which may be inherited from cell to cell independently of the nuclear genome." One might draw from this idea the possibility that areas of desmosome may split, stretch, or pinch in two or fuse but

Fig. 14. Scheme for origin of desmosomes through fragmentation. A representation of a cell surface at three times illustrating possible modes of fission. Sites could represent either actual desmosomes or precursor sites. Fragmentation of pre-existing structures, concomitant with continuing autocatalytic growth, could explain lack of "young" desmosomes in some tissues

not form *de novo* (see Fig. 14). This would explain the apparent lack of early developmental stages of desmosomes on cells of keratinizing epidermis, each cell of which must form dozens to hundreds of new desmosomes between each cell division. PATRIZI (1967a, b) also considers desmosomes to form only from permanently specialized areas of plasma membrane which retain their identity even after cells are dissociated and desmosomes are not visible; he notes that cell apposition is not enough to initiate desmosome formation, since in tissues many regions of closely apposed cell membranes remain free of desmosomes. An alternative possibility is that the presumed membrane heterogeneities are induced by the heterogeneities of a cell's microenvironments (see WEISS, 1957). This is attractive because it would relate desmosome formation to the tissue environment of a cell, a relation which is such a conspicuous aspect of desmosome location.

D. Conversion of Unspecialized Membrane

The similarity in basic composition of cell periphery at and between cell attachment zones has suggested that unspecialized membrane may be directly converted into contact membrane. Simple junctions provide more likely candidates for such transformation because these junctions seem to differ, morphologically, very little from normal membranes of opposed cells. Tight junctions might arise by the local removal of extracellular matrix and glycocalyx lining the two membranes (BENEDETTI and EMMELOT, 1968).

Tight junctions have essentially no stainable glycocalyx or mucopolysaccharide layer between the membranes (PEASE, 1966; RAMBOURG and LEBLOND, 1967). It cannot be presumed that they form at an embryonic stage at which no membrane coat is present because OVERTON (1969) has shown that in cells of chick embryos a well developed coat is present before tight junctions form.

If focal tight junctions originate through close apposition of two membranes, presumably one is not flat. The requirement for at least one of the surfaces to be pointed (i.e. have a very small radius of curvature in the contact region) appears essential in overcoming a variety of physico-chemical barriers (Pethica, 1961; Weiss, 1967; Curtis, 1967). There is evidence that in embryos, focal tight junctions arise mainly at the tips of narrow pseudopodia ("filopodia" or "microspikes") (Lesseps, 1963; Trelstad, Hay and Revel, 1967).

Minute regions of membrane contact have been observed in both embryonic and adult tissues (Trelstad, Hay, and Revel, 1967; Farquhar and Palade, 1963). It is likely that these represent the beginnings of tight junctions of larger extent such as zonulae occludentes. There are three ways in which an extension of focal tight junctions could occur. They could expand by 1. local increase in membrane area, 2. increase in the number of focal tight junctions until a zonula is essentially completely composed of contacting membranes, or 3. progressive tightening of membrane contacts at the periphery of a tight junction, in a zipper-like fashion (Schmitt, 1941). Zonulae of tight or close junctions commonly contain focal gaps, where membranes are not in contact. Such regions may be as small as a few millimicrons across. These appear as regions of non-contact which were entrapped during the process of zonula formation. They appear consistent with any of the three extension mechanisms mentioned above.

The force for the increase in size of incipient tight junctions could be the adhesiveness of the cells itself. One theory of morphogenetic movements and tissue

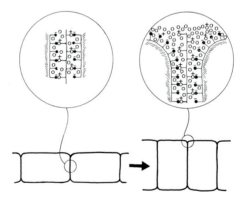

Fig. 15. Origin of contact surface by a "zipper" mechanism. Cells at left are in contact over a small area. Insert above schematizes an intercellular ionic molecular and monomolecular composition which favors adherence. In the abundance of such intercellular factors (right), free apical surface of the adjacent cells will be drawn together in "zipper" fashion, to form new contact surface. Redrawn after Schmitt (1941)

shape determination holds that the degree of mutual adhesiveness between adjacent epithelial cells determines the amount of contact surface between the cells. If adhesiveness increases during development, cells will become more columnar as illustrated in Fig. 15. This increase in contact has been described as a "zipper-like" phenomenon (Jacobson, 1962; Brown, Hamburger, and Schmitt, 1941; Mercer, 1961, p. 99; Schmitt, 1941; Gustafson, 1963).

Septate desmosomes (Fig. 16) could also arise by extracellular materials being partially eliminated or transformed. There is increasing evidence for a microscopic subunit structure of membranes, and the septate organization could simply reflect it (PEASE, 1962). The subunits would be aligned or crystallized between the two cells. This would allow an elementary means of explaining the reported variations in septate desmosome organization (in some, the septa appear as parallel lines in tangential section, while in others, they appear as hexagonal combs or discrete particles)

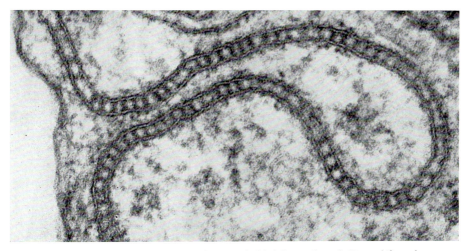

Fig. 16. Septate desmosome from epidermis of third instar larval *Drosophila melanogaster*. × 146,000. Courtesy of CLIFTON A. POODRY, University of California, Irvine

and dimensions (septal spacings range from 80 Å to 300 Å), since membrane substructure could vary from one cell type to another. This interpretation also provides a logical connection between septate junctions, gap junctions, (where an intercellular gap of 20—50 Å is visible) and tight junctions. All of these have hexagonal or square organizations as seen in tangential section (REVEL and KARNOVSKY, 1967; BENEDETTI and EMMELOT, 1965).

One recent study which extends this interpretation is that of WISSIG and GRANEY (1968) and GRANEY (1968). In ileal cells of the suckling rat, the outer lamella of the apical membrane is discontinuous (appearing "hyphenated" in cross section), with a particle of membrane coat on each "hyphen" (Fig. 17). In tangential section the membrane appears distinctly ridged; i.e., the membrane particles are aligned, probably in a rectangular grid. The ileal membrane thus closely resembles one membrane of a septate desmosome (Fig. 16), where some workers believe the outer lamella is hyphenated in register with the septa.

BULGER and TRUMP (1968) have reported that a variety of vertebrate membranes may be induced to form septate desmosome-type relations by treatments which are harmful to the tissues. These membranes, including both mitochondrial envelopes and plasma membranes, do not normally form such structures. This strongly supports the contention of PEASE (1962) and of BULGER and TRUMP (1968) that septate desmosomes represent a transformed state of ordinary membranes, with the septa representing aligned normal components of the membrane surface.

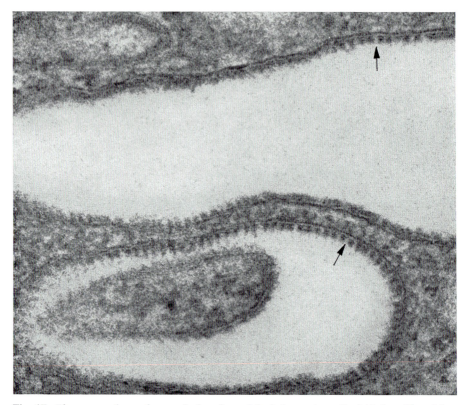

Fig. 17. Plasma membrane from apical surface of ileal absorptive cell of 14 day old suckling rat. At the points indicated by arrows, the plane is normal to the membrane and to rows of particles on its luminal face. At these points also, "hyphenation" of the outer unit membrane leaflet is apparent. × 190,000. Courtesy of Dr. S. L. Wissig; from J. Cell Biol. 39, 564 (1968)

E. Lateral Recruitment

The structure and history of desmosomes is highly suggestive of many components being compressed during development. Since most components are surface-associated materials, desmosomes might arise through progressive accumulation of materials which are recruited from adjacent portions of cell surface.

We have recently proposed a rather specific recruitment model which accounts for the concordant development of intercellular and extracellular components of two adjacent cells (Campbell, 1967) (see Fig. 18). The causative agent, in this model, is displacement of the cell membrane. If the membranes of two cells are adherent and mobile, their dynamics must be concerted. Movement of the membrane could affect subadjacent and supraadjacent material at precisely the same region of both cell peripheries.

Let us assume that there are two adherent cells whose membranes are not static but are moving laterally through the area of contact. This might occur if membrane were continuously deposited in the contact surface and removed from the free surfaces by pinocytosis, a process demonstrated by KAYE et al. (1962) (see Fig. 19). The resulting membrane displacement extends through a unique region (the junctional zone where the membranes diverge) and it is reasonable to suppose that membrane materials will be altered here. In particular, adhesive

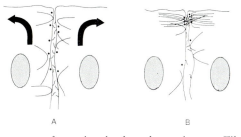

A   B

Fig. 18. Model of desmosome formation by lateral recruitment. Filaments and particles associated with membranes of adjacent epithelial cells (left) are considered to be compacted at the junctional region. One mechanism for such compaction could be systematic membrane displacement in directions indicated by arrows. Redrawn from CAMPBELL (1967)

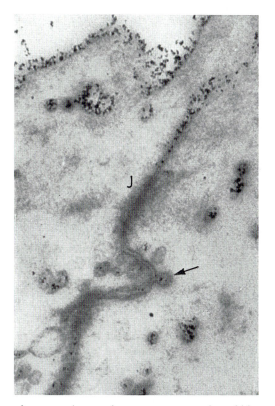

Fig. 19. Indication of systematic membrane movements in rabbit corneal epithelium. Membrane, stained with Thorotrast (dark granules, top) is withdrawn from free apical surface, transported past junctional complex (J) in the form of cytoplasmic vesicles, and redeposited (arrow) on lateral contact cell surface. × 48,000. Electron micrograph courtesy of Dr. GORDON I. KAYE, Columbia University (J. Cell Biol. 12, 481 [1962])

material entangled across the intercellular space will be unable to continue to follow the membranes as they diverge. Consequently, it will accumulate at the contact margin. Components of the membrane attached to the adhesive substance will also be impeded in their movements and cytoplasmic elements bound to them will accumulate directly under the extracellular accumulation. We have described elsewhere that this model of lateral recruitment is compatible with the substructure, morphology, and cellular location of junctions (CAMPBELL, 1967). For example, junctional complexes would be expected to form at margins of cell contact and this is their typical location. In epithelia they occur as a ring around the apex of each cell. They are also found at the edges of defined intercellular gaps such as the bile canaliculus (BLOOM and FAWCETT, 1962) and the neural canal (WATTERSON, 1965).

The crux of this hypothesis lies in the existence of systematic movements of membrane. If membrane is diplaced out of a region where its surface elements adhere to those of an adjacent cell some form of membrane alteration must occur. If some elements are less integral parts of the membrane than others, then selective accumulation is to be expected. From what we know about the dynamic structure of cells, extensive movements of the plasma membrane take place in diverse cell types (MARCUS, 1962; SCHAFFER, 1965). Direct observations on living cells reveal that their membranes are in continuous motion. From experimental studies MARCUS (1962) has shown that when cultured cells are infected with Newcastle disease virus hemagglutinating elements appear first at the tips of pseudopodia and then spread by systematic membrane displacement to other regions of the cell.

A variety of processes are known which locally affect the surface area of membrane; pinocytosis and vesicle formation at the expense of the plasma membrane, fusion of the membrane of cytoplasmic vesicles with the membrane of the cell, and possibly interstitial addition or dissolution of structural molecules of membrane. It would be surprising if all of these processes (plus others undoubtedly left to be discovered) exactly balanced each other in every small region or surface area. In fact these processes mentioned above typically are relegated to different surfaces of differentiated cells, enhancing the probability of systematic membrane movement coordinated with the overall organization of the cell. An excellent study of such movement is provided by KAYE et al. (1962) (see Fig. 19).

F. Developmental Models: Conclusions

Junction morphology is varied. No one mechanism can be expected to account for the development of all adhesive junctions, or even of the whole history of one junction. The models discussed above all have commendable features.

A remarkable feature of elaborate desmosomes is that materials external and internal to the cell membrane are organized in concert. The membrane either is able to transmit the signal for desmosome formation from one side to the other or itself acts as the organizing element. Moreover, a desmosome between two similar cells develops with a remarkable degree of symmetry. Either some precise feedback mechanism governs the elaboration of one cell surface in accordance with its complement or the causative factor is quantitatively similar at both cell peripheries.

The only cases where asymmetrical desmosomes have been reported are where the two participating cells are developmentally different. When the cells are very different, asymmetry is the rule. The most extreme example is the hemidesmosome (Fig. 21), where one half is acellular. Other examples include the synapse and desmosomes attaching different strata of keratinizing epithelia (Fig. 20). Desmosomes

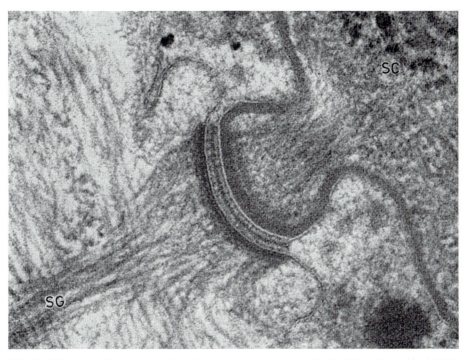

Fig. 20. "Composite", asymmetrical desmosome between cornified (SC) and granular (SG) cells at the base of the cornified layer in stratefied epidermis of frog skin. The two sides of the desmosome differ in geometry, appearance of cytoplasm, and thickness of plasma membrane. × 135,000. Courtesy of Dr. MARILYN G. FARQUHAR; from J. Cell Biol. 26, 263 (1965)

between cells of similar developmental history, such as different cell types in an epithelium, may be only subtly asymmetrical or apparently symmetrical. Desmosomes which are asymmetrical but in register betwen dissimilar cells suggest that cell coordination is adequate but that desmosome construction varies with the state of differentiation of a cell.

One should not understimate the variety of possible activities by which developing desmosomes and dynamic cell associations might interact. Epithelia are frequently under physical tension and this may in turn affect desmosome organization which resists such tension. For example, during post-mitotic radial growth of an epithelial cell the encircling junctional complex must be under circumferential tension which may well affect the disposition of the tangential desmosome fibers. Membrane displacements also must affect contact geometries and perhaps organizations. The phenomenon of "contact inhibition" of cell movement, by which contact involving minute regions of adjacent cells' surfaces immobilizes all nearby cell surface, indicates the profundity of cell surface interactions which must occur. These must at least influence cell junctions and might be determining factors for the generation and maintenance of junctions.

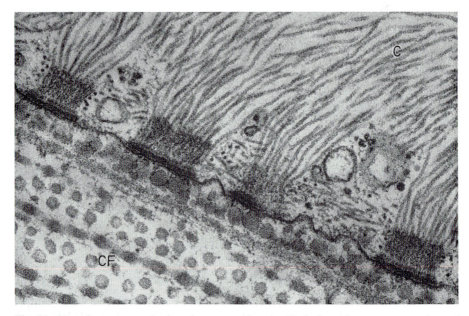

Fig. 21. Hemidesmosomes in larval newt epidermis. Each hemidesmosome consists of membrane, a dense subjacent plaque, tangential filaments, and perpendicular fibers which course through the cytoplasm (*C*). (*CF*), Collagen fibers in basement membrane of epidermis. Note that the extracellular region adjacent to the hemidesmosome is also organized locally. × 70,500. Courtesy of Douglas E. Kelly; from Kelly (1966)

VI. Continuity and Permanence of Desmosomes

A. Stability and Permanence

Desmosomes in general appear to be permanent cell structures. In most cells where they are systematically located, they are present regardless of the age of the animal. Desmosomes probably remain intact during cell division. Light microscopy indicates that the terminal bars of epithelial cells remain intact during mitosis (Sauer, 1937), as would seem necessary if the tissue is to remain coherent during growth. Persistence of desmosomes through mitosis has been investigated in a variety of cell types with the electron microscope (several chick epithelia, Allenspach and Roth, 1967; myocardium, Manasek, 1968b; amphibian epidermis, Buck and Krishan, 1965).

The widespread assumption that desmosomes are permanent structures has not gone unchallenged, however. Petry and his collaborators (Petry, Overbeck, and

VOGELL, 1961; PETRY, 1962) have investigated the cell contacts in successive layers of several renewing epithelia (such as the vaginal epithelium) and concluded that desmosomes increase and decrease in number, size, and compactness during the renewal cycle. They consider desmosomes to be temporary entities which continually change according to the cell's activities.

Certainly one source of the supposition that desmosomes are permanent is their physico-chemical stability. They survive purification procedures devised to

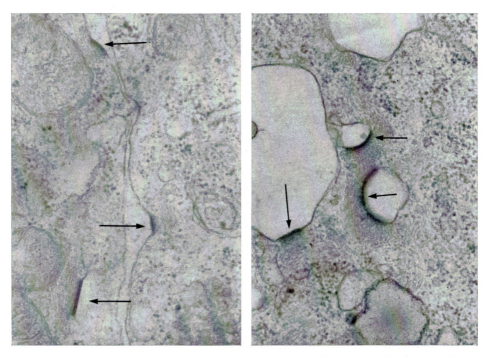

Fig. 22. Fate of desmosomes after tissue disaggregation. Chick *area pellucida* tissue following trypsin treatment. Arrows indicate desmosome halves at the cell surface, partially withdrawn from surface, or in vacuoles cut off from the surface. × 38,000. Courtesy of Dr. JANE OVERTON; from OVERTON (1968)

isolate plasma membranes, and often are more resistant than surrounding membrane (see BENEDETTI and EMMELOT, 1968; MILLINGTON, CRITCHLEY, and TOVELL, 1968). Tight junctions are the only portion of the hepatic plasma membrane not solubilized by 8% deoxycholate (BENEDETTI and EMMELOT, 1968). After cells have been separated by mechanical disruption of a tissue both membranes of a tight junction are found on one cell (MUIR, 1965), indicating that the junction is stronger than the membrane.

Desmosomes are also strikingly persistent in tissues following damage by hypotonic fluids (STEHBENS, 1966; PANNESE, 1968), detergents, temperature extremes (PEARSON, 1965), and enzymes (WEISS, 1957; KAHL and PEARSON, 1967a, b). The part of desmosomes most susceptible to chemical agents is the extracellular matrix.

In the absence of calcium ions, or in the presence of trypsin, or both, cells loosen from one another through symmetrical cleavage of the desmosome. These treatments affect only the extracellular component of desmosomes. The loosened cells frequently are viable and have normal-appearing "half-desmosomes" at their surfaces (OVERTON, 1962; HAYS, SINGER, and MALAMED, 1965; KAHL and PEARSON, 1967a) (Fig. 22). OVERTON (1968) studied the fate of these desmosomes. She found that shortly after cell disaggregation desmosome-like plaques were visible on the membranes of intracellular vesicles, indicating that the desmosome and the surrounding membrane had been withdrawn from the cell surface as a pinocytotic vesicle. These structures were recognizable for many hours and even preserved the flatness of the desmosome so that the vesicles were irregular in outline.

B. Disappearance

Despite the general permanence of cell attachments there are specific and important developmental and pathological circumstances during which they disappear:

1. Embryogenesis

Desmosomes disappear from certain cells during tissue differentiation especially when epithelial cells become nonepithelial. LYSER (1964) has described the withdrawal of neural tube cells from the epithelium lining the neural canal to become neuroblasts. The cells originally are bound into the epithelium by means of a junctional complex at the neural canal end (apex) of the cell. Immediately prior to withdrawal, the apical portion of the presumptive mesoblast becomes drawn out into a fine column still attached at the tip by the junctional complex. The process whereby contact with the neural canal was actually lost has not been described. Presumably either the junctional complex loosens and disappears, or else the cell breaks, leaving behind a minute, bound, apical portion. Other tissue transformations involving a similar histological process include the fragmentation of the vertebrate somite (the cells of which originally have a junctional complex around the apices facing the myocoel, TRELSTAD, HAY, and REVEL, 1967), ingressive movements during vertebrate gastrulation and neurulation (BALINSKY and WALTHER, 1961; PERRY and WADDINGTON, 1966) and primary mesenchyme formation in sea urchin embryos (GIBBINS, TILNEY, and PORTER, 1969; TILNEY and GIBBINS, 1969).

2. Tissue Wounding and Regeneration

Wound healing involves migration of epidermal cells which had been strongly attached to the basement membrane through hemidesmosomes. WEISS and FERRIS (1956) found that within minutes or hours of wounding the cells detach by a cleavage of the hemidesmosomes just external to the plasma membrane. The hemidesmosomal portion of the cell retains its structure throughout the cell migration but its eventual fate is unknown.

Regeneration by blastema formation involves the transformation of highly differentiated, adherent cells into masses of loose cells. Here again desmosome loss is indicated but the microscopic transitions have not been described.

3. Disease

The most abundantly documented pathological states involving loss of desmosomal attachment are skin diseases. Acantholysis is a condition, of multiple origin, in which the epidermis loses its compact histological structure and fluid spaces (blisters) appear. WILGRAM and CAULFIELD (1965) studying a variety of acantholytic diseases, reported graded series of images presumed to represent desmosome breakdown. The decrease in density of extracellular and intracellular membrane-associated plaques is closely correlated with a diminution of skeins of tonofilaments throughout the cell. The compact and regular filament bundles also give way to masses of poorly organized filaments filling the cell. These investigators suggest that in some cases desmosome degeneration occurs first, and that filament disarray (or poor development) is a result (see also MERCER, 1961, and WILGRAM, CAULFIELD and MADGIC, 1964); and that in other cases a fault in the formation of fibers impedes the normal maturation of desmosomes. However, in most of the cases they investigate, it is difficult to distinguish between lack of desmosome development and degeneration of desmosomes that had already formed. One curious and unexplained finding is that even in areas of skin showing advanced acantholytic degeneration, the hemidesmosomes between the basal cells and the basement lamella appear normal. Either the environment of these structures differs in important ways from that of the desmosomes between cells or the maintenance of the two types of attachments must depend upon different factors.

C. Mobility on Cell Surface

A problem related to the permanence of desmosomes is the extent to which desmosomes, once formed, are "fixed" to particular sites on the cell. It is commonly supposed that since they act as attachment devices desmosomes represent stationary points of the cell membrane. In an inactive epithelium there may be no reason to suppose otherwise. Yet, the structure and activities of certain cells indicate a degree of mobility to cell contacts. Several examples will be reviewed.

The bile canaliculus is a channel between adjacent cell surfaces, flanked on each side by a tight junctional complex. In the rat the canaliculi are typically lined by three or more cells so that their walls are tripartite as seen in cross section. WOOD (1965) has shown that they first develop as channels between only two cells, and that during late fetal life these channels rearrange so that more hepatocytes are involved. Junctional complexes are present throughout this period and probably must undergo changes in disposition. A similar situation during the development of sweat glands has been described by ELLIS (1968).

Merocrine secretion of cells involves fusion of cytoplasmic vesicles with the plasma membrane. This can result in enormous expansion of the secretory surface of the cell. During recovery from secretion this surface is reduced to its original size. Throughout the secretory cycle junctional complexes are found to be situated around the apex of the cell. This is an argument for the immobility of contacts, with the excess membrane being reabsorbed from the apex either by pinocytosis or dissolution (HOKIN, 1968). However, recovery could result from a shift of the extra membrane to other (lateral) regions of the cell periphery. If the junctional complexes were fixed to the membrane they would be displaced vis-a-vis their

19*

location in the tissue. It would be necessary for new complexes to be formed at successively higher levels to compensate for the movement. If membrane could slip through the junction in some fashion, the desmosome might retain its histological position although migrating relative to the surrounding membrane.

The idea that membrane can shift from one cell surface to another in some fashion is an old one (BROWN, HAMBURGER, and SCHMITT, 1941; SCHMITT, 1941) and should be given serious attention.

Many cells have desmosomes distributed irregularly or in bands along their lateral contact surfaces. There is no reason to view these as fixed points that membrane activities would not displace. Junctional complexes in the gastric mucosa undergo systematic movements during development, from the middle of the cell to the apex (OVERTON and SHOUP, 1964). Intercalated discs in cardiac muscle also arise from lateral desmosomes (FAWCETT and SELBY, 1958; MUIR, 1957). Maturation requires changes in cell shape and desmosome position so that the discs are at functional ends, not sides, of cells. It also appears that intercalated discs may fuse and fragment. Thus, desmosomes are not necessarily formed at the sites where they are observed, a further complication to understanding desmosome formation. It will almost certainly require more than straight electron microscopic observation to reveal the role of mobility of desmosomes in their formation. In an ultrathin section, problems of section orientation are so critical that it is difficult to determine the absence of a junctional complex.

The problem of what constitutes mobility of a desmosome is illustrated by the migration of cells from an epithelium. BALINSKY and WALTHER (1961) describe the migration of fixed epithelial cells from chick epithelial blastodisc at the primitive streak to become mesoblast cells. The cells first become flask shaped with the bulk of the cytoplasm at the basal end of the cell. (This is typical in vertebrate gastrulation.) The apical neck becomes increasingly drawn out leaving the junctional complex as a very narrow apical crown but still adherent to the epithelium. There are indications that the apical surface is progressively removed as pinocytotic vesicles. The final escape from the epithelium remains enigmatic. Should the desmosomes be considered to have migrated over the cell surface or did the cell slip out of a stationary junctional complex — and perhaps stationary membrane as well?

VII. Significance of the Origin and Continuity of Desmosomes

In closing, it is pertinent to comment on the relationship between the genes, the formation of desmosomes and cellular organization. The genetic implications of desmosomes are two-fold. On the one hand desmosomes are formed by cell activities which are ultimately determined by genes. On the other hand the location and specificity of cell junctions are key elements in the specificity of many histological and cytological characteristics such as tissue permeability, transmission of electrical signals, "recognition" of cells during embryogenesis, tissue differentiation and the shape, polarity and internal organization of cells.

Various authors have been tempted to place specific cell attachments as direct intermediates in the translation of genes into tissue functions. The essence of this viewpoint for histogenesis would be the following. In an aggregate of cells, genes for particular types of desmosomes are activated. The gene products are assembled into these desmosomes which cause the cells to adhere in a geometrical pattern so as to assume a particular tissue arrangement. However, one can make an equally strong case for an almost exact reversal of this presumed direction of cause and effect, since the formation of junctions may be a *result* of specific adhesion of dynamic cells. Also, there is more compelling evidence that cell contacts can signal a change in gene expression (e.g., contact inhibition) than that gene expression can signal a change in cell contact. Of course, desmosomal attachments unquestionably play a role in forming and maintaining specific characteristics of tissues. But, a simple cause and effect relationship oriented in a fashion that localized subcellular organization *causes* supracellular organization appears to be insufficient for sophisticated appreciation of specific tissue formation.

Cause and effect between desmosomes and intracellular organization also remains problematical. Various authors recognize desmosomes as key structures in organizing the cell. The most characteristic organizing effect of desmosome formation relates to the organization of fibrils and matrix between fibrils. Formation of the terminal web, incipient organization of myofilaments in cardiac muscle, and keratinization in skin may represent extensions of the organizing of fibers and matrix represented by desmosomes in less specialized cells.

Be that as it may, the involvement of cellular organization in formation of desmosomes appears to be no less important than the reciprocal role of desmosomes in generating cytological order. If desmosomes organize the fibers of keratinizing epithelium, the latter is essential in maintaining and perhaps locating desmosomes. Again, even if the position of terminal bars maintains the shape and polarity and distribution of fibers of columnar epithelial cells, it is probably the polarity of the cell and differential behavior at the two ends that causes desmosomes to form as rings perpendicular to the axis of the cell. Desmosomes may represent the most visible cytoplasmic element in the complex feedback relationship that causes cells and aggregates of cells to specialize. It is intriguing that we cannot yet sort out the direction of causality in the events of even this aspect of cellular differentiation.

References

ALLENSPACH, A. L., ROTH, L. E.: Structural variations during mitosis in the chick embryo. J. Cell Biol. **33**, 179—196 (1967).

BALINSKY, B. I., WALTHER, H.: The immigration of presumptive mesoblast from the primitive streak in the chick as studied with the electron microscope. Acta Embryol. Morph. Exp. **4**, 261—283 (1961).

BARTOSZEWICZ, W., BARRNETT, R. J.: Fine structural localization of nucleoside phosphatase activity in the urinary bladder of the toad. J. Ultrastruct. Res. **10**, 599—609 (1964).

BENEDETTI, E. L., EMMELOT, P.: Hexagonal array of subunits in tight junctions separated from isolated rat liver plasma membranes. J. Cell Biol. **38**, 15—24 (1968).

BERGER, E. R.: Subsurface membranes in paired cone photoreceptor inner segments of adult and neonatal *Lebistes* retinae. J. Ultrastruct. Res. **17**, 220—232 (1967).

BLOOM, W., FAWCETT, D. W.: A Textbook of Histology, 720 pp., Philadelphia: Saunders 1962.

BLÜMCKE, S., RODE, J., NIEDORF, R.: Formation of the basement membrane during regeneration of the corneal epithelium. Z. Zellforsch. **93**, 84—92 (1969).

BRIGHTMAN, M. W., PALAY, S. L.: The fine structure of ependyma in the brain of the rat. J. Cell Biol. **19**, 415—439 (1963).

Brightman, M. W., Reese, T. S.: Junctions between intimately apposed cell membranes in the vertebrate brain. J. Cell Biol. **40**, 648—677 (1969).

Brown, M. G., Hamburger, V., Schmitt, F. O.: Density studies on amphibian embryos with special reference to the mechanism of organizer action. J. exp. Zool. **88**, 353—372 (1941).

Buck, R. C., Krishan, A.: Site of membrane growth during cleavage of amphibian epithelial cells. Exp. Cell Res. **38**, 426—428 (1965).

Bulger, R. E., Trump, B. F.: Occurrence of repeating septate subunits between apposed cellular membranes. Exp. Cell Res. **51**, 587—594 (1968).

Campbell, R. D.: Desmosome formation: An hypothesis of membrane accumulation. Proc. nat. Acad. Sci. (Wash.) **58**, 1422—1429 (1967).

Challice, C. E., Edwards, G. A.: Some observations on the intercalated disc. Proc. Europ. Reg. Conf. Electron Microscopy, Delft **2**, 774—777 (1960).

Chapman, G. B., Dawson, A. B.: Fine structure of the larval anuran epidermis, with special reference to the figures of Eberth. J. biophys. biochem. Cytol. **10**, 425—435 (1961).

Copeland, E.: Septate desmosomes and juxtaposition membranes. J. Cell Biol. **31**, 24 A (1966).

Curtis, A. S. G.: The cell surface: Its molecular role in morphogenesis, 405 pp. London: Academic Press 1967.

Dan, K.: Cyto-embryology of echinoderms and amphibia. Intern. Rev. Cytol. **9**, 321—367 (1960).

Deane, H. W., Wurzelmann, S.: Mitochondrial-desmosome complexes in maturing columnar epithelia in organs of the male reproductive tract. J. Cell Biol. **27**, 131 A (1965a).

— — Electron microscopic observations on the postnatal differentiation of the seminal vesicle epithelium of the laboratory mouse. Amer. J. Anat. **117**, 91—133 (1965b).

Easty, G. C., Mercer, E. H.: An electron microscope study of model tissues formed by the agglutination of erythrocytes. Exp. Cell Res. **28**, 215—227 (1962).

Ellis, R. A.: Eccrine sweat glands: electron microscopy, cytochemistry and anatomy. In: Handbuch der Haut- und Geschlechtskrankheiten, I/1, 224—266 (1968).

Essner, E., Novikoff, A. B., Masek, B.: Adenosinetriphosphatase and 5-nucleotidase activities in the plasma membrane of liver cells as revealed by electron microscopy. J. biophys. biochem. Cytol. **4**, 711—716 (1958).

Farquhar, M. G., Palade, G. E.: Junctional complexes in various epithelia. J. Cell Biol. **17**, 375—412 (1963).

— — Functional organization of amphibian skin. Proc. nat. Acad. Sci. (Wash.) **51**, 569—577 (1964).

— — Adenosine triphosphatase localization in amphibian epidermis. J. Cell Biol. **30**, 359—379 (1966).

Fawcett, D. W.: An Atlas of Fine Structure, pp. 365—382. New York: Saunders 1966.

— Selby, C. C.: Observations on the fine structure of the turtle atrium. J. biophys. biochem. Cytol. **4**, 63—72 (1958).

Gibbins, J. R., Tilney, L. G., Porter, K. R.: Microtubules in the formation and development of the primary mesenchyme in *Arbacia punctulata*. I. The distribution of microtubules. J. Cell Biol. **41**, 201—226 (1969).

Gorgas, K.: Über Fibrillärstrukturen im Nebennierenmark von Haus- und Wildmeerschweinchen (*Cavia aperea* f. porcellus L. und *Cavia aperea tschudii* Fitzinger). Z. Zellforsch. **87**, 377—388 (1968).

Graney, D.: Ultrastructure of the apical plasma membrane of intestinal lining cells. Anat. Rec. **148**, 373—374 (Abstract) (1964).

Graney, D. O.: The uptake of ferritin by ileal absorptive cells in suckling rats. An electron microscope study. Amer. J. Anat. **123**, 227—254 (1968).

Grimley, P. A., Edwards, G. A.: The ultrastructure of cardiac desmosomes in the toad and their relationship to the intercalated disc. J. biophys. biochem. Cytol. **8**, 305—318 (1960).

Gustafson, T.: Cellular mechanisms in the morphogenesis of the sea urchin embryo. Exp. Cell Res. **32**, 570—589 (1963).

GUSTAFSON, T., WOLPERT, L.: Cellular movement and contact in sea urchin morphogenesis. Biol. Rev. **42**, 442—498 (1967).

HAYS, R. M., SINGER, B., MALAMED, S.: The effect of calcium withdrawal on the structure and function of the toad bladder. J. Cell Biol. **25**, 195—208 (1965).

HICKS, R. M.: The function of the golgi complex in transitional epithelium. Synthesis of the thick cell membrane. J. Cell Biol. **30**, 623—643 (1966).

HOKIN, L. E.: Dynamic aspects of phospholipids during protein secretion. Intern. Rev. Cytol. **23**, 187—208 (1968).

HORSTMANN, E., KNOOP, A.: Elektronenmikroskopische Studien an der Epidermis. I. Rattenpfote. Z. Zellforsch. **47**, 348—362 (1958).

JACOBSON, C.-O.: Cell migration in the neural plate and the process of neurulation in the axolotl larva. Zool. Bidrag (Uppsala), **35**, 433—449 (1962).

JURAND, A.: Ultrastructural aspects of early development of the fore-limb buds in the chick and the mouse. Proc. roy. Soc. Ser. B. Biol. Sci. **162**, 387—405 (1965).

KAHL, F. R., PEARSON, R. W.: Ultrastructural studies of experimental vesiculation. I. Papain. J. Invest. Dermatol. **49**, 43—60 (1967a).

— — Ultrastructural studies of experimental vesiculation. II. Collagenase. J. Invest. Dermatol. **49**, 616—631 (1967b).

KALLENBACH, E., CLERMONT, Y., LEBLOND, C. P.: The cell web in the ameloblasts of the rat incisor. Anat. Rec. **153**, 55—70 (1965).

KARRER, H. E.: Cell interconnections in normal human cervical epithelium. J. biophys. biochem. Cytol. **7**, 181—183 (1960).

KAVANAU, J. L.: Structure and function in biological membranes. Vols. I, II. 760 pp. San Francisco: Holden Day 1965.

KAYE, G. I., PAPPAS, G. D.: Studies on the ciliary epithelium and zonule. III. The fine structure of the rabbit ciliary epithelium in relation to the localization of ATPase activity. J. Microscopie **4**, 497—508 (1965).

— PAPPAS, G. D., DONN, A., MALLETT, N.: Studies on the cornea II. The uptake and transport of colloidal particles by the living rabbit cornea *in vitro*. J. Cell Biol. **12**, 481—501 (1962).

— WHEELER, H. O., WHITLOCK, R. T., LANE, N.: Fluid transport in the rabbit gallbladder. A combined physiological and electron microscopic study. J. Cell Biol. **30**, 237—268 (1966).

KELLY, D. E.: Fine structure of desmosomes, hemidesmosomes, and an adepidermal globular layer in developing newt epidermis. J. Cell Biol. **28**, 51—72 (1966).

— Myofibrillogenesis and Z-band differentiation. Anat. Rec. **163**, 403—425 (1969).

— LUFT, J. H.: Fine structure, development, and classification of desmosomes and related attachment mechanisms. Sixth Intern. Congr. Electron Microscopy, Kyoto, 401—402 (1966).

LEBLOND, C. P., PUCHTLER, H., CLERMONT, Y.: Structures corresponding to terminal bars and terminal web in many types of cells. Nature (Lond.) **186**, 784—788 (1960).

LESSEPS, R. J.: Cell surface projections: Their role in the aggregation of embryonic chick cells as revealed by electron microscopy. J. exp. Zool. **153**, 171—182 (1963).

LEVI, C., PORTE, A.: Ultrastructure de l'endostyle de l'ascidie *Microcosmus claudicans* Savigny. Z. Zellforsch. **62**, 293—309 (1964).

LISTGARTEN, M. A.: The ultrastructure of human gingival epithelium. Amer. J. Anat. **114**, 49—69 (1964).

LOEWENSTEIN, W. R.: Permeability of membrane junctions. Ann. N. Y. Acad. Sci. **137**, 441—472 (1966).

LYSER, K. M.: Early differentiation of motor neuroblasts in the chick embryo as studied by electron microscopy. I. General aspects. Develop. Biol. **10**, 433—466 (1964).

MANASEK, F. J.: Embryonic development of the heart. I. A light and electron microscopic study of myocardial development in the early chick embryo. J. Morph. **125**, 329—365 (1968a).

— Mitosis in developing cardiac muscle. J. Cell Biol. **37**, 191—196 (1968b).

MARCUS, P. I.: Dynamics of surface modification in myxovirus-infected cells. Cold Spr. Har. Symp. quant. Biol. **27**, 351—365 (1962).

Mercer, E. H.: Keratin and keratinization. An essay in molecular biology. 316 pp. New York: Pergammon 1961.
— Protein synthesis and epidermal differentiation. In: The Epidermis (Ed.: William Montagna and Walter C. Lobitz), pp. 161—178. New York: Academic Press 1964.
— Intercellular adhesion and histogenesis. In: Organogenesis (Ed.: Robert L. Dehaan and Heinrich Ursprung), pp. 29—53. New York: Holt, Rinehart and Winston 1965.
Milhaud, M., Pappas, G. D.: Postsynaptic bodies in the habenula and interpeduncular nuclei of the cat. J. Cell Biol. 30, 437—441 (1966).
Millington, P. F., Critchley, D. R., Tovell, P. W. A.: A study on the isolation of intestinal brush borders in saline. Z. Zellforsch. 87, 401—408 (1968).
Muir, A. R.: An electron microscope study of the embryology of the intercalated disc in the heart of the rabbit. J. biophys. biochem. Cytol. 3, 193—202 (1957).
— Further observations on the cellular structure of cardiac muscle. J. Anat. 99, 27—46 (1965).
— Peters, A.: Quintuple-layered membrane junctions at terminal bars between endothelial cells. J. Cell Biol. 12, 443—448 (1962).
Odland, G. F., Reed, T. H.: Epidermis. In: Ultrastructure of normal and abnormal skin (Ed.: A. S. Zelickson), pp. 54—75. Philadelphia: Lea and Febiger 1967.
Olsson, R.: The relationship between ciliary rootlets and other cell structures. J. Cell Biol. 15, 596—599 (1962).
Overton, J.: Desmosome development in normal and reassociating cells in the early chick blastoderm. Devel. Biol. 4, 532—548 (1962).
— The fate of desmosomes in trypsinized tissue. J. exp. Zool. 168, 203—214 (1968).
— A fibrillar intercellular material between reaggregating embryonic chick cells. J. Cell Biol. 40, 136—143 (1969).
— Shoup, J.: Fine structure of cell surface specializations in the maturing duodenal mucosa of the chick. J. Cell Biol. 21, 75—85 (1964).
Pannese, E.: Temporary junctions between neuroblasts in the developing spinal ganglia of the domestic fowl. J. Ultrastruct. Res. 21, 233—250 (1968).
Patrizi, G.: Ricomparsa di desmosomi tra cellule amniotiche umane isolate cultivate in vitro. Sperimentale 117, 189—203 (1967a).
— Desmosomes in tissue cultures: Their reconstruction after trypsinization. J. Cell Biol. 35, 182A (Abstract) (1967b).
Paweletz, N.: Elektronenmikroskopische Untersuchungen von Aggregaten dissoziierter Tumorzellen. Virchows Arch. Abt. B Zellpathol. 2, 114—124 (1969).
Pearson, R. W.: Response of human epidermis to graded thermal stress. Arch. Environ. Hth 11, 498—507 (1965).
Pease, D. C.: Demonstration of a highly ordered pattern upon a mitochondrial surface. J. Cell Biol. 15, 385—389 (1962).
— Polysaccharides associated with the exterior surface of epithelial cells: Kidney, intestine, brain. J. Ultrastruct. Res. 15, 555—588 (1966).
Perry, M. M., Waddington, C. H.: Ultrastructure of the blastopore cells in the newt. J. Embryol. Exp. Morphol. 15, 317—330 (1966).
Pethica, B. A.: The physical chemistry of cell adhesion. Exp. Cell Res. Suppl. 8, 123—140 (1961).
Petry, G.: Desmosomen. Dtsch. med. Wschr. 87, 1012—1014 (1962).
— Overbeck, L., Vogell, W.: Sind Desmosomen statische oder temporäre Zellverbindungen? Naturwissenschaften 48, 166—167 (1961).
Pflugfelder, O., Schubert, G.: Elektronenmikroskopische Untersuchungen an der Haut von Larven- und Metamorphosestadien von Xenopus laevis nach Kaliumperchloratbehandlung. Z. Zellforsch. 67, 96—112 (1965).
Potts, M.: The attachment phase of ovoimplantation. Amer. J. Obstet. Gynec. 96, 1122—1128 (1966).
Puchtler, H., Leblond, C. P.: Histochemical analysis of cell membranes and associated structures as seen in the intestinal epithelium. Amer. J. Anat. 102, 1—31 (1958).
Rambourg, A., Leblond, C. P.: Electron microscope observations on the carbohydrate-rich cell coat present at the surface of cells in the rat. J. Cell Biol. 32, 27—53 (1967).

RAMBOURG, A., HERNANDEZ, W., LEBLOND, C. P.: Detection of complex carbohydrates in the golgi apparatus of rat cells. J. Cell Biol. **40**, 395—414 (1969).

RASH, J. E., SHAY, J. W., BIESELE, J. J.: Urea extraction of Z bands, Intercalated Disks, and Desmosomes. J. ultrastruct. Res. **24**, 181—189 (1968).

REVEL, J. P., KARNOVSKY, M. J.: Hexagonal array of subunits in intercellular junctions of the mouse heart and liver. J. Cell Biol. **33**, C7—C12 (1967).

ROGERS, D. C.: An electron microscope study of the parathyroid gland of the frog (*Rana clemitans*). J. Ultrastruct. Res. **13**, 478—499 (1965).

ROSENBLUTH, J.: Subsurface cisterns and their relationship to the neuronal plasma membrane. J. Cell Biol. **13**, 405—421 (1962).

— Ultrastructure of somatic muscle cells in *Ascaris lumbricoides*. II. Intermuscular junctions, neuromuscular junctions, and glycogen stores. J. Cell Biol. **26**, 579—591 (1965).

SANDBORN, E., KOEN, P. F., MCNABB, J. D., MOORE, G.: Cytoplasmic microtubules in mammalian cells. J. Ultrastruct. Res. **11**, 123—138 (1964).

— SZEBERENYI, A., MESSIER, P.É., BOIS, P.: A new membrane model derived from a study of filaments, microtubules and membranes. Rev. Canad. Biol. **24**, 243—276 (1965).

SAUER, F. C.: Mitosis in the neural tube. J. comp. Neurol. **62**, 377—405 (1935).

— Some factors in the morphogenesis of vertebrate embryonic epithelia. J. Morph. **61**, 563—579 (1937).

SCHAFFER, J.: Das Epithelgewebe. In: Handbuch der mikroskopischen Anatomie des Menschen (Ed.: W. V. MÖLLENDORFF), Bd. II/II, 28—46. Berlin: Springer 1927.

SCHMIDT, A. J.: Cellular biology of vertebrate regeneration and repair. 420 pp. Chicago: University of Chicago Press 1968.

SCHMITT, F. O.: Some protein patterns in cells. Growth, Third Growth Symposium **5**, 1—20 (1941).

SCHNEIDER, K. C.: Lehrbuch der vergleichenden Histologie der Tiere, 988 pp. Jena: Fischer 1902.

— Histologisches Praktikum der Tiere. Jena: Fischer 1908.

SEDAR, A. W., FORTE, J. G.: Effects of calcium depletion on the junctional complex between oxyntic cells of gastric glands. J. Cell Biol. **22**, 173—188 (1964).

SHAFFER, B. M.: Mechanical control of the manufacture and resorption of cell surface in collective amoebae. J. theoret. Biol. **8**, 27—40 (1965).

SJÖSTRAND, F. S., ANDERSSON-CEDERGREN, E., DEWEY, M. M.: The ultrastructure of the intercalated discs of frog, mouse, and guinea pig cardiac muscle. J. Ultrastruct. Res. **1**, 271—287 (1958).

SNELL, R. S.: The fate of epidermal desmosomes in mammalian skin. Z. Zellforsch. **66**, 471—487 (1965).

— The fate of epidermal desmosomes in mammalian skin. Anat. Rec. **154**, 425 (Abstract) (1966).

SPIRO, D., HAGOPIAN, M.: On the assemblage of myofibrils. Symp. Intern. Soc. Cell Biol. **6**, 71—78 (1967).

STEHBENS, W. E.: The basal attachment of endothelial cells. J. Ultrastruct. Res. **15**, 389—399 (1966).

STERNLIEB, I.: Mitochondrion-desmosome complexes in human hepatocytes. Z. Zellforsch. **93**, 249—253 (1968).

TILNEY, L. G., GIBBINS, J. R.: Microtubules in the formation and development of the primary mesenchyme in *Arbacia punctulata*. II. An experimental analysis of their role in development and maintenance of cell shape. J. Cell Biol. **41**, 227—250 (1969).

TRELSTAD, R. L., HAY, E. D., REVEL, J. P.: Cell contact during early morphogenesis in the chick embryo. Devel. Biol. **16**, 78—106 (1967).

— REVEL, J. P., HAY, E. D.: Tight junctions between cells in the early chick embryo as visualized with the electron microscope. J. Cell Biol. **31**, C6—C10 (1966).

WARTIOVAARA, J.: Cell contacts in relation to cytodifferentiation in metanephrogenic mesenchyme in vitro. Ann. Med. exp. Biol. Fenn. **44**, 469—503 (1966).

WATTERSON, R. L.: Structure and mitotic behavior of the early neural tube. In: Organogenesis (Ed.: R. L. DEHAAN and H. URSPRUNG), pp. 129—159. New York: Holt, Rinehart and Winston 1965.

Weis, P.: Confronting subsurface cisternae in chick embryo spinal ganglia. J. Cell Biol. **39**, 485—488 (1968).

Weiss, L.: The cell periphery, metastasis and other contact phenomena. 388 pp. Amsterdam: North-Holland Publ. Comp. 1967.

Weiss, P.: Perspectives in the field of morphogenesis. Quart. Rev. Biol. **25**, 177—198 (1950).

— Cell contact. Intern. Rev. Cytol. **7**, 391—423 (1957).

— Ferris, W.: The basement lamella of amphibian skin its reconstruction after wounding. J. biophys. biochem. Cytol. **2** (Suppl.), 275—282 (1956).

Wilgram, G., Caulfield, J. B., Madgic, E. B.: A possible role of the desmosome in the process of keratinization. In: The Epidermis (Ed.: W. Montagna and W. Lobitz), pp. 275—301. New York: Academic Press 1964.

Wilgram, G. F., Caulfield, J. B., Madgic, E. B.: An electron microscopic study of genetic errors in keratinization in man. In: Biology of Skin and Hair Growth (Ed.: A. Lyne and B. Short), pp. 251—266. Sidney: Angus and Robertson 1965.

Wissig, S. L., Graney, D. O.: Membrane modifications in the apical endocytic complex of the ileal epithelial cells. J. Cell Biol. **39**, 564—579 (1968).

Wolpert, L., Mercer, E. H.: An electron microscope study of the development of the blastula of the sea urchin embryo and its radial polarity. Exp. Cell Res. **30**, 280—300 (1963).

Wood, R. L.: Intercellular attachment in the epithelium of hydra as revealed by electron microscopy. J. biophys. biochem. Cytol. **6**, 343—352 (1959).

— An electron microscope study of developing bile canaliculi in the rat. Anat. Rec. **151**, 507—529 (1965).

On Relationships between Endosymbiosis and the Origin of Plastids and Mitochondria

Eberhard Schnepf and R. Malcolm Brown, Jr.

Lehrstuhl für Zellenlehre der Universität, Heidelberg,

and Department of Botany, University of North Carolina,
Chapel Hill, North Carolina

I. Introduction

In 1883 Schimper wrote in a footnote: „Sollte es sich definitiv bestätigen, dass die Plastiden in den Eizellen nicht neu gebildet werden, so würde ihre Beziehung zu dem sie enthaltenden Organismus einigermassen an eine Symbiose erinnern. Möglicherweise verdanken die grünen Pflanzen wirklich einer Vereinigung eines farblosen Organismus mit einem von Chlorophyll gleichmässig tingierten ihren Ursprung".

A similar hypothesis on the evolution of the cells of higher organisms from endosymbiontic associations of different "bioblasts" was expressed by Altmann (1890).

In the past, ideas on the evolution of plastids and mitochondria were based on a poor knowledge of the organelles themselves. Therefore, the hypothesis of Schimper and Altmann was accepted (e.g. by Mereschkowsky, 1905), rejected (Buchner, 1953; Klein and Cronquist, 1967) though convincing facts for the interpretations were lacking (Geitler, 1923), or it was ignored. Presently, there are many new lines of evidence which allow us to view these interpretations in a new context. Obviously, it is extremely difficult to present direct evidence for the evolution of plastids and mitochondria. The following remark depicts the current situation: "Indeed, it is hard to see how it would be possible to be sure of the correctness of a hypothesis like that ... But even if in the long run it should turn out that the hypothesis of the exogenous origin is a March hare, the chase should be stimulating and illuminating" (*Anonymous*, 1967). The idea of exogenous origin already has influenced several new proposals on the phylogeny and the classification of higher organisms (Sagan, 1967; Margulis, 1968; Whittaker, 1969).

All ideas on the origin of plastids and mitochondria must take into account that they are absent in prokaryotes and that they never arise *de novo* but by division. This last statement is the result of many morphological, genetic and biochemical studies which are summarized in this volume by Stubbe and Baxter.

Only very few authors presently question this position and as time passes, even fewer authors will doubt the validity of this point. Those who continue to doubt take into consideration only isolated observations (which may well be misinterpretations) and not the abundance of coherent contrary findings. Furthermore, it seems well established that plastids and mitochondria are semi-autonomous organelles. The genetic independence of plastids was clearly given by Renner (1934) and his colleagues (Schötz, 1967; Kirk and Tilney-Basset, 1967). The genetic properties of mitochondria were detected later (Wagner, 1969).

Plastidal and mitochondrial semiautonomy are based on different and specific DNA and protein synthesizing capacities. Any hypothesis which tries to describe the evolution of plastids and mitochondria has to explain, first, the origin and the peculiarities of these specific biochemical systems and, second, a compartmentation which is unique among the cell organelles. In contrast to all other cell organelles, plastids and mitochondria are composed of two compartments which are enclosed by two membrane systems which completely separate them from the cytoplasm.

One explanation of this compartmentalization has been provided by Robertson (1964) who stated that a cell's own pseudopodium is phagocytized and then transformed into a mitochondrion. If such a process ever occurred, it could not have been a recent one, contrary to the explanations of Robertson (1964), which do not take into account the different characteristics of the protein synthesizing systems of the mitochondrion and the cytoplasm (which were not yet known when Robertson published his article).

Another explanation, based on the ideas of Schimper (1883) and Altmann (1890), suggests that plastids and mitochondria have evolved during a long process from prokaryotic organisms which lived as endosymbionts in a host cell. This hypothesis is favored by the similarities of plastids and mitochondria with prokaryotic organisms as well as by the study of recent forms of endosymbiosis (Schnepf, 1966a).

It is not necessary here to give a full account of all the studies which demonstrate the ability of plastids and mitochondria to code for and to synthesize a part of their own proteins. The number of papers in this area alone exceeds several hundred.

The semiautonomy of plastids and mitochondria is based on their: 1. specific DNA (which replicates independently and differs in the base composition, size, and structure from nuclear DNA); and 2. specific DNA- and RNA-polymerases, specific ribosomes, tRNA's and amino acid-activating enzymes (Nass, Nass, and Afzelius, 1965; Parthier and Wollgiehn, 1966; Tuppy and Wintersberger, 1966; Schötz, 1967; Scott and Smillie, 1967; Roodyn and Wilkie, 1968; Nass and Buck, 1969; Meyer and Simpson, 1969; Rabinowitz et al., 1969).

On the other hand, special considerations should be given to possible relationships between the protein synthesizing systems of plastids and mitochondria, and those of the prokaryotes. There are increasing numbers of studies which point out certain similarities among these systems on the one hand and differences to those in the nucleo-cytoplasm of the eukaryotic cell on the other. Surveys on this topic have been given by Swift (1965), Echlin (1966a), Heil (1968), and Roodyn and Wilkie (1968).

II. Similarities between Plastids, Mitochondria, and Prokaryotes

The following compilation of similarities between plastids, mitochondria, and prokaryotes includes features of different value which range from size and morphological similarities to biochemical peculiarities. Of secondary importance is the similarity of size and form. The derivation of structural correspondence follows in the next chapter. More fundamental are the biochemical relationships, especially in the protein synthesizing system.

The DNA of plastids and mitochondria is double stranded and has physical properties and a mode of replication like bacterial or cyanophycean DNA (EDELMAN et al., 1967; KIRSCHNER, WOLSTENHOLME, and GROSS, 1968; WERZ and KELLNER, 1968). Like bacterial DNA (KIRSCHNER, WOLSTENHOLME, and GROSS, 1968), mitochondrial DNA often occurs in circular form (KROON et al., 1966; AVERS et al., 1968; PIKÓ et. al., 1968).

The plastid and mitochondrial DNA has fixing and staining properties similar to those of prokaryotes (RIS, 1962; NASS and NASS, 1963; NASS, NASS, and AFZELIUS, 1965; BISALPUTRA and BISALPUTRA, 1967). This may be connected with the fact that both kinds of DNA are not associated with histone (SWIFT, 1965) as was demonstrated by STEINERT (1965) in the kinetonuclei of trypanosomes which are modified mitochondria. In contrast, the nuclei of eukaryotic cells contain DNA-histone-complexes with the exception of the Dinophyceae (NASS, NASS, and AFZELIUS, 1965; SWIFT, 1965; KUBAI and RIS, 1969). The latter reveal structures in electron microscopic studies of thin sections comparable to the "nucleoids" of prokaryotes, plastids, and mitochondria. Like bacterial DNA, mitochondrial DNA is associated with the membrane (NASS, 1969).

Of special significance are results from the analysis of the nearest neighbor frequencies of dinucleotide sequences. The nuclear DNA of the slime mold, *Physarum*, has like that of other higher organisms a linear arrangement frequency of cytosine-phosphate-guanine much less than the value for a random arrangement of nucleotides. On the other hand, this frequency is near the random value in *Physarum* mitochondrial DNA, a relationship commonly observed in bacterial DNA (CUMMINS, RUSCH, and EVANS, 1967). The mitochondrial ($=$ kinetonuclear) DNA in Trypanomatidae reacts like bacterial DNA on treatment with ethidium bromide and not like nuclear DNA (STEINERT, 1969).

The similarities between blue-green algal and plastidic DNA and between DNA of prokaryotes and organelles of eukaryotes are summarized by EDELMAN et al. (1967). These authors point out the apparent unimodality of both in a CsCl equilibrium density gradient and note that the base frequencies of different DNA's favor the idea of an evolution *via* endosymbiontic prokaryotes.

Even more conspicuous relationships between prokaryotes, plastids, and mitochondria are evident through a comparison of the ribosome in which its size (70 S) as well as the size of its subunits is similar. These ribosomes differ from cytoplasmic ribosomes in their size (80 S) and their subunits (DURE, EPLER, and BARNETT, 1967; KÜNTZEL and NOLL, 1967; STUTZ and NOLL, 1967; SVETAILO, PHILIPPOVICH, and SISSAKIAN, 1967; HOOBER and BLOBEL, 1969). Prokaryotic, plastidal and mitochondrial rRNA have nearly identical sedimentation behavior (STUTZ and NOLL, 1967). Their sensibility to Mg^{++} concentration is the same and differs from that of cytoplasmic ribosomes (Boardman, Francki, and Wildman, 1966; Hoober and Blobel, 1969).

It is important that most inhibitors of translation such as chloramphenicol or cycloheximide have different effects on the protein synthesis of prokaryotes, plastids and mitochondria on the one hand and the cytoplasm of the eukaryotic cell on the other (Anderson and Smillie, 1966; Linnane and Stewart, 1967; Smillie et al., 1967; Haslam et al., 1968; Ellis, 1969; Hoober and Blobel, 1969).

A special attribute of the 70 S ribosomes of prokaryotes, plastids, and mitochondria is the initiation of the protein synthesis by N-formylmethionine-tRNA. The 80 S type ribosomes of eukaryotic cytoplasm have another mechanism of initiation (Schwartz et al., 1967; Smith and Marcker, 1968; Galper and Darnell, 1969).

In general, the protein synthesizing apparatus of plastids and mitochondria is more similar to that of prokaryotes than to that of the nuclear-cytoplasmic matrix of the eukaryotic cell (Küntzel, 1969). It should be expected that, with increasing knowledge, further relations will be detected.

The differences between the free living prokaryotes and the plastids and mitochondria should not be neglected. The organelles certainly must demand proteins which are coded by nuclear genes. This is demonstrated by genetic experiments which show Mendelian inheritance of certain plastidal and mitochondrial defects while other defects follow a non-Mendelian inheritance and therefore are expected to be localized within the organelles themselves (see Gibor and Granick, 1964; Stubbe, 1966; Kirk and Tilney-Basset, 1967; Wagner, 1969, for references).

Calculations on the DNA content of the organelles likewise demonstrate their *semi*autonomy. In mitochondria a DNA strand usually is about 5 μ long (Sinclair et al., 1967). Provided that the strands are identical they cannot be expected to code for more than about 60 polypeptide chains of a molecular weight of 15 000 (Tewari et al., 1966) which is not enough to build up the complete organelle. These considerations agree with experimental data that both mitochondrial and non-mitochondrial systems synthesize mitochondrial proteins (Kadenbach, 1967; Clark-Walker and Linnane, 1967; Küntzel, 1969). On the other hand, mitochondrial products, such as the adrenocorticotropic hormone in the adrenal cortex, are not restricted to mitochondria (Garren and Crocco, 1967).

For plastids, the same may be valid (Gibor and Granick, 1964; Parthier and Wollgiehn, 1966; Tewari and Wildmann, 1968), though it is of interest to note that the total DNA content of *Euglena gracilis* chloroplasts and the blue-green alga *Anabaena variabilis* are of the same order (Craig, Leach, and Carr, 1969). These calculations are rendered more difficult because of the "polyploidy" of plastids (Hermann, 1969) and also of mitochondria (Nass, 1969). It is thought that plastids synthesize more own proteins than do mitochondria (Smillie et al., 1967).

III. Comparison of Plastids and Mitochondria with the Recent "Endosymbionts"

The semiautonomy of plastids and mitochondria implies that they are firmly established parts of the eukaryotic cell (vanden Driessche, 1967). So far, it has been impossible to culture isolated plastids or mitochondria though several important biochemical processes occur in the isolated state (DNA synthesis: SPENCER and WHITFELD, 1967; PARSONS and SIMPSON, 1967; KAROL and SIMPSON, 1968; SCOTT, SHAH, and SMILLIE, 1968; RNA synthesis: LUCK and REICH, 1964; KIRK, 1964; SCHWEIGER and BERGER, 1964; protein synthesis: NEUPERT, BRDICZKA, and BÜCHER, 1967; MARGULIES, GANTT, and PARENTI, 1968; MARGULIES and PARENTI, 1968). Likewise, the capacity of isolated plastids and mitochondria to perform assimilation or the respiratory pathway is well known; however, the culturing and growing in the *in vitro* state never has been achieved.

Perhaps one day it will be possible to culture these organelles in an artificial medium which would have to contain nuclear gene coded components, for the nuclear DNA seems to be involved in the biosynthesis of some plastidal and mito-chondrial proteins (SHERMAN et al., 1966; KADENBACH, 1967; SCHWEIGER, MASTER, and WERZ, 1967; GROSS, MCCOY, and GILMORE, 1968; TEWARI and WILDMAN, 1968). This requirement is shown furthermore by the Mendelian inheritance of several mitochondrial and plastidal features. In fact, plastids of green algae (presumably *Caulerpa* or *Codium*), phagocytized by the cells lining the tubules of the digestive diverticulum of different sacoglossan opisthobranch gastropods, survive for a long time, and in one case they remained functional up to 6 weeks (KAWAGUTI and YAMASU, 1965; TRENCH, 1969).

In two cases (yeast: TUPPY and WILDNER, 1965; *Neurospora:* DIACUMAKOS, GARNJOBST, and TATUM, 1965) a transfer of mitochondria from one cell into another, followed by a multiplication of the transferred organelles, presumably was possible; however, there have been some difficulties in the interpretation of these results (see JAKOB in the discussion of the paper of TUPPY and WINTERSBERGER, 1966).

In many respects it is clear that the plastids and mitochondria resemble pro-karyotic endosymbionts. There are many accounts of different stages of endo-symbiosis, in which the endosymbionts substitute for mitochondria or plastids.

Pelomyxa palustris is an ameba without mitochondria (LEINER and BHOWMICK, 1967). In this case it may be possible that symbiontic bacteria perform the respiration for the whole cell (LEINER et al., 1968).

In *Geosiphon* (VON WETTSTEIN, 1915; KNAPP, 1933), a symbiontic association of a phycomycete and the blue-green alga *Nostoc*, a loose association occurs, and both organisms can live separately. It is relatively easy to survey the situation here (SCHNEPF, 1964). The *Nostoc* cells are taken up presumably in a phagocytosis-like process (KNAPP, 1933), and they live as intact filaments in special vesicles which are derivatives of the phagocytotic vesicles (Fig. 1). The vesicle membrane apparently originates from the plasmalemma. Every *Nostoc* cell has its own cell wall and a completely normal plasma. The thylakoid arrangement is reticulate. Sometimes thylakoidal membranes fuse with the cyanophycean plasmalemma, and the intrathylakoidal space is then connected with the extraplasmatic space.

More information concerning the origin of plastids and mitochondria comes from organisms like *Glaucocystis* and *Cyanophora* which are considered apoplastidic unicellular algae with endosymbiontic blue-green algae. Free living members of these "syncyanoses" (PASCHER, 1914) are not known.

Glaucocystis has a rigid cellulose cell wall and two reduced flagella. Figs. 2 and 3 show its organization. *Cyanophora* is more similar to a true flagellate because it has normal flagella and a pellicula.

The host cell classification of these two organisms is uncertain. The most recent proposal has been to classify them with primitive dinoflagellates (Mignot, Joyon, and Pringsheim, 1969). This suggestion is not completely satisfactory because

Fig. 1. Outer part of a *Nostoc* cell endosymbiontic in *Geosiphon*. Plasmalemma of the cyanelle N, plasmalemma of the host cell P, tonoplast of the host cell T, cell wall of the cyanelle with different layers W. Arrows label connections between the plasmalemma and the thylakoidal membranes of the *Nostoc* cell. KMnO₄, 60,000 ×

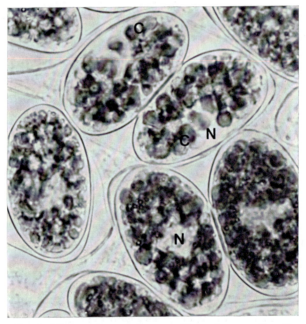

Fig. 2. Living cells of *Glaucocystis nostochinearum*. Note the cyanelles C and the nucleus region N. Bright field, 1,200 ×

Glaucocystis and *Cyanophora* do not have the dinokaryon (a unique and most characteristic nucleus of the *Dinophyceae* with DNA free of histones and, therefore, with chromosomes which resemble prokaryotic nucleoids; Giesbrecht, 1965; Kubai and Ris, 1969).

The "cyanelles" of *Glaucocystis* and *Cyanophora* have not been cultivated separately (Geitler, 1923; Pringsheim, 1958, 1963; Echlin, 1967; Edelman et al., 1967; Hall and Claus, 1967). Both organisms are autotrophic. The host cell obviously shares the photosynthetic products of the cyanelles.

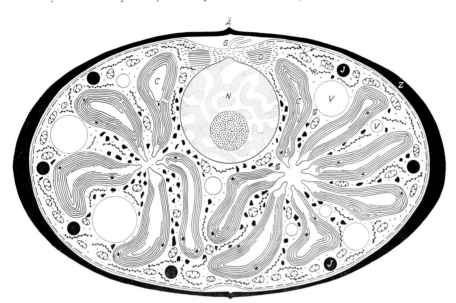

Fig. 3. Scheme of the organization of *Glaucocystis*. Equator *Ä*, cyanelle *C*, dictyosome *D*, flagellum *G*, osmiophilic inclusion body in the host cell *J*, lacuna *L*, mitochondrion *M*, nucleus *N*, endoplasmic reticulum *R*, starch grain *S*, vacuole *V*, cell wall *Z*. From Schnepf, Koch, and Deichgräber, 1966

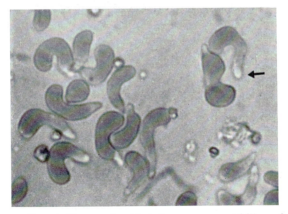

Fig. 4. Isolated cyanelles of *Glaucocystis geitleri*. Arrow: pyrenoid-like polar body. 2,400 ×

The question is now raised whether these cyanelles (Fig. 4) are indeed *autonomous* cyanophycean cells or whether they could be considered as special forms of chloroplasts (Pringsheim, 1958). There are at least three possibilities of thoughts on this subject:

1. Conventional thought has been directed toward the cyanophycean nature of the cyanelles. Strong support for this idea has come from Hall and Claus. These authors even went so far as to give a taxonomic description of the cyanelles (*Skuja-pelta nuda* for *Glaucocystis*: Hall and Claus, 1967; and *Cyanocyta korschikoffiana* for *Cyanophora*: Hall and Claus, 1963).

2. We propose a radical alternative to consider the cyanelles as chloroplasts, and *Glaucocystis* and *Cyanophora* as primitive rhodophycean algae. Skuja (1954), in establishing the division "Glaucophyta" which includes *Glaucocystis* and similar

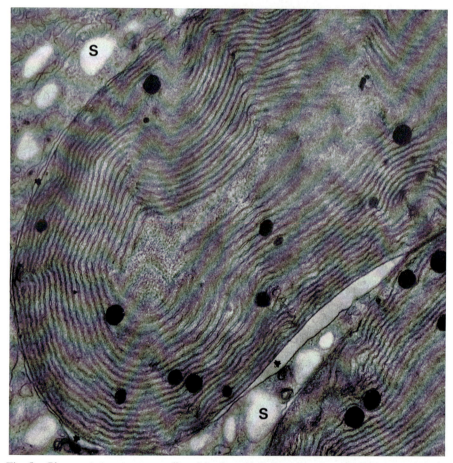

Fig. 5. *Glaucocystis incrassata* cyanelle with plastoglobuli and concentrically arranged thylakoids. Starch grains in the host cytoplasma S, enlarged spaces between the two membranes of the envelope of the cyanelle labeled by asteriscs. OsO_4, 40,000 ×

endocyanoses, already had mentioned this possibility. Certain similarities between the cyanelles and the chloroplasts of rhodophycean algae were listed by Schnepf (1966a, b). Pigment analyses by Chapman (1966) point in the same direction though the relations are not expressed literally.

3. In addition, we propose a third suggestion that these cyanelles are semi-autonomous endosymbionts developing to chloroplasts, representing not the true ancestors of rhodophycean chloroplasts (which could have arisen in a similar way) but forms of parallel evolution.

There are arguments for and against each of these 3 ideas. To understand better the organization of *Glaucocystis* and *Cyanophora* for ideas on the origin of plastids and mitochondria, it is necessary to give a brief description of their cyanelles (HALL and CLAUS, 1963, 1967; LEFORT, 1965; SCHNEPF, 1966a, 1966b; SCHNEPF, KOCH, and DEICHGRÄBER, 1966; LEFORT and POUPHILE, 1967; BOURDU and LEFORT, 1967; ECHLIN, 1967; MIGNOT, JOYON, and PRINGSHEIM, 1969).

The following features are taken into consideration:

1. *Wall* – A typical cyanophycean cell wall is absent (Fig. 5, 9; HALL and CLAUS, 1963; SCHNEPF, KOCH, and DEICHGRÄBER, 1966).

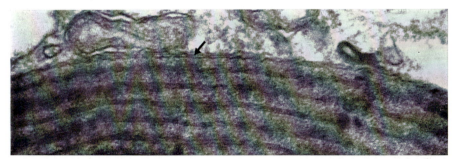

Fig. 6. *Glaucocystis nostochinerarum.* The wall remnant between the two membranes of the envelope of the cyanelle is labeled by an arrow. Glutaraldehyde/OsO$_4$, 120,000 ×

2. *Diaminopimelic acid* – This substance, a typical constituent of the cell wall of bacteria and cyanophycean algae, is absent (HOLM-HANSEN, PRASAD, and LEWIN, 1965).

3. *"Wall remnant"* – A thin (about 6 nm) non-convoluted electron dense layer (perhaps corresponding to the innermost layer of the *Nostoc* cell wall shown in Fig. 1) sometimes is seen between the two unit membranes enclosing the cyanelles (Fig. 6, 10) (SCHNEPF, 1966a; SCHNEPF, KOCH, and DEICHGRÄBER, 1966; HALL and CLAUS, 1963). Sometimes in *Glaucocystis* these membranes are separated by an enlarged space (Fig. 5).

4. *Thylakoid arrangement* – The thylakoids are separate and arranged concentrically (Fig. 5, 9).

5. *Phycobilisomes* – Granules with a diameter of about 35 nm, attached to the thylakoids, and containing phycobilins (GANTT, 1968) are present (Fig. 8; LEFORT, 1965; LEFORT and POUPHILE, 1967; ECHLIN, 1967)[1].

6. *Plastoglobuli* – Electron-dense interthylakoidal granules are present (Fig. 5, 9; HALL and CLAUS, 1963; SCHNEPF, 1966b).

7. *Inclusion bodies* – Inclusion bodies typical for cyanophycean algae like poly-glucoside granules, polyhedral bodies, cyanophycin granules, and polyphosphate

1 Phycobilosomes have been observed only in red algae (Fig. 11) but not in free living blue-green algae (with perhaps one exception, GANTT and CONTI, 1966).

20*

granules (Lang, 1968) are absent (*Glaucocystis:* Schnepf, 1966b) with the possible exception that polyphosphate granules may be present in *Cyanophora* (Hall and Claus, 1963).

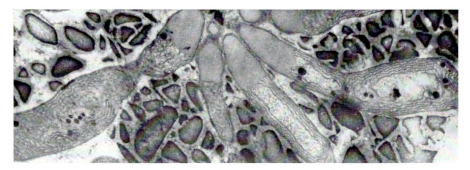

Fig. 7. *Glaucocystis geitleri*. Several cyanelles lying together with their pyrenoid-like polar bodies. Electron dense starch grains in the host cytoplasm. OsO_4, 13,000 ×

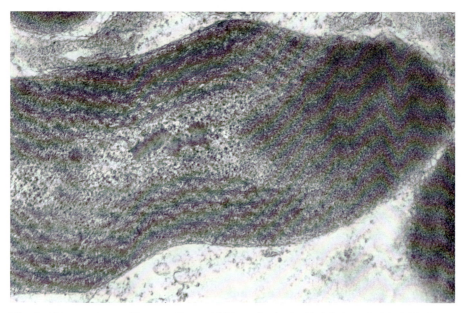

Fig. 8. *Glaucocystis nostochinearum*. Pyrenoid-like polar body (right), interthylakoidal phycobilisomes. Glutaraldehyde/OsO_4, 64,000 ×

8. *Pigments* – C- and R-phycocyanin are present (common also in free living blue-green and most red algae[2]. The common cyanophycean xanthophylls, echinenone, and myxoxanthophyll, are absent (Chapman, 1966; Kleinig, personal communication, 1969).

---

2 For exceptions see Chapman (1966).

9. *DNA* – a) The staining properties with acridine orange are similar to chloroplasts and not to the centroplasm of free living cyanophyceae. b) The G-C – ratio of *Cyanophora* DNA is similar to that of chroococcalian cyanophyceae (EDELMAN et al., 1967).

10. *"Pyrenoid"* – A thylakoid and ribosome free body in the polar region in the *Glaucocystis* cyanelles (Fig. 7, 8) resembles a pyrenoid (SCHNEPF, 1966b). Its

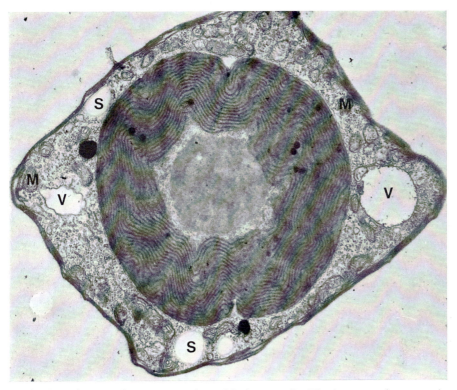

Fig. 9. *Cyanophora paradoxa*, cross section with the cyanelle. The host cytoplasm contains mitochondria *M*, starch grains *S* and vacuoles *V*. OsO₄, 19,000 ×

lattice-like striation (ECHLIN, 1967) has a correspondence to the crystalline structure of the pyrenoids from the diatom *Achnanthes* (HOLDSWORTH, 1968). Free-living blue-green algae never have pyrenoids (BROWN et al., 1967).

11. *Starch* – Starch grains giving a wine-red reaction with iodine typical of floridean starch are found in the cytoplasm of the "host cells", in *Glaucocystis* concentrated near the pyrenoids (Fig. 7, 9; SCHNEPF, 1966b). The starch of red algae is, likewise, extraplastidic.

Comparing the structural and biochemical features with the three considerations of the state of the "cyanelles" the following conclusions can be drawn:

Consideration (1) – The cyanophycean nature of the cyanelle – The following features are in favour: 3, 9b, against: 1, 2, 5, 7, 8, 9a, 10, 11, or neutral: 4, 6.

Consideration (2) — The rhodophycean chloroplast nature of the cyanelle — The following features are in favour: 1, 2, 5, 7, 8, 9a, 10, 11, against: 3, and neutral: 4, 6. The structure of the "host cell", at least in *Glaucocystis*, is not so peculiar that these organisms could not be considered as red algae. For conventional thinking it is strange that there are flagella. But why shouldn't primitive rhodophycean algae have flagella? It is not yet proven that the Rhodophyta as "Aconta" (in the sense of Christensen, 1962) have a lower position than the "Contophora". It may well be that in phylogeny most of the living red algae have lost flagellation.

Fig. 10. *Cyanophora paradoxa*. Wall remnant W between the two membranes *A* and *B* of the envelope of the cyanelle. OsO₄, 83,000 ×

Consideration (3) — The intermediate nature of the cyanelle — All points discussed here could be considered to support this idea with the following explanation: results of a reduction are features 1, 2, 7, 8, 9 and results of a new development are features 5, 10, 11. The taxonomic position of the host cell remains uncertain.

If one assumes the cyanelles are on the way to becoming chloroplasts, the following homologies can be stated:

Cyanelle	Plastid
membrane of the phagocytotic vesicle	outer membrane of the plastidal envelope
plasmalemma of the organelle	inner membrane of the plastidal envelope
matrix of the cyanelle (with DNA, ribosomes etc.)	matrix of the plastid (with DNA, ribosomes etc.)
invagination of the *Nostoc*-plasmalemma in *Geosiphon*	invagination of the inner membrane of the plastidal envelope
thylakoids	thylakoids
derivative of the phagocytotic vesicle with remnants of the cell wall of the cyanelle	space between outer and inner plastidal membrane; no cell wall remnant

Similar comparisons are possible with mitochondria although endosymbiontic bacteria with reduced walls are not known. This correspondence is further strengthened by some biochemical data. The outer membrane of mitochondria resembles in cholesterol content (PARSONS and YANO, 1967) and in enzymatic properties (SOTTOCASA et al., 1967) the membranes of the endoplasmic reticulum. The inner mitochondrial membrane is unique among cell membranes and resembles more a bacterial plasma membrane (PARSONS and YANO, 1967). Similar conclusions are drawn from the finding of NEUPERT, BRDICZKA, and BÜCHER (1967) and CLARK-WALKER and LINNANE (1967). Here chloramphenicol (which interferes with the protein synthesis on 70 S ribosomes) inhibits only the formation of cristae but does not influence the formation of the outer mitochondrial membrane.

IV. Discussion

At present, it is impossible to come to a definitive decision between the different suggestions on the state of the *Glaucocystis* and *Cyanophora* cyanelles. Nevertheless, there are so many similarities between the *Glaucocystis* cyanelles and rhodophycean chloroplasts (Fig. 11) that it is difficult to discern them easily. We feel the necessity to find good criteria for both cyanelles and chloroplasts. The existence or absence of a wall or at least a wall remnant and the degree of autonomy (complete or incomplete) should be useful in this respect.

Basic questions on the origin of plastids and mitochondria are the following: Do the similarities necessarily lead to the conclusion that plastids and mitochondria have evolved from endosymbiontic organisms, are they accidental, or were they conditioned by the special function of these cell parts? Do these similarities reflect homologies or analogies?

Again, it should be pointed out that all deductions must be circumstantial. At present, there seem to be almost no severe objections against the endosymbiont hypothesis. (A possible exception results from studies of DU BUY and RILEY, 1967, who found a considerable hybridisation between nuclear and mitochondrial DNA, for contradicting observations in other objects see DAWID, 1965, 1966). GIBOR (1967) argues that there are some difficulties in that the organelles of different species differ in their DNA. This fact should require that different symbionts enter different cells and all proceed to evolve in an identical pattern. One can object that during evolution the genetic information of both members of the endosymbiosis can evolve. With respect to the long time in evolution it is not surprising that such differences exist.

Surely, the origin of plastids and mitochondria from a part of the cell with satellite DNA which is separated from the cytoplasm by a process similar to that postulated by ROBERTSON (1964) cannot be excluded. Such a self-phagocytizing process should have occurred at the prokaryotic level (because of the prokaryotic properties of the protein synthesizing apparatus of the organelles).

In spite of these arguments and alternatives, most criteria seem to satisfy the endosymbiont hypothesis which may be the most useful basis for further studies which will ultimately perhaps prove its correctness or incorrectness.

SAGAN (1967) has proposed some criteria for suggesting that an organelle arose as an endosymbiont. They are:

1. The organelle must have once been able to replicate DNA and its own protein synthesizing apparatus.

2. The daughter cell must receive at least one copy of the genome of the endo-symbiont.

3. One would not expect to find "intermediate intracellular stages" of the organelle in a given organism. It would either be present or absent.

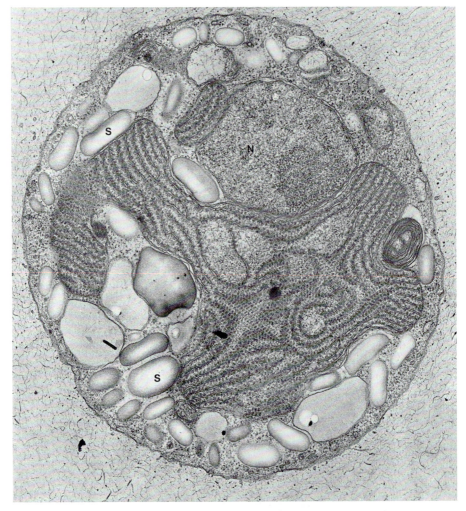

Fig. 11. Section of *Porphyridium cruentum* showing the interthylakoidal phycobilisomes, nucleus N, starch S. Glutaraldehyde/OsO$_4$, 23,000 ×. Kindly supplied by Dr. E. Gantt (J. Cell Biol. 26, 365)

4. If the symbiont is lost, there would be a concomitant loss of all metabolic characteristics coded by the genome of the symbiont.

5. The symbiont will show non-Mendelian genetics (no classical segregation and recombination with nuclear genes).

6. It should be possible to find "naturally occurring counterparts" among organisms that exist at the present time. ROODYN and WILKIE (1968) in reviewing these criteria admit that they may be insufficient.

"Nevertheless, it is most striking that most of these criteria appear to be met to a greater or lesser extent by mitochondria. Perhaps it is for the reader to judge . . . how far, for example, criteria 3, 4, and 6 are met (the "natural counterpart" in criterion 6 would, of course, be an aerobic bacterium). Criteria 1, 2, and 5 seem to the authors, at least, to be fully satisfied." (ROODYN and WILKIE, 1968, p. 58).

Centrioles and basal bodies, whose DNA content is dubious (RANDALL and DISBREY, 1965; HOFFMAN, 1965; SMITH-SONNEBORN and PLAUT, 1967; PYNE, 1968) and which are not self-duplicating bodies (DINGLE and FULTON, 1966) do not meet these criteria in the same way as do plastids and mitochondria. Furthermore, they are not separated from the ground plasma by an envelope of two membranes. Therefore, it seems unlikely to us that they arose from endosymbionts. Likewise, the nucleoplasm is unified with the ground plasma by the pores in the nuclear envelope (SCHNEPF, 1966a).

The evolution of plastids and mitochondria from free living prokaryotes would be a very long process which includes the adaptation of the endosymbionts to their host cell and *vice versa*. The special environment would allow the endosymbionts to give up parts of their genetic information, not only such genes which code the formation of the cell wall but also those which regulate fundamental biosynthetic processes. The host cell, on the other hand, can restrict its own energy producing apparatus. Certainly, this adaptation is very old and has been assumed to originate in the Precambrian, 1 billion or more years ago (ECHLIN, 1966b; SAGAN, 1967).

SAGAN (1967, see ROODYN and WILKIE, 1968) postulates the following different stages of evolution:

1. Evolution of prokaryotic cells about 3—5 billion years ago when earth's atmosphere was reducing in nature.

2. Evolution of two types of prokaryotic cells: Primitive respirers using, perhaps, nitrate and sulphate as electron acceptors ("anaerobic respirers") and primitive photosynthesizers which enrich oxygen in the atmosphere.

3. This enabled the anaerobic respirers to become aerobic respirers.

4. With increasing oxygen content in the atmosphere the cells had to become adapted to the aerobic environment. The appearance of ozone in the atmosphere cut off ultraviolet radiation and hence prevented the "abiogenic" formation of organic matter. All forms of life thus became dependent on photosynthesis. Heterotrophs were forced to eat organic matter directly or indirectly from photosynthetic or chemiautotrophic organisms.

5. This process was improved by the evolution of ingestion mechanisms. Anaerobic heterotrophic cells ingest aerobic respirers. A symbiontic relationship results. This may have allowed the cell to develop complex intracellular membranes (in the meaning of the present authors the development of intracellular membranes as well as the development of true flagellation might have primarily evolved as a consequence of ingestion processes which demands the evolution of larger cells).

6. Mitochondria were present in the primitive eukaryote ancestor when plastids were first acquired by symbiosis. Thus, eukaryotes evolved from prokaryotes not only by mutational steps but by endosymbiosis.

To speculate further it can be postulated that the assembly of different "prokaryotes" to form an eukaryote, must have evolved special, more complicated ways of regulation.

It is an open question, at least in the evolution of the plastids, whether such an assembly occurred only once or repeatedly. There might have been several

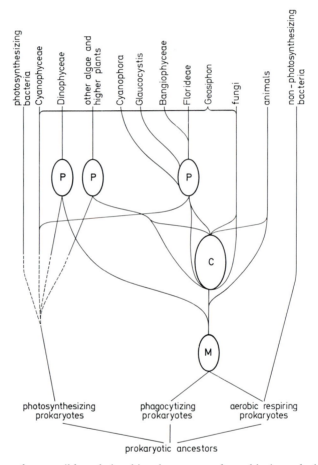

Fig. 12. Scheme for possible relationships between endosymbiosis and the origin of plastids and mitochondria. *M* origin of mitochondria via endosymbiotic bacteria. *P* origin of plastids via endosymbiotic photosynthesizing prokaryotes. *C* development of higher chromosomes

different cases of invasion rendering the origin of plastids in different phyla poly-phyletic (Schiff and Epstein, 1965). The recent forms of endocyanosis would thus be illustrations of the former evolution. For example, ancestors of rhodophycean plastids should have been blue-green algae. The plastids of other phyla could have arisen as well from photosynthesizing bacteria, but the fact that bacteria do not possess chlorophyll a and plastoquinone, in contrast to blue-green algae and plastids, suggests that this pathway would be less probable. Furthermore, it should be noted that plastids as well as blue-green algae usually are larger than mitochondria and bacteria.

Assuming a monophyletic evolution of plastids the position of *Glaucocystis* and *Cyanophora* is reflected again. Are they living fossils, relics of former developmental stages, and thus ancestors of the Rhodophyta? The possibility that red algae arose from flagellates cannot be excluded.

Fig. 12 represents an abbreviated scheme of one of the concepts on the origin of higher organisms with special emphasis on the *Cyanophora-Glaucocystis*-Rhodophyta-line. The ellipses are intended to show the origin of organelles and the development of higher chromosomes in a multistep process requiring a long time of evolution. The position of the individual steps in this scheme does not reflect the real time in which they occurred. Due to their lower degree of autonomy, mitochondria should have evolved earlier than plastids.

Several other points should be annotated: The *Geosiphon* line represents a present attempt of endosymbiosis. The scheme does not show the possible development of fungi by a loss of plastids and, conversely, that algae could have arisen from the symbiontic association of a fungal host and a photosynthesizing prokaryote.

The present evidence for this scheme seems to be strongest with the rhodophycean line of development. It seems to be obvious that the ancestors of the rhodophycean plastids should have been blue-green algae. It is not possible at present to specify the ancestors of other types of plastids as well as the ancestors of mitochondria.

It is hoped that future investigations, on amino-acid sequences of proteins (MARGOLIASH, 1963) or with new techniques and new ideas, will enlighten our knowledge of the origin of cells and organelles and give us a better understanding to life itself.

Note. The first author's electron microscopic work has been supported by grants from the Deutsche Forschungsgemeinschaft, Bad Godesberg, Germany.

References

ALTMANN, R.: Die Elementarorganismen und ihre Beziehung zu den Zellen. Leipzig: Veit & Comp. 1890.

ANDERSON, L. A., SMILLIE, R. M.: Binding of chloramphenicol by ribosomes from chloroplasts. Biochem. biophys. Res. Commun. **23**, 535—539 (1966).

Anonymous: Where plastids come from. Nature (Lond.) **215**, 1320—1321 (1967).

AVERS, C. J., BILLHEIMER, F. E., HOFFMANN, H.-P., PAULI, R. M.: Circularity of yeast mitochondrial DNA. Proc. nat. Acad. Sci. (Wash.) **61**, 90—97 (1968).

BISALPUTRA, T., BISALPUTRA, A.-A.: The occurrence of DNA fibrils in chloroplasts of *Laurencia spectabilis*. J. Ultrastruct. Res. **17**, 14—22 (1967).

BOARDMAN, N. K., FRANCKI, R. I. B., WILDMAN, S. G.: Protein synthesis by cell-free extracts of tobacco leaves. III. Comparison of the physical properties and protein synthesizing activities of 70 S chloroplast and 80 S cytoplasmic ribosomes. J. molec. Biol. **17**, 470—489 (1966).

Bourdu, R., Lefort, M.: Structure fine, observée en cryodécapage, des lamelles photo-synthetiques des Cyanophycées endosymbiotiques: *Glaucocystis nostochinearum* Itzigs., et *Cyanophora paradoxa* Korschikoff. C. R. Acad. Sci (Paris), Sér. D, **265**, 37—40 (1967).

Brown, R. M., Arnott, H. J., Bisalputra, T., Hoffman, L. R.: The pyrenoid: its structure, distribution and function. J. Phycol. **3** (Supp.), 5—7 (1967).

Buchner, P.: Endosymbiose der Tiere mit pflanzlichen Mikroorganismen. Basel-Stuttgart: Birkhäuser 1953.

du Buy, H. G., Riley, F. L.: Hybridization between the nuclear and kinetoplast DNA's of *Leishmania enriettii* and between nuclear and mitochondrial DNA's of mouse liver. Proc. nat. Acad. Sci. (Wash.) **57**, 790—797 (1967).

Chapman, D. J.: The pigments of the symbiotic algae (cyanomes) of *Cyanophora paradoxa* and *Glaucocystis nostochinearum* and two Rhodophyceae, *Porphyridium aerugineum* and *Asterocytis ramosa*. Arch. Mikrobiol. **55**, 17—25 (1966).

Christensen, T.: Alger. In: T. W. Böcher, M. Lange, and T. Sørensen: Botanik, Bd. II: Syst. Bot., Nr. 2, København: Munksgaard 1962.

Clark-Walker, G. D., Linnane, A. W.: The biogenesis of mitochondria in *Saccharomyces cerevisiae*. A comparison between cytoplasmic respiratory-deficient mutant yeast and chloramphenicol-inhibited wild type cells. J. Cell Biol. **34**, 1—14 (1967).

Craig, I. W., Leach, C. K., Carr, N. G.: Studies with deoxyribonucleic acid from blue-green algae. Arch. Mikrobiol. **65**, 218—227 (1969).

Cummins, J. E., Rusch, H. P., Evans, T. E.: Nearest neighbor frequencies and the phylogenetic origin of mitochondrial DNA in *Physarum polycephalum*. J. molec. Biol. **23**, 281—284 (1967).

Dawid, I. B.: Deoxyribonucleic acid in amphibian eggs. J. molec. Biol. **12**, 581—599 (1965).

— Evidence for the mitochondrial origin of frog egg cytoplasmic DNA. Proc. nat. Acad. Sci. (Wash.) **56**, 269—276 (1966).

Diacumakos, E. G., Garnjobst, L., Tatum, E. L.: A cytoplasmic character in *Neurospora crassa*. The role of nuclei and mitochondria. J. Cell Biol. **26**, 427—443 (1965).

Dingle, A. D., Fulton, C.: Development of the flagellar apparatus of *Naegleria*. J. Cell Biol. **31**, 43—54 (1966).

Driessche, vanden T.: The nuclear control of the chloroplasts circadian rhythms. Sci. Prog. (Oxf.) **55**, 293—303 (1967).

Dure, L. S., Epler, J. L., Barnett, W. E.: Sedimentation properties of mitochondrial and cytoplasmic ribosomal DNA's from *Neurospora*. Proc. nat. Acad. Sci. (Wash.) **58**, 1883—1887 (1967).

Echlin, P.: The cyanophytic origin of higher plant chloroplasts. Brit. phycol. Bull. **3**, 150—151 (1966a).

— Origins of photosynthesis. Sci. J. April 1966, 2—7 (1966b).

— The biology of *Glaucocystis nostochinearum*. I. The morphology and fine structure. Brit. phycol. Bull. **3**, 225—239 (1967).

Edelman, M., Swinton, D., Schiff, J. A., Epstein, H. T., Zeldin, B.: Deoxyribonucleic acid of the blue-green algae (Cyanophyta). Bact. Rev. **31**, 315—331 (1967).

Ellis, R. J.: Chloroplast ribosomes: stereospecificity of inhibition by chloramphenicol. Science **163**, 477—478 (1969).

Galper, J. B., Darnell, J. E.: The presence of N-formylmethionyl-tRNA in HeLa cell mitochondria. Biochem. biophys. Res. Commun. **34**, 205—214 (1969).

Gantt, E.: An electron microscope study of phycoerythrin. J. Cell Biol. **39**, 49a (1968).

— Conti, S. F.: Granules associated with the chloroplast lamellae of *Porphyridium cruentum*. J. Cell Biol. **29**, 423—434 (1966).

Garren, L. D., Crocco, R. M.: Amino acid incorporation by mitochondria of the adrenal cortex: The effect of chloramphenicol. Biochem. biophys. Res. Commun. **26**, 722—729 (1967).

Geitler, L.: Der Zellbau von *Glaucocystis Nostochinearum* und *Gloeochaete Wittrockiana* und die Chromatophoren-Symbiosetheorie von Mereschkowsky. Arch. Protistenk. **47**, 1—24 (1923).

GIBOR, A.: Inheritance of cytoplasmic organelles. In K. B. WARREN: Formation and fate of cell organelles. P. 305—316. New York-London: Academic Press 1967.

— GRANICK, S.: Plastids and mitochondria: inheritable systems. Science **145**, 890—897 (1964).

GIESBRECHT, P.: Über das Ordnungsprinzip in den Chromosomen von Dinoflagellaten und Bakterien. Zbl. Bakt. Abt. 1. Orig. **196**, 516—519 (1965).

GROSS, S. R., MCCOY, M. T., GILMORE, E. B.: Evidence for the involvement of a nuclear gene in the production of the mitochondrial leucyl-tRNA synthetase of *Neurospora*. Proc. nat. Acad. Sci. (Wash.) **61**, 253—260 (1968).

HALL, W. T., CLAUS, G.: Ultrastructural studies on the blue-green algal symbiont in *Cyanophora paradoxa* Korschikoff. J. Cell Biol. **19**, 551—563 (1963).

— — Ultrastructural studies on the cyanelles of *Glaucocystis nostochinearum* Itzigsohn. J. Phycol. **3**, 37—51 (1967).

HASLAM, J. M., DAVEY, P. J., LINNANE, A. W., ATKINSON, M. R.: Differentiation in vitro by phenanthrene alkaloids of yeast mitochondrial protein synthesis from ribosomal systems of both yeast and bacteria. Biochem. biophys. Res. Commun. **33**, 368—373 (1968).

HEIL, K.: Stammen Chloroplasten und Mitochondrien von Bakterien ab? Naturwiss. Rdsch. **21**, 69—70 (1968).

HERRMANN, R. G.: Are chloroplasts polyploid? Exp. Cell Res. **55**, 414—416 (1969).

HOFFMAN, E. J.: The nucleic acids of basal bodies isolated from *Tetrahymena pyriformis*. J. Cell Biol. **25** (2, 1), 217—228 (1965).

HOLDSWORTH, R. H.: The presence of a crystalline matrix in pyrenoids of the diatom, *Achnanthes brevipes*. J. Cell Biol. **37**, 831—837 (1968).

HOLM-HANSEN, O., PRASAD, R., LEWIN, R. A.: Occurrence of α, ε-diaminopimelic acid in algae and flexibacteria. Phycologia **5**, 1—14 (1965).

HOOBER, J. K., BLOBEL, G.: Characterization of the chloroplastic and cytoplasmic ribosomes of *Chlamydomonas reinhardi*. J. molec. Biol. **41**, 121—138 (1969).

KADENBACH, B.: Synthesis of mitochondrial proteins: Demonstration of a transfer of proteins from microsomes into mitochondria. Biochim. biophys. Acta (Amst.) **134**, 430—442 (1967).

KAROL, M. H., SIMPSON, M. V.: DNA biosynthesis by isolated mitochondria: A replicative rather than a repair process. Science **162**, 470—473 (1968).

KAWAGUTI, S., YAMASU, T.: Electron microscopy on the symbiosis between an elysioid gastropod and chloroplasts of a green alga. Biol. J. Okayama Univ. **11**, 57—65 (1965).

KIRK, J. T. O.: Studies on RNA synthesis in chloroplast preparations. Biochem. biophys. Res. Commun. **16**, 233—238 (1964).

— TILNEY-BASSETT, R. A. E.: The plastids. Their chemistry, structure, growth and inheritance. London-San Francisco: W. H. Freeman & Comp. 1967.

KIRSCHNER, R. H., WOLSTENHOLME, D. R., GROSS, N. J.: Replicating molecules of circular mitochondrial DNA. Proc. nat. Acad. Sci. (Wash.) **60**, 1466—1472 (1968).

KLEIN, R. M., CRONQUIST, A.: A consideration of the evolutionary and taxonomic significance of some biochemical, micromorphological, and physiological characters in the thallophytes. Quart. Rev. Biol. **42**, 105—296 (1967).

KNAPP, E.: Über *Geosiphon pyriforme* Fr. Wettst., eine intrazelluläre Pilz-Algen-Symbiose. Ber. dtsch. bot. Ges. **51**, 210—216 (1933).

KROON, A. M., BORST, P., BRUGGEN, VAN E. F. J., RUTTENBERG, G. J. C. M.: Mitochondrial DNA from sheep heart. Proc. nat. Acad. Sci. (Wash.) **56**, 1836—1843 (1966).

KUBAI, D. F., RIS, H.: Division in the dinoflagellate *Gyrodinium cohnii* (Schiller). A new type of nuclear reproduction. J. Cell Biol. **40**, 508—528 (1969).

KÜNTZEL, H.: Proteins of mitochondrial and cytoplasmic ribosomes from *Neurospora crassa*. Nature (Lond.) **222**, 142—146 (1969).

— NOLL, H.: Mitochondrial and cytoplasmic polysomes from *Neurospora crassa*. Nature (Lond.) **215**, 1340—1345 (1967).

LANG, N. J.: The fine structure of blue-green algae. Ann. Rev. Microbiol. **22**, 15—46 (1968).

Lefort, M.: Sur le chromatoplasme d'une Cyanophycée endosymbiotique, *Glaucocystis nostochinearum* Itzigs. C. R. Acad. Sci. (Paris) **261**, 233—236 (1965).
— Pouphile, M.: Données cytochimiques sur l'organisation structurale du chromatoplasma de *Glaucocystis nostochinearum*. C. R. Soc. Biol. **161**, 992—994 (1967).
Leiner, M., Bhowmick, D. K.: Über *Pelomyxa palustris* Greeff. Z. mikr.-anat. Forsch. **77**, 529—552 (1967).
— Schweikhardt, F., Blaschke, G., König, K., Fischer, M.: Die Gärung und Atmung von *Pelomyxa palustris* Greeff. Biol. Zbl. **87**, 567— 591 (1968).
Linnane, A. W., Stewart, P. R.: The inhibition of chlorophyll formation in *Euglena* by antibiotics which inhibit bacterial and mitochondrial protein synthesis. Biochem. biophys. Res. Commun. **27**, 511—516 (1967).
Luck, D. J. L., Reich, E.: DNA in mitochondria of *Neurospora crassa*. Proc. nat. Acad. Sci. (Wash.) **52**, 931—938 (1964).
Margoliash, E.: Primary structure and evolution of cytochrome c. Proc. nat. Acad. Sci. (Wash.) **50**, 672—679 (1963).
Margulies, M. M., Gantt, E., Parenti, F.: In vitro protein synthesis by plastids of *Phaseolus vulgaris*. II. The probable relation between ribonuclease insensitive amino acid incorporation and the presence of intact chloroplasts. Plant Physiol. **43**, 495—503 (1968).
— Parenti, F.: In vitro protein synthesis by plastids of *Phaseolus vulgaris*. III. Formation of lamellar and soluble chloroplast protein. Plant Physiol. **43**, 504—514 (1968).
Margulis, L.: Evolutionary criteria in thallophytes: A radical alternative. Science **161**, 1020—1022 (1968).
Mereschkowsky, C.: Über Natur und Ursprung der Chromatophoren im Pflanzenreiche. Biol. Cbl. **25**, 593—604, 689—690 (1905).
Meyer, R. R., Simpson, M. V.: DNA biosynthesis in mitochondria: Differential inhibition of mitochondrial and nuclear DNA polymerases by the mutagenic dyes ethidium bromide and acriflavin. Biochem. biophys. Res. Commun. **34**, 238—244 (1969).
Mignot, J. P., Joyon, L., Pringsheim, E. G.: Quelques particularités structurales de *Cyanophora paradoxa* Korsch., protozoaire flagellé. J. Protozool. **16**, 138—145 (1969).
Nass, M. M. K.: Mitochondrial DNA. J. molec. Biol. **42**, 521—528 (1969).
— Buck, C. A.: Comparative hybridization of mitochondrial and cytoplasmic aminoacyl transfer RNA with mitochondrial DNA from rat liver. Proc. nat. Acad. Sci. (Wash.) **62**, 506—513 (1969).
— Nass, S., Afzelius, B. A.: The general occurrence of mitochondrial DNA. Exp. Cell Res. **37**, 516—539 (1965).
Nass, S., Nass, M. M. K.: Intramitochondrial fibers with DNA characteristics. II. Enzymatic and other hydrolytic treatments. J. Cell Biol. **19**, 613—629 (1963).
Neupert, W., Brdiczka, D., Bücher, T.: Incorporation of amino acids into the outer and inner membrane of isolated rat liver mitochondria. Biochem. biophys. Res. Commun. **27**, 488—493 (1967).
Parsons, D. F., Yano, Y.: The cholesterol content of the outer and inner membranes of guinea-pig liver mitochondria. Biochim. biophys. Acta (Amst.) **135**, 362—364 (1967).
Parsons, P., Simpson, M. V.: Biosynthesis of DNA by isolated mitochondria: incorporation of thymidine triphosphate-2-C^{14}. Science **155**, 91—93 (1967).
Parthier, B., Wollgiehn, R.: Nucleinsäuren und Proteinsynthese in Plastiden. In P. Sitte: Funktionelle und morphologische Organisation der Zelle. 3. wiss. Konf. Ges. Dtsch. Naturf. u. Ärzte: Probleme der biologischen Reduplikation. P. 244—270. Berlin-Heidelberg-New York: Springer 1966.
Pascher, A.: Über Symbiosen von Spaltpilzen und Flagellaten mit Blaualgen. Ber. dtsch. botan. Ges. **32**, 339—352 (1914).
Pikó, L., Blair, D. G., Tyler, A., Vinograd, J.: Cytoplasmic DNA in the unfertilized sea urchin egg: Physical properties of circular mitochondrial DNA and the occurrence of catenated forms. Proc. nat. Acad. Sci. (Wash.) **59**, 838—845 (1968).
Pringsheim, E. G.: Organismen mit blaugrünen Assimilatoren. Studies in Plant Physiology (Praha), S. 165—184 (1958).
— Farblose Algen. Ein Beitrag zur Evolutionsforschung. Stuttgart: Fischer 1963.

PYNE, M. C. K.: Sur l'absence d'incorporation de la thymidine tritiée dans les cinétosomes de *Tetrahymena pyriformis* (Ciliés holotriches). C. R. Acad. Sci. (Paris), Sér. D, **267**, 755—757 (1968).

RABINOWITZ, M., GETZ, G. S., CASEY, J., SWIFT, H.: Synthesis of mitochondrial and nuclear DNA in anerobically grown yeast during the development of mitochondrial function in response to oxygen. J. molec. Biol. **41**, 381—400 (1969).

RANDALL, J., DISBREY, C.: Evidence for the presence of DNA at basal body sites in *Tetrahymena pyriformis*. Proc. roy. Soc. (Lond.) B, **162**, 473—491 (1965).

RENNER, O.: Die pflanzlichen Plastiden als selbständige Elemente der genetischen Konstitution. Ber. sächs. Akad. Wiss., math. phys. Kl., **86**, 214—266 (1934).

RIS, H.: Ultrastructure of certain self-dependent cytoplasmic organelles. 5. Int. Congr. Electr. Micr. Philadelphia, XX 1, 1962.

ROBERTSON, J. D.: Unit membranes: A review with recent new studies of experimental alterations and a new subunit structure in synaptic membranes. In M. LOCKE: Cellular membranes in development. P. 1—81. New York-London: Academic Press 1964.

ROODYN, D. B., WILKIE, D.: The biogenesis of mitochondria. London: Methuen & Co. Ltd. 1968.

SAGAN, L.: On the origin of mitosing cells. J. theor. Biol. **14**, 225—274 (1967).

SCHIFF, J. A., EPSTEIN, H. T.: The continuity of the chloroplast in *Euglena*. In M. LOCKE: Reproduction: Molecular, subcellular and cellular. P. 131—189. New York-London: Academic Press 1965.

SCHIMPER, A. F. W.: Über die Entwicklung der Chlorophyllkörner und Farbkörner (1. Teil). Bot. Z. **41**, 105—114 (1883).

SCHNEPF, E.: Zur Feinstruktur von *Geosiphon pyriforme*. Ein Versuch zur Deutung cytoplasmatischer Membranen und Kompartimente. Arch. Mikrobiol. **49**, 112—131 (1964).

— Organellen-Reduplikation und Zellkompartimentierung. In: P. SITTE: Funktionelle und morphologische Organisation der Zelle. 3. wiss. Konf. Ges. dtsch. Naturf. u. Ärzte: Probleme der biologischen Reduplikation. S. 372—393. Berlin-Heidelberg-New York: Springer 1966a.

— Die Kompartimentierung der Zelle in morphologischer Sicht. Biol. Rdsch. **4**, 259—275 (1966b).

— KOCH, W., DEICHGRÄBER, G.: Zur Cytologie und taxonomischen Einordnung von *Glaucocystis*. Arch. Mikrobiol. **55**, 149—174 (1966).

SCHÖTZ, F.: Extrachromosomale Vererbung. Ber. dtsch. bot. Ges. **80**, 523—538 (1967).

SCHWARTZ, J. H., MEYER, R., EISENSTADT, J. M., BRAWERMAN, G.: Involvement of N-formylmethionine in initiation of protein synthesis in cell-free extracts of *Euglena gracilis*. J. molec. Biol. **25**, 571—574 (1967).

SCHWEIGER, H. G., BERGER, S.: DNA-dependent RNA synthesis in chloroplasts of *Acetabularia*. Biochim. biophys. Acta (Amst.) **87**, 533—535 (1964).

— MASTER, R. W. P., WERZ, G.: Nuclear control of a cytoplasmic enzyme in *Acetabularia*. Nature (Lond.) **216**, 554—557 (1967).

SCOTT, N. S., SHAH, V. C., SMILLIE, R. M.: Synthesis of chloroplast DNA in isolated chloroplasts. J. Cell Biol. **38**, 151—157 (1968).

— SMILLIE, R. M.: Evidence for the direction of chloroplast ribosomal RNA synthesis by chloroplast DNA. Biochem. biophys. Res. Commun. **28**, 598—603 (1967).

SHERMAN, F., STEWART, J. W., MARGOLIASH, E., PARKER, J., CAMPBELL, W.: The structural gene for yeast cytochrome c. Proc. nat. Acad. Sci. (Wash.) **55**, 1498—1504 (1966).

SINCLAIR, J. H., STEVENS, B. J., GROSS, N., RABINOWITZ, M.: The constant size of circular mitochondrial DNA in several organisms and different organs. Biochim. biophys. Acta (Amst.) **145**, 528—531 (1967).

SKUJA, H.: Glaucophyta. In A. ENGLER: Syllabus der Pflanzenfamilien. 21. Aufl. (H. MELCHIOR und E. WEDERMANN), Bd. I, pp. 56—57. Berlin: Borntraeger 1954.

SMILLIE, R. M., GRAHAM, D., DWYER, M. R., GRIEVE, A., TOBIN, N. F.: Evidence for the synthesis in vivo of proteins of the Calvin cycle and of the photosynthetic electron-transfer pathway on chloroplast ribosomes. Biochem. biophys. Res. Commun. **28**, 604—610 (1967).

Smith, A. E., Marcker, K. A.: N-formylmethionyl transfer DNA in mitochondria from yeast and rat liver. J. molec. Biol. **38**, 241—243 (1968).

Smith-Sonneborn, J., Plaut, W.: Evidence for the presence of DNA in the pellicle of *Paramecium*. J. Cell Sci. **2**, 225—234 (1967).

Sottocasa, G. L., Kuylenstierna, B., Ernster, L., Bergstrand, A.: An electron-transport system associated with the outer membrane of liver mitochondria. A biochemical and morphological study. J. Cell Biol. **32**, 415—438 (1967).

Spencer, D., Whitfeld, P. R.: DNA synthesis in isolated chloroplasts. Biochem. biophys. Res. Commun. **28**, 538—542 (1967).

Steinert, M.: L'absence d'histone dans le kinétonucléus des trypanosomes. Étude cytochimique. Exp. Cell Res. **39**, 69—73 (1965).

— Specific loss of kinetoplastic DNA in Trypanomatidae treated with ethidium bromide. Exp. Cell Res. **55**, 248—252 (1969).

Stubbe, W.: Die Plastiden als Erbträger. In: Funktionelle und morphologische Organisation der Zelle (Ed.: P. Sitte). 3. wiss. Konf. Ges. Dtsch. Naturf. u. Ärzte: Probleme der biologischen Reduplikation, S. 273—286. Berlin-Heidelberg-New York: Springer 1966.

Stutz, E., Noll, H.: Characterization of cytoplasmic and chloroplast polysomes in plants: Evidence for three classes of ribosomal RNA in nature. Proc. nat. Acad. Sci. (Wash.) **57**, 774—781 (1967).

Svetailo, E. N., Philippovich, I. I., Sissakian, N. M.: Differences in sedimentation properties of chloroplast and cytoplasmic ribosomes from pea seedlings. J. molec. Biol. **24**, 405—415 (1967).

Swift, H.: Nucleic acids of mitochondria and chloroplasts. Amer. Naturalist **49**, 201—227 (1965).

Tewari, K. K., Vötsch, W., Mahler, H. R., Mackler, B.: Biochemical correlations of respiratory deficiency. VI. Mitochondrial DNA. J. molec. Biol. **20**, 453—481 (1966).

— Wildman, S. G.: Function of chloroplast DNA, I. Hybridization studies involving nuclear and chloroplast DNA with RNA from cytoplasmic (80 S) and chloroplast (70 S) ribosomes. Proc. nat. Acad. Sci. (Wash.) **59**, 569—576 (1968).

Trench, R. K.: Chloroplasts as functional endosymbionts in the mollusc *Tridachia crispata* (Bergh), (Opisthobranchia, Sacoglossa). Nature (Lond.) **222**, 1071—1072 (1969).

Tuppy, H., Wildner, G.: Cytoplasmic transformation: Mitochondria of wild-type baker's yeast restoring respiratory capacity in the respiratory deficient "petite" mutant. Biochem. biophys. Res. Commun. **20**, 733—738 (1965).

— Wintersberger, E.: Mitochondrien als Träger genetischer Information. In: Funktionelle und morphologische Organisation der Zelle (Ed.: P. Sitte). 3. Wiss. Konf. Ges. Dtsch. Naturf. u. Ärzte: Probleme der biologischen Reduplikation, P. 323—335. Berlin-Heidelberg-New York: Springer 1966.

Wagner, R. P.: Genetics and phenogenetics of mitochondria. Science **163**, 1026—1031 (1969).

Werz, G., Kellner, G.: Molecular characteristics of chloroplast DNA of *Acetabularia* cells. J. Ultrastruct. Res. **24**, 109—115 (1968).

Wettstein, F. v.: *Geosiphon* Fr. Wettst., eine neue, interessante Siphonee. Österr. bot. Z. **65**, 145—156 (1915).

Whittaker, R. H.: New concepts of kingdoms of organisms. Science **163**, 150—160 (1969).

Note Added in Proof. Since the time this manuscript went to press, a number of important papers on this subject have been published; some of them are added to complete the article.

Ashwell, M. A., Work, T. S.: The functional characterization of ribosomes from rat liver in mitochondria. Biochem. biophys. Res. Commun. **39**, 204—211 (1970).

Attardi, B., Attardi, G.: Sedimentation properties of RNA species homologous to mitochondrial DNA in HeLa cells. Nature (Lond.) **224**, 1079—1083 (1969).

BENDICH, A. J., McCARTHY, B. J.: Ribosomal RNA homologies among distantly related organisms. Proc. Nat. Acad. Sci. **65**, 349—356 (1970).

BISALPUTRA, T., BURTON, H.: The ultrastructure of chloroplast of a brown alga, *Sphacelaria* sp. II. Association between the chloroplast DNA and the photosynthetic lamellae. J. Ultrastruct. Res. **29**, 224—235 (1969).

BORST, P., KROON, A. M.: Mitochondrial DNA: physicochemical properties, replication, and genetic function. Int. Rev. Cytol. **26**, 107—190 (1969).

BROWN, T. E., RICHARDSON, F. L., PATTHOFF, D. E., WOODRUFF, M. N.: A physiological and ultrastructural characterization of *Glaucosphaera vacuolata* with some comparative data for *Cyanophora paradoxa*. 11. Int. Bot. Congr., p. 24 (1969).

CLAUSS, H., LÜTTKE, A., HELLMANN, F., REINERT, J.: Chloroplastenvermehrung in kernlosen Teilstücken von *Acetabularia mediterranea* und *Acetabularia cliftonii* und ihre Abhängigkeit von inneren Faktoren. Protoplasma **69**, 313—329 (1970).

EDELMAN, M., VERMA, I. M., LITTAUER, U. Z.: Mitochondrial ribosomal RNA from *Aspergillus nidulans:* characterization of a novel molecular species. J. Mol. Biol. **49**, 67—83 (1970).

HENDLER, R. W., BURGESS, A. H., SCHARFF, R.: Respiration and protein synthesis in *Escherichia coli* membrane-envelope fragments. I. Oxidative activities with soluble substrates. J. Cell Biol. **42**, 715—732 (1969).

HERRMANN, F., BAUER-STÄB, G.: Lamellarproteine mutierter Plastiden der Plastommutante *albomaculata*-1 von *Antirrhinum majus* L. Flora, Abt. A **160**, 391—393 (1969).

HERRMANN, R. G., KOWALLIK, K. V.: Multiple amounts of DNA related to the size of chloroplasts. II. Comparison of electron-microscopic and autoradiographic data. Protoplasma **69**, 365—372 (1970).

KNIGHT, E., JR., SUGIYAMA, T.: Transfer RNA: a comparison by gel electrophoresis of the tRNA in HeLa cytoplasm, HeLa mitochondrial fraction, and *E. coli*. Proc. Nat. Acad. Sci. **63**, 1383—1388 (1969).

LEEDALE, G. F.: Observations on endonuclear bacteria in euglenoid flagellates. Österr. bot. Z. **116**, 279—294 (1969).

MEYER, R. R., SIMPSON, M. V.: DNA biosynthesis in mitochondria: differential inhibition of mitochondrial and nuclear DNA polymerases by the mutagenic dyes ethidium bromide and acriflavin. Biochem. biophys. Res. Commun. **34**, 238—244 (1969).

NASS, M. M. K.: Mitochondrial DNA. I. Intramitochondrial distribution and structural relations of single- and double-length circular DNA. J. mol. Biol. **42**, 521—528 (1969).

— Mitochondrial DNA: Advances, problems and goals. Science **165**, 25—35 (1969).

— Uptake of isolated chloroplasts by mammalian cells. Science **165**, 1128—1131 (1969).

— BUCK, C. A.: Comparative hybridization of mitochondrial and cytoplasmic aminoacyl transfer RNA with mitochondrial DNA from rat liver. Proc. Nat. Acad. Sci. **62**, 506—513 (1969).

NASS, S.: The significance of the structural and functional similarities of bacteria and mitochondria. Int. Rev. Cytol. **25**, 55—129 (1969).

SCHENK, H. E. A.: Nachweis einer lysozymempfindlichen Stützmembran der Endocyanellen von *Cyanophora Paradoxa* Korschikoff. Z. Naturforsch. **25b**, 656 (1970).

SCHNAITMAN, C. A.: Comparison of rat liver mitochondrial and microsomal membrane proteins. Proc. Nat. Acad. Sci. **63**, 412—419 (1969).

SCHWEYEN, R., KAUDEWITZ, F.: Protein synthesis by yeast mitochondria in vivo. Quantitative estimate of mitochondrially governed synthesis of mitochondrial proteins. Biochem. biophys. Res. Commun. **38**, 728—735 (1970).

SHMERLING, Zh. G.: The effect of rifamycin of RNA synthesis in the rat liver mitochondria. Biochem. biophys. Res. Commun. **37**, 965—969 (1969).

SURZYCKI, S. J.: Genetic functions of the chloroplast of *Chlamydomonas reinhardi:* effect of rifampin on chloroplast DNA-dependent RNA polymerase. Proc. Nat. Acad. Sci. **63**, 1327—1334 (1969).

TAYLOR, D. L.: Chloroplasts as symbiotic organelles. Int. Rev. Cytol. **27**, 29—64 (1970).

Taylor, F. J. R., Blackbourn, D. J., Blackbourn. J.: Ultrastructure of the chloroplasts and associated structures within the marine ciliate *Mesodinium rubrum* (Lohmann). Nature **224**, 819—821 (1969).

Trench, R. K., Greene, R. W., Bystrom, B. G.: Chloroplasts as functional organelles in animal tissues. J. Cell Biol. **42**, 404—417 (1969).

Wilson, J. F.: Mitochondrial transplantation studies in *Neurospora*. 11. Int. Bot. Congr., p. 240 (1969).

Wood, D. D., Luck, D. J. L.: Hybridization of mitochondrial ribosomal RNA. J. mol. Biol. **41**, 211—224 (1969).

Woodcock, C. L. F., Bogorad, L.: Evidence for variation in the quantity of DNA among plastids of *Acetabularia*. J. Cell Biol. **44**, 361—375 (1970).

Cell Organelles
and the Differentiation of Somatic Plant Cells

F. A. L. CLOWES

Botany School, Oxford University

I. Compartments and Autonomy

Organelles enable a diversity of chemical pathways to occur simultaneously within a single cell. The division of a cell into compartments must confer an enormous advantage upon organisms and this must be true however the organelles have been acquired in the course of evolution. It is possible to believe that chloroplasts are blue-green algae that have been captured and enslaved by larger cells and that, in subsequent evolution, the alga has become specialized only for photosynthesis and related reactions and the rest of the cell has lost any other photosynthetic apparatus if it possessed any originally. This is a theory going back more than 60 years that has revived recently with the discovery that plastid ribosomes differ from cytoplasmic ribosomes of the same cell in size and in their RNA and that they resemble those of blue-green algae and bacteria (STUTZ and NOLL, 1967). The partial autonomy and specific DNA of plastids conforms with the theory, although we have to explain how the nucleus comes to contain some of the genes affecting chloroplasts and how photosynthesis comes to be regulated by the nucleus although the plastids appear to have all the apparatus necessary for the process. I shall discuss later the roles of nuclear and plastid genes. Nuclear control of photosynthesis is shown by experiments on *Acetabularia* by SCHWEIGER and SCHWEIGER (1965) in which the nucleus and the plastids were given inherent rhythms out of phase with each other and evolution of oxygen followed the nuclear rhythm, not that of the plastids.

An interesting side light on this problem is provided by recent work on the unicellular algae that live symbiotically inside animal cells. This work is reviewed by SMITH, MUSCATINE, and LEWIS (1969). The algae give carbohydrates to the animals, which become infected by the algae either systemically or by ingestion and the algal cells lose their flagella and eyespots, if they had them, and sometimes their walls as well. In certain molluscs the algal symbionts are actually chloroplasts derived from aseptate algae of the Bryopsidophyceae on which the animals feed. Some of these chloroplasts are capable of functioning for at least six weeks when the animal is starved. At least one animal can take in several species of algae as symbionts, but the one found in nature promotes the growth of the host better

21*

than the others and the animal will eliminate existing symbionts in its preference for this species.

A similar theory of endosymbiosis may also be true of mitochondria and bacteria, flagella and motile parasites. We may imagine that an organism which had enslaved either algae or bacteria would become more efficient in its cell chemistry than one in which all reactions occur within a single compartment. This would, of course, be equally true if, instead, organelles evolved by the specialization of regions of the protoplasm, by the differentiation of a bounding membrane for respiration as in the special membranes of anaerobic yeasts, or by the adaptation of a vacuole by increasing the complexity of its bounding membrane as may occur in the onto-genetic origin of plastids and mitochondria. But the possession of 70 S ribosomes by both these organelles as opposed to the 80 S ribosomes of the cytoplasm of eukaryotic cells suggests a symbiotic origin and may allow for two quite separate protein-synthetic systems within a cell. These are inhibited by different drugs and their diversity may be advantageous to a cell (see also Baxter's contribution in this volume).

II. The Differences between Cells in Their Organelles

The problem of differentiation is that all the cells of an individual (with a few exceptions) inherit all the genes of the precursor zygote or spore, but they come to differ from one another. The problem of organelles and their relation to differentiation in plants is this: all cells possess mitochondria, Golgi bodies, endoplasmic reticulum, ribosomes, vacuoles, microtubules; most plant cells also possess plastids. Yet there are differences between cells which must come about largely by the activity of the organelles. How do these differences arise? Five general points emerge:

1. It is the cytoplasm that decides whether the nucleus shall synthesise DNA or RNA, whether the cell shall divide or differentiate. Many kinds of experiments including nuclear transplantation and fusion of unlike cells of animals show this to be so.

2. There are differences between cells in the numbers of particular organelles, but there is not much systematic data in relation to cells of different kinds. Many cytoplasmic organelles increase their numbers proportionately with increase in cell size. This is true of the almost unvacuolate cells of root caps (Fig. 1) where the number of mitochondria and Golgi and the area of endoplasmic reticulum keep pace with expanding cytoplasmic volume (Juniper and Clowes, 1965).

3. There are differences in the development of the organelles. This is the most marked for plastids where even the light microscope shows that some tissues have chloroplasts, others amyloplasts and others proplastids. The electron microscope reveals differences in the size of the Golgi and their output of vesicles and also in the degree of development of the cristae of the mitochondria in different tissues and sometimes in the same cell. The massive slab-like mitochondria of insect flight muscle are conspicuous examples of an organelle changed in size according to the kind of cell it occupies. Other sorts of mitochondrial heterogeneity have been reported for plants (Avers and King, 1960) and for animals and yeasts.

A special problem here concerns those organisms that produce cilia or flagella in some of their cells. In some animals the centrioles present in all the cells seem to play a part in producing the precursors of cilia and flagella. They do this either by breaking up into a number of component blepharoplasts which move to the surface of the cell and act as basal granules or by simply migrating to the surface and becoming a basal granule from which the flagellum grows. In higher plants

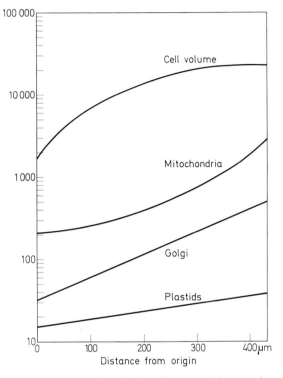

Fig. 1. Numbers of organelles per cell and cell volume (μm^3) in maturing root cap cells of Zea mays. Mitochondria and Golgi keep pace with expanding volume; plastids do not. Data from JUNIPER and CLOWES (1965)

centrioles do not exist until the time approaches for the production of a flagellate cell – within the antheridium in *Marsilea* (MIZUKAMI and GALL, 1966). Yet, on the analogy of animals, it is expected that blepharoplasts do not arise *de novo* in the cell lineage leading up to sperms, but must come from a pre-existing organelle which multiplies in step with mitosis. It may be that techniques are still inadequate to reveal such an organelle in plants or it may be that its role has not yet been recognised. SAGAN (1967) believes that the basal granules of flagella are derived from symbiotic motile organisms, perhaps like Spirochaetes. These may have later evolved to become the centromeres of chromosomes and the centrioles of spindles as well as the organisers of flagella. Origin from a parasite would explain partial autonomy and self-replication as in plastids and mitochondria. Centrioles have their own double-stranded DNA like plastids and mitochondria, but they are usually inherited patroclinally.

4. There are differences in the positions occupied by some organelles within cells. In multicellular plants this phenomenon is usually related to asymmetrical cell division. The pre-prophase band of microtubules that lies in a ring around the cell periphery at the level of the equator of the nucleus is asymmetrical when the next cell division is asymmetrical as in the production of the guard cells of stomata and trichoblasts in certain roots (Pickett-Heaps and Northcote, 1966a, b). This special band of microtubules appears before the onset of mitosis and possibly orientates the nucleus ready for division, but we do not know whether the microtubules are created specially for this purpose or whether the ordinary microtubules which occur all the time all over the cell's periphery are moved to form the band.

5. The most prominent differences between plant cells are in the structure of their walls. There is now evidence that wall synthesis is partly effected by the activity of various cytoplasmic organelles. The Golgi, microtubules and endoplasmic reticulum all appear to be concerned in wall synthesis and so their activity and positions must be a causal factor in the differentiation of cells and it is almost certainly true that we have not yet discovered all the significant facts about this.

III. Replication and Inheritance

The partial autonomy of organelles presents interesting problems in the study of differentiation. The evidence for autonomy of certain organelles is presented in the other chapters of this volume and is summarized by Clowes and Juniper (1968). Briefly, just as all cells are derived from other cells, so many organelles are derived from other organelles of the same kind. For plastids, the genetic evidence for continuity through generations of cells is strong. The non-Mendelian inheritance of plastids and the way in which mutant plastids sort out in the development of plants show that they replicate over many cell generations and are also passed on from one plant to another in sexual reproduction. Whether there is ever a stage from which all plastids are eliminated and then created *de novo* is disputed. Bell and Mühlethaler (1964) suggest that this occurs in the eggs of *Pteridium*. Certainly some extraordinary process goes on in these cells and some people believe that a purge of all of the self-replicating parts of a cell is likely once in a life cycle so that new parts could be specified by the new nuclear DNA of the zygote or spore. However, the evidence for this in relation to organelles is small. Bell and Mühlethaler hold that, in *Pteridium*, new membranes evaginate from the nuclear envelope and form plastids and mitochondria in the egg. Once formed in this way, they presumably reproduce themselves to populate the somatic cells. Many people doubt whether the membrane-bound vesicles seen in *Pteridium* eggs are plastids and mitochondria and others doubt whether any similar cleaning up process ever occurs in other plants (Diers, 1966; Jensen, 1965). If it does, it may be merely an example of what is called dedifferentiation, a process that occurs whenever a differentiated cell becomes meristematic. Dedifferentiation is easily recognised because the cell changes its affinity for dyes probably by increasing the amount of RNA in the cytoplasm. Other processes also occur during dedifferentiation. In *Splachnum* both plastids and mitochondria produce little buds which probably separate and grow into new organelles like their parent. This enables a rapid increase in the numbers of organelles to occur when cell division is accelerated (MALTZAHN and MÜHLETHALER, 1962). A period of cell division appears always to intervene when a differentiated cell changes to some other course of differentiation. A common example of this is in the pericycle of roots when lateral roots are initiated. This usually, though not invariably, occurs in parenchyma outside the apical meristem of the mother root. DNA is synthesised in a few contiguous cells, the nuclei swell, RNA increases in the cytoplasm and then mitosis occurs to form a primordium. There is thus a reversal in ageing which probably becomes more difficult the older the tissue (STANGE, 1965). A point about dedifferentiation in somatic cells that is not known yet is whether only the mature organelles themselves proliferate or whether a reservoir of their precursors is always present and provides for such contingencies. In this connexion it has been shown by CAPORALI (1959) that two sorts of amyloplasts develop in the roots of *Lens*. One sort develops early and the other much later. MARINOS (1967) believes that in dormancy the plastids present in meristems are potentially multifunctional, being able to store a variety of substances – starch, lipids, nucleic acids, proteins or phytoferritin in potato tuber buds, for example. But in general a cell contains only one kind of plastid except after mutation.

In *Pteridium*, BELL, FREY-WYSSLING and MÜHLETHALER (1966) show that the prolific evagination of the nuclear membrane starts before fertilization and ends before the male nucleus becomes genetically active. So, in this case, it is not the new DNA of the zygote that is specifying the plastids and mitochondria, but a maternal nucleus, albeit after meiosis. It is possible therefore to reconcile BELL's theory with matriclinal inheritance, should this be necessary.

It may be that plastids repress the nuclear genes concerned with plastid production and only on the rare occasions when plastids are completely eliminated do the nuclear genes become derepressed. So many of the structural genes which affect plastid development are known to reside in the nucleus: the evidence comes from experimental breeding that distinguishes between Mendelian and cytoplasmic inheritance. Therefore it is tempting to assume that the function of the organelle DNA is to supply the regulator genes. As KIRK (1966) has pointed out, the fact that "removal" of chloroplast DNA by streptomycin or ultra-violet rays leads to cessation of synthesis of many plastid compounds does not disprove the regulatory theory about plastid DNA. The chloroplasts may well lose their ability to make ribosomal RNA by the treatment although the nucleus is still providing messenger RNA for the chloroplasts. Also it is likely that treatments do not remove plastid DNA, but probably cause the genetic code to be read wrongly. The alternative theory that the organelles carry the structural genes and the nucleus the regulator genes seemed to KIRK to be less likely because the nuclear gene mutations affecting plastids are usually recessive to the wild type and they should be dominant if the mutation merely caused the production of a faulty repressor. However, this criticism assumes the regulators to be of the negative type. There is now evidence for positive regulators i.e. genes that switch on rather than off and so the argument against nuclear regulators of organelle structural genes is now invalid.

On the analogy of mitochondrial systems, it is possible that the plastid proteins are synthesised according to the instructions of the plastid DNA and that the rest is under nuclear control. The use of inhibitors of bacterial protein synthesis to inhibit specifically the 70 S ribosomes of higher organisms has elucidated this problem and that of the synthesis of the organelle ribosomes themselves. In *Neurospora*, where the ribosomal proteins are mostly different in the mitochondrial (73 S) ribosomes compared with the cytoplasmic (77 S) ribosomes, Küntzel (1969) has shown that the proteins of both kinds of ribosomes are synthesised by cytoplasmic ribosomes and probably coded for by the nuclear DNA. This means that the mitochondrial ribosomal proteins must be passed into the mitochondria from outside although the RNA of the mitochondrial ribosomes is made inside the mitochondria and specified by mitochondrial DNA. We also know that some of the mitochondrial enzymes are specified by the nuclear DNA and that the structural genes for the flagellar apparatus in *Chlamydomonas* are nuclear (Randall, 1969).

Clearly the zygote or spore inherits more than the nucleus given to it by syngamy or meiosis and the ability of plastids and some other organelles to replicate themselves makes this an important point genetically.

Plastids, like everything else in the cell, come under the influence of the nucleus. In fact some chloroplasts fail to develop in the presence of certain nuclei. Nuclear genes can produce mutations in plastids, but, once changed, the mutant plastid produces more mutant plastids of the same kind even when the original nucleus that caused the mutation is replaced by one lacking the mutated gene. Mutations can also occur directly in the plastids themselves without a nuclear gene intervening. It may be supposed that the plastid DNA is responsible for such mutations.

There is good evidence also for the self-replication of mitochondria in fungi and we may assume that the same happens in higher plants. It is not so easy to obtain the necessary information for a rigid proof because higher plants are absolutely dependent on mitochondria for respiration. Alternative systems are available for anaerobic respiration in yeasts and so mutant mitochondria can survive there. Similarly cells containing mutant chloroplasts without chlorophyll survive only because they are supported by neighbouring green cells as in variegated plants or because they become holozoic as in *Euglena*. It is assumed that mitochondrial DNA, which is different from both the nuclear and plastid DNA's, is partly responsible for self-replication, probably through specifying the mitochondrial ribosomes. Cytoplasmic mutants of yeast are known with characteristically modified mitochondria (Yotsuyanagi, 1962).

Centrioles and blepharoplasts also appear to reproduce themselves in animals though it is not clear exactly what happens in higher plants. It is also likely that Golgi bodies can divide. They certainly keep pace in numbers with meristematic activity and growth of cells and it sometimes looks as if they are dividing into two, either by separation of cisternae or by fission across the diameter of a dictyosome (Clowes and Juniper, 1968, Figs. 53, 60, and 62; Morré, Mollenhauer, and Bracker, this volume). There is little evidence that any of the other organelles are self-replicating and the Golgi bodies are probably not autonomous.

The significance for differentiation of the partial autonomy of organelles lies in the balance between the nucleus and the organelles in determining what kind of organelles are produced in a particular cell. If the nucleus is wholly in control of organelle development, then the sort of organelle produced depends on the repression of the nuclear genes and presumably there is some feedback mechanism from the cytoplasm, or even the existing organelles, instructing the nucleus to produce the appropriate message or not. If the organelle's own DNA is in control of development, presumably some similar feedback system acts directly on the organelle or *via* the nucleus. Plastids have about 300 μm or 10^{-15} g of DNA per plastid. This comes to about 2 million nucleotides and, if the DNA is double-stranded and "haploid" and there are three bases to a codon, there is enough for 1 500 genes. This figure is reduced if the DNA strands are duplicated as appears likely from some electron micrographs or if the DNA is "polytene". 1 500 seems to be adequate for all that goes on in a plastid, but it has to be admitted that ideas on structural and regulator genes in plants are entirely conjectural still and some of the 2 million nucleotides are used for coding for ribosomal and transfer RNA and also perhaps for ribosomal proteins. Mitochondrial DNA has a more limited maximal coding capacity, enough for about 30 proteins (ROODYN and WILKIE, 1968). It may specify the RNA and proteins for mitochondrial ribosomes and the nuclear DNA may specify most of the other mitochondrial proteins. This assumes that messenger-RNA can enter the mitochondria, but certainly the formation of mito-chondria is regulated by nuclear genes as well as by the mitochondria themselves.

Centrioles also contain DNA (about 2×10^{-16} g). This has been proved for the basal granules concerned with producing cilia in *Paramecium* and *Tetrahymena* (RANDALL and DISBREY, 1965) and is likely for those concerned with mitosis even in those ciliates whose centrioles appear not to be related ontogenetically with the basal granules.

IV. How Differences between Cells Arise

Differences between cells in their content of organelles can arise by somatic mutations, either nuclear or cytoplasmic. Obviously this is not a normal method of differentiation, but if the mutation is cytoplasmic and if it does not affect the proliferation of the organelle, it could give rise to heterogeneous cells which, in a non-meristematic tissue, might increase the versatility of the cell and, in a meriste-matic tissue, give rise by sorting out to mosaics and chimeras, some of which become fairly stable because of the segregation of the layers in angiosperm shoot apices by the method of cell division (CLOWES, 1961; NEILSON-JONES, 1969). Others are fairly permanent in the absence of discrete germ layers because of the high frequency of branching at the shoot apex which never allows the shoot system to become populated wholly either by the mutant cells or by the normal cells. The normal controls of differentiation are imposed on chimeras. For example, in *Pelargonium zonale* cultivar Freak of Nature, the epidermal cells of leaves are genetically capable of producing green plastids and the other cells of the leaves are genetically capable of producing white plastids only. But the epidermis does not produce chloroplasts because the outer layer of cells in this species never does so (except in guard cells). We know that its cells are potentially green because it proliferates periclinally at the margins of the chimerical leaves to produce green internal tissue.

An interesting point about these genetical changes is the influence they have on other cells without the mutation. In chimeras with cells whose chloroplasts are genetically white, the genetically green tissue nearby is also sometimes white. The phenomenon is called "bleaching" and has been known for some time. An example of it occurs in a yellow and green striped *Hemerocallis* that was shown by THIELKE (1955) to be a mericlinal chimera with a genetically green epidermis. But the guard cells, instead of all having green plastids, have green ones only over the green strips of mesophyll and non-green ones over the yellow strips.

A similar interaction which results in a kind of intercellular complementation has been described by STUBBE (1958) for *Oenothera albivelutina*, a hybrid between *O. suaveolens* and *O. lamarckiana*. This plant may have yellow-green plastids derived from *O. suaveolens* and pale plastids derived from *O. lamarckiana*. But individuals also have cells with green plastids unlike those of either parent. STUBBE showed that the green plastids are those of *O. lamarckiana* which had greened under the influence of the yellow-green *O. suaveolens* plastids. The green cells exist only above and below tissue containing *suaveolens* plastids. As with "bleaching" the effect is transmitted to overlying and underlying cells, not to adjacent cells to the side. It looks as if the *suaveolens* plastids can provide something essential for the greening of the otherwise pale *lamarckiana* plastids.

More important for differentiation are those differences between cells that arise from gradients set up in cells. Gradients occur in cells for many reasons. For free egg cells it has been shown that light, temperature, pH, oxygen and the point of insertion of the sperm can all initiate polarity. In higher plants the egg is not free and a gradient is imposed by the maternal tissue. But, even if we start with a homogeneous cell, mitosis itself imposes a gradient on a cell lineage. BROWN (1958) points out that the half-spindles of the first mitosis give the daughters a polarity to the cytoplasm so that, of two sister grand-daughter cells, one contains the remnants of the pole of the first spindle and the other contains the remnants of the equator. The spindle remnants themselves may be unimportant, but the organelles it excludes or attracts may be. Further divisions in parallel planes enhance the differences between the ends of a file of cells derived from a single cell in this way. This source of heterogeneity has never been worked on in sufficient detail for us to know how it affects organelles except in the most obvious examples — cells that divide asymmetrically. Among these, AVERS (1963) showed that there was a polar distribution of endoplasmic reticulum in the precursor of a trichoblast and its non-hair-forming sister cell in *Phleum* roots. It may again be of no special significance that one cell receives more endoplasmic reticulum than another, but the endoplasmic reticulum may have displaced some of the self-replicating organelles so that these may account for or enhance the difference between the two daughter cells. Another example of a cytoplasmic gradient leading up to an asymmetrical division occurs in angiosperm zygotes. JENSEN (1965) has shown that the inner of the two daughter cells of the two-celled filament which results from the first division of the zygote is provided with more plastids, larger mitochondria and more RNA than the cell nearest to the micropyle. This polarity is increased in the subsequent development, the inner cell giving rise to most of the embryo and its sister often to only the basal cell and suspensor. The differentiation between the generative cell and the vegetative cell of a pollen grain also follows an unequal distribution of RNA (LA COUR, 1949). Several other examples of asymmetry within cells being followed by different fates of the daughter cells are known. STEBBINS

(1966) explains the origin of such asymmetrical cells in terms of their reaction to several hormones flowing from the apex and from the base of the organ containing the cell. The differential effect, he believes, is enhanced by an impedence to the flow of hormones operated by the meristematic cells themselves and by precociously thickened cell walls. The fate of the cell is decided by its state of maturity at the time it comes under these opposing influences and this depends on the state of activation and repression of the genes.

Plasmodesmata are not evenly distributed around cells. Preferential growth in one direction and cell division across the axis of elongation ensure that transverse walls have more plasmodesmata than longitudinal walls. JUNIPER and BARLOW (1969) think that this asymmetry may impose gradients in tissues with polar growth on the supposition that plasmodesmata are paths of conduction.

Even where there is no asymmetry in the environment to initiate heterogeneity in a homogeneous initial cell, heterogeneity may arise by the initiation of instability in the way suggested by TURING (1952) on purely theoretical grounds. TURING showed that the random movements of molecules of two substances reacting together in an initially homogeneous system would inevitably lead to inhomogeneity which would build up in the way that standing waves build up in an oscillatory circuit to form a pattern to fit the dimensions of the system. In the cultures of free cells used by STEWARD (1966) and others the formation of clumps of cells or embryo-like organs is presumably preceded by the division of the initial cell in a plane determined by the onset of some such internal inhomogeneity.

V. Differentiated Cells and Their Organelles

A. Plastids

We know more about plastids in relation to differentiation than about any other organelle because plastids are easily seen and many of the differences between them are easily seen, too. This results in different names being used — chloroplast, chromoplast, amyloplast, leucoplast, proteinoplast, elaioplast etc. The genetic relationship between all these is, however, obscure. In the algae and some of the other lower plants the chloroplasts reproduce themselves by fission and do not stem from proplastids. But in higher plants, chloroplasts do not exist in the embryos or meristems and probably they proliferate only at the final stages of cell differentiation as in leaf mesophyll or in extraordinary circumstances. Precursors of chloroplasts, called proplastids, occur in meristematic cells of shoots and it is assumed that the proliferation of these accounts for most of the plant's quota of chloroplasts. It is assumed, too, that proplastids are the precursors of all other kinds of plastids though these may go through a different differentiated state first. What we do not know is whether all proplastids are capable of giving rise to any kind of plastid or whether there are genetically distinct "races" of proplastids. At one time, shortly after the introduction of the electron microscope, it was widely believed that there were common precursors of both plastids and mitochondria. The reason for this was that it was impossible to distinguish between early stages of the two and this view has become less popular now because electron microscopists recognise earlier stages by sight and because the genetical evidence from *Euglena* suggests otherwise.

In strains of *Euglena* it is possible to stop for ever the production of chloroplasts by treatment with streptomycin, yet mitochondria are presumably produced to cope with respiration in the colourless future generations.

Proplastids are known to differentiate as amyloplasts, chloroplasts and chromoplasts. It is possible that they also give rise to proteinoplasts and elaioplasts, but the evidence is not adequate. Chloroplasts can give rise to amyloplasts and *vice versa*. They can also change into chromoplasts, as can amyloplasts. It is possible that chromoplasts can change to chloroplasts and that chloroplasts can change to elaioplasts. On this evidence, it looks as if proplastids are totipotent for all kinds of plastids except perhaps proteinoplasts and elaioplasts. The only plastids known to be able to divide are the proplastids and chloroplasts, but the others have not received enough attention for us to be sure if they proliferate or not. Against the totipotency view is the fact that some kinds of plastids, those in the roots of some species for example, do not change to chloroplasts when light is supplied although root plastids of other species do turn green. It may merely be that the right conditions for conversion are absent in some plants.

Certainly a cell seems normally to have only plastids of one kind. We do not find cells with both chloroplasts and amyloplasts and it looks as if all the proplastids of a cell are usually converted to one sort of plastid simultaneously.

A slightly different situation exists for centrioles, for, although transplanting tracts of cilia proves that the basal granules are genetically totipotent (NANNEY and RUDZINSKA, 1960), they produce different patterns of cilia in different parts of the same ciliate cell.

Chloroplasts and the proplastids that give rise to them often contain small starch grains even before the thylakoid system develops so that the course of differentiation to chloroplasts or amyloplasts may not be determined until a fairly late stage. The glucose subunits presumably can come from outside the plastids in both cases though the starch synthesising enzymes must be present within the plastids. Although chlorophyll itself is not usually formed in the dark, it is not only light that initiates the start of differentiation to chloroplasts. In the dark the vesicles attached to the inner membrane of a chloroplast envelope aggregate into tubes in a regular array and form the prolamellar bodies. This never happens in developing amyloplasts. In the light no prolamellar body is produced in chloroplasts and as little as two minutes of light can disperse a prolamellar body already formed. Protochlorophyll occurs in prolamellar bodies and these may be the earliest sites of chlorophyll synthesis when their tubes are converted into the vesicles of the thylakoid system (VIRGIN, KAHN, and WETTSTEIN, 1963; BOARDMAN and ANDERSON, 1964). There may also be some focus for the production of thylakoids and chlorophyll in the absence of a prolamellar body and in some species a prolamellar body is produced in both light and dark.

Another example of diversification from similar proplastids occurs in the leaves of *Saccharum* where there are dimorphic chloroplasts. Those in the bundle sheath have no grana, but are larger and accumulate a lot of starch. Those in the mesophyll have normal grana, but little starch. LAETSCH and PRICE (1969) have shown that diversification occurs fairly late in the maturation of the organelles and that both sorts initially produce grana. The chloroplasts in the bundle sheath begin to lose their grana by reducing the number of constituent thylakoids at the time when the plastid contains about 75% of its maximum chlorophyll. There are prominent plasmodesmata between the two tissues and this suggests that the two kinds of cells specialise by producing carbohydrate at one site and storing it at another and this is done by modifying the chloroplast development.

The plastid ribosomes contain RNA of base composition different from that of the cytoplasmic ribosomes (BRAWERMAN, 1962) and are presumably responsible for the production of the plastid proteins. If this is so, it seems likely that some kind of repression and derepression of the plastid DNA genes occurs analogous to that of nuclear genes to determine the course of differentiation of the plastids. It is, of course, not known whether this occurs at the transcriptional or translational level. MORTON and RAISON (1963) reported that actinomycin D, which inhibits nuclear DNA-dependent RNA synthesis, does not inhibit incorporation of amino acids into the proteins of proteinoplasts in the endosperm of *Triticum*. These proteinoplasts, like other plastids, contain their own ribosomes, but the transcription of the plastid DNA, if there is any here, may be protected from the drug by the plastid.

Within the main classes of plastids there are further subdivisions which are related to differentiation of the enclosing cells. Among amyloplasts, for example, there are differences in behaviour according to the tissue. In the pollen of *Zea* the amyloplasts have a single starch grain (LARSON, 1965) whereas in the roots of *Zea* each amyloplast has several grains. Those amyloplasts that appear to act as statoliths in geotropically sensitive regions of the plant differ from other amyloplasts. They exist in statenchyma such as root caps and root endodermises even in species that do not otherwise make starch such as *Allium* and *Iris*. In *Allium* the starch stains differently from normal starch. Statoliths withstand starvation of the plant longer than other amyloplasts (AUDUS, 1962). In this connexion it is worth noting that THIMANN and CURRY (1961) have proposed that another kind of plastid may behave in relation to light as amyloplasts do in relation to gravity. They think that proplastids containing carotenoids in the tips of shoots may migrate to the wall furthest from a light source and there, perhaps by enhancing the flow of auxin through that wall, cause the positive phototropic response of the shoot.

Some of the differences between cells in their plastid complement can be ascribed to environmental differences. It is known, for example, that nitrogenous fertilizers affect the number of chloroplasts per cell and light, as already mentioned, affects the development of chloroplasts – so much so that the chloroplasts of dark-grown plants are called etioplasts. Under some circumstances chloroplasts and amyloplasts are converted into chromoplasts, plastids containing pigments other than chlorophyll. In this process the lamellae usually disappear and carotenoids and fats accumulate. In some tissues the carotenoids form fibrils which distort the plastid. These conversions take place in special tissues such as petal mesophyll and fruits and are controlled by the nuclear genes of the plant and are selected for roles in attracting animals. But similar conversions also occur of chloroplasts in plants deficient of certain minerals (THOMSON, WEIER, and DREVER, 1964) and in the autumn coloration of leaves.

An interesting point in connexion with tissue specificity of organelles is that when cells of the gastrodermis of *Hydra viridis* dedifferentiate to form ectodermal cells after excision they throw out the algal symbionts that they formerly contained (HAYNES and BURNETT, 1963).

The greening of plastids has already been mentioned in connexion with genetic and physiological differences between cells. There is still much that we do not understand about this — why genetically competent cells of the epidermis often do not produce chloroplasts, for instance. They contain competent nuclei and proplastids that are capable of becoming chloroplasts in inner tissues of a leaf. They receive illumination adequate for thylakoid formation and chlorophyll synthesis.

Sunderland and Wells (1968) have worked on this problem on cultured callus derived from the endosperm of *Oxalis*. The amyloplasts of this tissue are virescent. In light, thylakoids are formed between the starch grains and they proliferate into grana as the starch disappears until they fill the whole enlarged plastid. Carotenoids are formed in the amyloplasts, but chlorophyll is not produced except when the grana develop. An interesting point is that cell division and growth decline as the plastids green and Sunderland and Wells consider that this is due to the diversion of nitrogen from general protoplasmic synthesis to thylakoid production. Now in leaf ontogeny the epidermal cells, although they stop dividing before the mesophyll, go on expanding after the mesophyll and thus it may be that their expansion is at the expense of chloroplast development.

Another point that comes from this work is that auxin treatments of the callus suppress thylakoid development, but not that of the amyloplast villi which probably bear the starch-synthesising enzymes. Previously it had been assumed that the photosynthetic lamellae were derived in light by invagination of the inner membrane of the plastid envelope exactly as the villi that occur in amyloplasts arise. This may still be true, and the fact that there is a difference in the evocation of the two kinds of membranes draws attention to some differences observed in the mode of formation of the membranes. So far as we know the inner membrane of the envelope always invaginates in both amyloplasts and chloroplasts. In the latter it may form the starch grains that often precede thylakoid formation. I have already mentioned the role of prolamellar bodies in forming the thylakoids in plastids developed in the dark when exposed to light. Israel and Steward (1967) hold that there is always a pre-thylakoid body from which the photosynthetic lamellae are made separately from the invaginations from the envelope. Their material was cultured cells of *Daucus* and in these the pre-thylakoid body, even when formed in the dark is different in construction from the prolamellar body described by Gunning (1965) from etioplasts of *Avena*. It is possible that neither of these precursors of thylakoids is derived from the envelope and so the difference between amyloplasts and chloroplasts is more fundamental than it was formerly supposed to be.

We now have three factors that influence the course of differentiation of proplastids to chloroplasts or amyloplasts — light, auxin, and intracellular competition for substrates. We do not yet know enough about any of them to say how they might affect the differentiation of cells in nature.

There is a number of other cases known where the plastids are specialized in relation to the differentiation of cells. One of these is the resin-secreting cell. These cells occur in many conifers often forming an epithelium around a canal. They contain many plastids with poorly developed internal membranes in contact with starch grains. A special feature is that they are invested by sheets of endoplasmic reticulum on which the ribosomes are attached only on the side away from the plastid (WOODING and NORTHCOTE, 1965a). Similarly sheathed plastids also occur in the companions cells of the phloem of *Acer* (WOODING and NORTHCOTE, 1965b). It looks as if sheathed plastids of this type are associated with cells that produce and export various substances, though the Golgi bodies are responsible for secreting carbohydrates in other kinds of cells. But ESAU and CRONSHAW (1968) apparently found no sheathing in the companion cells of *Cucurbita*.

The plastids of sieve tubes sometimes survive the changes that occur in the protoplasm of these cells during maturation though they commonly appear less dense than those of neighbouring cells. In monocotyledons they acquire proteinaceous bodies with a crystalline array of 5—6 nm subunits and they may even retain starch (BEHNKE, 1969). NORTHCOTE (1969) considers that the sheathing of plastids in the degenerating cytoplasm of xylem and phloem cells by endoplasmic reticulum screens the organelles from general autolysis so that the maturation of these cells goes stepwise. The ER is later withdrawn and the plastids then lose their contents except for their starch grains.

B. Golgi Bodies

After the plastids we know most about the Golgi in relation to differentiation. These bodies seem to be concerned with wall synthesis and with the secretion of carbohydrates. They have different functions and a different organization in animals.

The Golgi bodies of plants are discrete dictyosomes each consisting of a stack of flat, fenestrate cisternae. They occur in the somatic cells of all higher plants except for mature sieve tubes. The differences between cells in respect of the Golgi are in number and size of the dictyosomes and in their activity in budding off vesicles from the periphery of the cisternae. The Golgi, like the plastids, increase in number per cell at the summit of shoot apices when a plant is given a photoinductive cycle to promote flowering (GIFFORD and STEWART, 1965). This increase precedes growth and division of the cells.

One of the striking features of differentiation in root cap cells is the hypertrophy of the Golgi near the periphery of the cap. This has been described for *Zea* and *Triticum* particularly well (MOLLENHAUER, WHALEY, and LEECH, 1961; MOLLENHAUER and WHALEY, 1963; WHALEY, MOLLENHAUER and KEPHART, 1962; NORTHCOTE and PICKETT-HEAPS, 1966). In *Zea*, in the quiescent centre of root tips, where the least specialized cells occur, the cisternae of the Golgi are small and flat and they produce few vesicles. About half way (200 μm) from the quiescent centre to the tip of the cap the cisternae have increased diameters and many small vesicles come off their rims. At the periphery of the cap, but before the cells are sloughed off, the area of the cisternae increases further to six or seven times the area in the quiescent centre and the rims swell considerably so that the vesicles they cut off in greater numbers are much bigger. In cells about to be sloughed off the tip of the root the Golgi are smaller and produce small vesicles as in the partially differentiated cap cells. The hypertrophy is correlated with the secretion of polysaccharide to the outside of the cell carried in the vesicles which pass through the plasmalemma. NORTHCOTE and PICKETT-HEAPS (1966) showed that, in *Triticum* root caps, glucose is taken up by the Golgi and there converted to galactose, galacturonic acid, arabinose etc. These substances are then exported probably as insoluble pectic substances in the vesicles and then incorporated into the cell wall and the slime that surrounds mature cap cells.

This activity of the Golgi at the periphery of the root cap probably involves a change in the functions of the organelles possibly triggered off by a change in the availability of substrates. At the time it occurs the amyloplasts are breaking down and so the starch they contain probably becomes available and this may account for the sudden hypertrophy of the Golgi. Until this occurs the amyloplasts may compete for carbohydrate with the Golgi (Juniper and Roberts, 1966).

Echlin and Godwin (1968b) have shown that the Golgi are concerned with the laying down of thick callose walls in the pollen mother cells and pollen grains of *Helleborus*. Electron-dense material moves across the cytoplasm in Golgi vesicles as the wall thickens. Another instructive example of differentiation in relation to the Golgi also occurs in pollen. In the pollen grain, the vegetative nucleus becomes associated with more dictyosomes each of which has more cisternae and generates more vesicles than the generative nucleus. This difference is probably related to the vegetative nucleus's role in controlling pollen tube wall synthesis (Larson, 1963, 1965; Rosen et al., 1964; Dashek and Rosen, 1966). There is also a temporal change in the Golgi, for, when a pollen grain germinates, the Golgi stop producing ordinary small vesicles 42 nm across and start producing large vesicles 150 nm across possibly by whole cisternae on one face of the dictyosome partitioning themselves into vesicles. These vesicles migrate to the plasmalemma, sometimes after fusing with other vesicles in the cytoplasm, and there they hand over their contents to the wall especially near the tip of the pollen tube where the growth occurs. It is possible here that the vesicles contain precursors for proteins as well as for polysaccharides.

The Golgi are also implicated in root hair development and Bonnett and Newcomb (1966) think that they may produce two different kinds of vesicles, one with a chambered coat containing proteins and the other with a smooth envelope containing carbohydrates. The separation of the two components appears to occur in the large smooth vesicles just as they are budded off from the cisternae when the chambered vesicles are budded off from the smooth vesicles. Both kinds of vesicles deliver their loads to the wall of the growing hair.

In both pollen tubes and root hairs the walls have already been formed when the Golgi vesicles are conspicuously concerned with wall building and the role of the Golgi is probably related to rapid wall extension. It is also likely that the Golgi are concerned with the formation of new walls at the cell plate after mitosis, but this does not affect differentiation. An aspect of Golgi activity that does bear upon differentiation is their role in lignification. Golgi and their vesicles increase in numbers in tracheary elements of the xylem compared with xylem parenchyma and presumably they help to thicken the wall, but little is known about their role yet.

Golgi are also concerned with the activities of some special cells – some of the glands of insectivorous plants and the haustoria of flowering plant parasites. In both of these, the cells with enhanced Golgi activity secrete substances – a sugary "*Fangschleim*" in the stalked glands of *Drosophyllum* (Schnepf, 1966) and enzymes at the tips of the haustorial hyphae of *Cuscuta* (Bonneville and Voeller, 1963). In *Dionaea*, where there is a digestive cycle lasting 7–10 days, the Golgi of resting gland cells make small vesicles. When the glands have been stimulated, the Golgi make two kinds of large vesicles at a stage in the cycle that suggests that they are concerned with the replenishing of the used store of proteolytic enzymes and is related to the appearance of polysomes (SCHWAB, SIMMONS and SCALA, 1969).

One of the most remarkable enterprises of the Golgi is the prefabrication of the elaborate scales that occur on the outside of the walls and flagella of certain flagellates (MANTON and ETTL, 1965; MANTON and PARK, 1965; GREEN and JEN-NINGS, 1967). In *Mesostigma*, for example, three different kinds of scales are made inside the cell and are transported in Golgi vesicles to the exterior and placed in the appropriate position on the surface of the cell.

A somewhat similar fate has been suggested for the Ubisch bodies in stamens though they are not apparently created by the Golgi. Ubisch bodies are formed as spheroidal vesicles within tapetal cells, pass through the plasmalemma and reach the thecal fluid surrounding the pollen cells during their maturation. They acquire the resistant sporopollenin from the thecal fluid and form aggregates at the same time as the exine of the pollen grains also acquires its sporopollenin. It is tempting to conclude that the elaborately sculptured exine is made up of aggregated Ubisch bodies, but there is no real evidence for this view (GODWIN, 1968; ECHLIN and GODWIN, 1968a).

C. Mitochondria

All plant cells contain mitochondria except for the blue-green algae and anaerobic fungi. They are necessary in each cell because of their role in respiration and therefore their presence is not specially related to differentiation except in so far as they are modified in size, numbers or structure for the different respiratory needs of the cell. The large mitochondria of muscle are clearly related to the large demand for energy by these cells. Similar, though less conspicuous differences occur in plants. In secondary xylem the ray cells have bigger mitochondria than the tracheids – 0.6 μm in diameter compared with 0.3–0.5 μm (CRONSHAW and WARDROP, 1964).

There are also considerable differences in numbers per cell often related to cell size. The flagellate alga, *Micromonas*, has only one mitochondrion, the large amoeba, *Chaos*, has 500,000. The glandular cells of nectaries have more mitochondria than neighbouring cells (FREY-WYSSLING and MÜHLETHALER, 1965). In cells that have localized demands for ATP the mitochondria may be concentrated in a region of the cell — at the bases of flagella, near synapses in nerve cells. The internal membranes may also be elaborated in relation to differentiation. The total length of the cristae doubles during the maturation of *Arum* spadix cells (SIMON and CHAPMAN, 1961). In root tips of *Zea* the total length of cristae seen in sections varies from 1.3 μm in the quiescent centre to 2.3 μm in the statocytes of the cap and 1.7 μm in the elongated cells of the cortex (CLOWES and JUNIPER, 1964). The generative cell in pollen has fewer mitochondria with fewer cristae than the vegetative cell (LARSON, 1963). There are many other instances of apparent differences between cells in the numbers and in the development of the cristae. Some of these are subjective estimates and may be misleading. There is also a suggestion that the matrix of the organelles varies in density with cell differentiation.

Other Organelles

We know less about other organelles in relation to differentiation, but it is worth noting some of the correlations. The endoplasmic reticulum, though perhaps not strictly an organelle, shows some differentiation into rough (studded with ribosomes) and smooth (free of ribosomes), tubular and sheet-like. The significance of these differences is not fully known. Dormant cambium of *Fraxinus* has smooth ER and meristematic cambium has rough ER (Srivastava and O'Brien, 1966), but, in apical meristems of roots, the actively dividing cells have ER with fewer ribosomes than differentiating cells. The tips of pollen tubes have smooth ER which is tube-like here, but the tips of haustorial hyphae of *Cuscuta* have rough ER. Protein-secreting cells in plants are not always rich in rough ER, which is always sheet-like, as they are in animals. In the cotyledons of developing seeds of *Vicia faba* the ratio of free to membrane-bound ribosomes is high during the period of rapid cell division and then more membrane-bound ribosomes are produced during the accumulation of food reserves. Payne and Boulter (1969) consider that the two groups of ribosomes specialise in synthesising different sorts of proteins. Probably the polysomes on the endoplasmic reticulum produce the storage protein, but during maturation of the seeds they are detached and appear free in the cytoplasm. There is some evidence to suggest that the ER is concerned with movement of substances – carbohydrate in companion cells and resin or its precursors into resin canals (Wooding and Northcote, 1965a, b) for example and it may be that, even when secretion is effected by the Golgi vesicles, the ER supplies the enzymes or substrates to the Golgi bodies. The ER also plays a role in the sculpturing of thick cell walls made of both lignified cellulose and of callose as in xylem vessels and pollen grains (Pickett-Heaps and Northcote, 1966a; Angold, 1967; Echlin and Godwin, 1968b). It may also temporarily protect organelles from degradation while the rest of the cytoplasm undergoes drastic changes.

The possible role of the microtubules in determining mitotic spindle orientation or the plane of a dividing wall has already been mentioned. They also seem to play a role in aligning the microfibrils in the bands of secondary thickening in xylem vessels and tracheids (Hepler and Newcomb, 1964; Wooding and Northcote, 1964) and possibly also in aligning Golgi vesicles in the construction of the cell plate after mitosis.

Lysosomes, vacuoles, Golgi vesicles, pinocytosis vesicles, microbodies and spherosomes look rather alike in plant cells and the information available about their relation to differentiation in plants is still meagre. Lysosomes are identified in animals by their possession of hydrolases, particularly acid phosphatase, but this enzyme has been reported in spherosomes and mitochondria in plants. No doubt many organelles with different batteries of enzymes have evolved for specialized cells such as the peroxisomes discovered by de Duve (1969).

No essay on differentiation could be complete without mentioning nucleoli although they are not always treated as organelles. It was partly the fact that nucleoli enlarge in cells that become meristematic and are large in embryonic tissues that started the theory that RNA is concerned with protein synthesis, a theory that has grown to the whole subject of molecular genetics. It may be useful to consider the nucleolus as the manifestation of the activity of a gene, the nucleolar organizer, just as some people regard the puffs on polytene chromosomes or the loops on lampbrush chromosomes, as indicators of gene activity. Nucleoli are also larger than normal in rapidly growing malignant cells and in protein secreting cells, at any rate in animals, as in silk glands. The correlation between nucleolar size and protein synthesis almost certainly results from the role of the nucleolus in assembling ribosomes and possibly also in synthesising nuclear proteins. Many early stages of embryos lack nucleoli and if nucleolate nuclei are transplanted to animal eggs they initially lose their nucleoli. Early stages of embryos appear to have sufficient maternal ribosomes for their needs and they do not synthesise ribosomal RNA until later on. When it does start to be produced, it is at different ages in different kinds of cells (GURDON and WOODLAND, 1969) and is thus related to differentiation.

References

ANGOLD, R. E.: The ontogeny and fine structure of the pollen grain of *Endymion nonscriptus*. Rev. Palaeobotan. Palynol. **3**, 205—212 (1967).

AUDUS, L. J.: The mechanism of the perception of gravity by plants. Symp. Soc. exp. Biol. **16**, 196—226 (1962).

AVERS, C. J.: Fine structure studies of *Phleum* root meristem cells. II. Mitotic asymmetry and cellular differentiation. Amer. J. Botany **50**, 140—148 (1963).

— KING, E. E.: Histochemical evidence of intracellular enzymatic heterogeneity of plant mitochondria. Amer. J. Botany **47**, 220—225 (1960).

BEHNKE, H. D.: Die Siebröhren-Plastiden der Monokotyledonen. Planta **84**, 174—184 (1969).

BELL, P. R., FREY-WYSSLING, A., MÜHLETHALER, K.: Evidence for the discontinuity of plastids in the sexual reproduction of a plant. J. Ultrastruct. Res. **15**, 108—121 (1966).

— MÜHLETHALER, K.: The degeneration and reappearance of mitochondria in the egg cells of a plant. J. Cell Biol. **20**, 235—248 (1964).

BOARDMAN, N. K., ANDERSON, J. M.: Studies on the greening of dark-grown bean plants. Aust. J. Biol. Sci. **17**, 86—92 (1964).

BONNETT, H. T., NEWCOMB, E. H.: Coated vesicles and other cytoplasmic components of growing root hairs of radish. Protoplasma **62**, 59—75 (1966).

BONNEVILLE, M. A., VOELLER, B. R.: A new cytoplasmic component of plant cells. J. Cell Biol. **18**, 703—708 (1963).

BRAWERMAN, G.: A specific species of ribosomes associated with the chloroplasts of *Euglena gracilis*. Biochem. biophys. Acta (Amst.) **61**, 313—315 (1962).

BROWN, R.: Cellular basis for the induction of morphological structures. Nature (Lond.) **181**, 1546—1547 (1958).

CAPORALI, L.: Recherches sur les infrastructures des cellules radiculaires de *Lens culinaris* et particulièrement sur l'évolutions des leucoplastes. Ann. Sci. nat. Botan. **11** sér 20, 215—247 (1959).

CLOWES, F. A. L.: Apical meristems. Oxford: Blackwell Scientific Publications 1961.

— JUNIPER, B. E.: The fine structure of the quiescent centre and neighbouring tissues in root meristems. J. exp. Bot. **15**, 622—630 (1964).

— — Plant Cells. Oxford: Blackwell Scientific Publications (1968).

CRONSHAW, J., WARDROP, A. B.: The organization of cytoplasm in differentiating xylem. Aust. J. Botany **12**, 15—23 (1964).

DASHEK, W. V., ROSEN, W. G.: Electron microscopical localization of chemical components in the growth zone of lily pollen tubes. Protoplasma **61**, 192—204 (1966).

DE DUVE, C.: The peroxisome: a new cytoplasmic organelle. Proc. Roy. Soc. B173, 71—83 (1969).

DIERS, L.: On the plastids, mitochondria, and other cell constituents during oogenesis of a plant. J. Cell Biol. **28**, 527—543 (1966).

Echlin, P., Godwin, H.: The ultrastructure and ontogeny of pollen in *Helleborus foetidus*. I. The development of the tapetum and Ubisch bodies. J. Cell Sci. **3**, 161—174 (1968a).
— — The ultrastructure and ontogeny of pollen in *Helleborus foetidus*. II. Pollen grain development through the callose special wall stage. J. Cell Sci. **3**, 175—186 (1968b).
Esau, K., Cronshaw, J.: Plastids and mitochondria in the phloem of *Cucurbita*. Can. J. Botany **46**, 877—880 (1968).
Frey-Wyssling, A., Mühlethaler, K.: Ultrastructural Plant Cytology. Amsterdam: Elsevier 1965.
Gifford, E. M., Stewart, K. D.: Ultrastructure of vegetative and reproductive apices of *Chenopodium album*. Science **149**, 75—77 (1965).
Godwin, H.: The origin of the exine. New Phytol. **67**, 667—676 (1968).
Green, J. C., Jennings, D. H.: A physical and chemical investigation of the scales produced by the Golgi apparatus within and found on the surface of the cells of *Chrysochromulina chiton*. J. exp. Bot. **18**, 359—370 (1967).
Gunning, B. E. S.: The greening process in plastids. I. The structure of the prolamellar body. Protoplasma **60**, 111—130 (1965).
Gurdon, J. B., Woodland, H. R.: The influence of the cytoplasm on the nucleus during cell differentiation with special reference to RNA synthesis during amphibian cleavage. Proc. R. Soc. B173, 99—111 (1969).
Haynes, J., Burnett, A.: Dedifferentiation and redifferentiation of cells in *Hydra viridis*. Science **142**, 1481—1483 (1963).
Hepler, P. K., Newcomb, E. H.: Microtubules and fibrils in the cytoplasm of *Coleus* cells undergoing secondary wall deposition. J. Cell Biol. **20**, 529—534 (1964).
Israel, H. W., Steward, F. C.: The fine structure and development of plastids in cultured cells of *Daucus carota*. Ann. Bot. **31**, 1—18 (1967).
Jensen, W. A.: The ultrastructure of the egg and central cell of cotton. Amer. J. Bot. **52**, 781—797 (1965).
Juniper, B. E., Barlow, P. W.: The distribution of plasmodesmata in the root tip of maize. Planta **89**, 352—360 (1969).
— Clowes, F. A. L.: Cytoplasmic organelles and cell growth in root caps. Nature **208**, 864—865 (1965).
— Roberts, R. M.: Polysaccharide synthesis and the fine structure of root cap cells. J. R. Microsc. Soc. **85**, 63—72 (1966).
Kirk, J. T. O.: Nature and function of chloroplast DNA. 'Biochemistry of Chloroplasts' (Ed.: Goodwin), 319—340. London-New York: Academic Press 1966.
Küntzel, H.: Proteins of mitochondrial and cytoplasmic ribosomes from *Neurospora crassa*. Nature (Lond.) **222**, 142—146 (1969).
La Cour, L. F.: Nuclear differentiation in the pollen grain. Heredity **3**, 319 (1949).
Laetsch, W. M., Price, I.: Development of the dimorphic chloroplasts of sugar cane. Amer. J. Botany **56**, 77—87 (1969).
Larson, D. A.: Cytoplasmic dimorphism within pollen grains. Nature (Lond.) **200**, 911—912 (1963).
— Fine structural changes in the cytoplasm of germinating pollen. Amer. J. Botany **52**, 139—154 (1965).
Maltzahn, K. von, Mühlethaler, K.: Observations on division of mitochondria in dedifferentiating cells of *Splachnum ampullaceum*. Experientia (Basel) **18**, 315—316 (1962).
Manton, I., Ettl, H.: Observations on the fine structure of *Mesostigma viride*. J. Linn. Soc. (Bot.) **59**, 175—184 (1965).
— Park, M.: Observations on the fine structure of two species of *Platymonas* with special reference to flagellar scales and the mode of origin of the theca. J. mar. Biol. Ass. U. K. **45**, 743—754 (1965).
Marinos, N. G.: Multifunctional plastids in the meristematic region of potato tuber buds. J. Ultrastruct. Res. **17**, 91—113 (1967).
Mizukami, I., Gall, J.: Centriole replication. II. Sperm formation in the fern, *Marsilea* and the cycad, *Zamia*. J. Cell Biol. **29**, 97—111 (1966).
Mollenhauer, H. H., Whaley, W. G.: An observation on the functioning of the Golgi apparatus. J. Cell Biol. **17**, 222—225 (1963).

MOLLENHAUER, H. H., LEECH, J. H.: A function of the Golgi apparatus in outer rootcap cells. J. Ultrastruct. Res. **5**, 193—200 (1961).

MORTON, R. K., RAISON, J. K.: A complete intracellular unit for incorporation of amino acid into storage utilizing adenosine triphosphate generated from phytate. Nature (Lond.) **200**, 429—433 (1963).

NANNEY, D. L., RUDZINSKA, M. A.: In: The Cell (Ed.: J. BRACHET and J. MIRSKY) **4**, 109—150. New York-London: Academic Press 1960.

NEILSON-JONES, W.: Plant Chimeras. London Methuen 1969.

NORTHCOTE, D. H.: Fine structure of cytoplasm in relation to synthesis and secretion in plant cells. Proc. roy. Soc. B173, 21—30 (1969).

— PICKETT-HEAPS, J. D.: A function of the Golgi apparatus in polysaccharide synthesis and transport in the root-cap cells of wheat. Biochem. J, **98**, 159—167 (1966).

PAYNE, P. I., BOULTER, D.: Free and membrane bound ribosomes of the cotyledons of *Vicia faba*. Planta **84**, 263—271 (1969).

PICKETT-HEAPS, J. D., NORTHCOTE, D. H.: Relationship of cellular organelles to the formation and development of the plant cell wall. J. exp. Bot. **17**, 20—26 (1966a).

— — Cell division in the formation of the stomatal complex of the young leaves of wheat. J. Cell Sci. **1**, 121—128 (1966b).

RANDALL, J.: The flagellar apparatus as a model organelle for the study of growth and morphopoiesis. Proc. roy. Soc. B173, 31—62 (1969).

— DISBREY, C.: Evidence for the presence of DNA at basal body sites in *Tetrahymena pyriformis*. Proc. roy. Soc. B **162**, 473—491 (1965).

ROODYN, D. B., WILKIE, D.: The Biogenesis of Mitochondria. London: Methuen 1968.

ROSEN, W. G., GAWLICK, S. R., DASHEK, W. V., SIEGESMUND, K. A.: Fine structure and cytochemistry of *Lilium* pollen tubes. Amer. J. Botany **51**, 61—71 (1964).

SAGAN, L.: On the origin of mitosing cells. J. theor. Biol. **14**, 225—274 (1967).

SCHNEPF, E.: Feinbau und Funktion pflanzlicher Drüsen. Umschau Wiss. Technik **16**, 522—527 (1966).

SCHWAB, D. W., SIMMONS, E., SCALA, J.: Fine structure changes during function of the digestive gland of Venus's flytrap. Amer. J. Botany **56**, 88—100 (1969).

SCHWEIGER, H. G., SCHWEIGER, E.: The role of the nucleus in a cytoplasmic diurnal rhythm. Circadian Clocks (Ed.: ASCHOFF), pp. 195—197. Amsterdam: North-Holland 1965.

SIMON, E. W., CHAPMAN, J. A.: The development of mitochondria in *Arum spadix*. J. exp. Bot. **12**, 414—420 (1961).

SMITH, D., MUSCATINE, L., LEWIS, D.: Carbohydrate movement from autotrophs to heterotrophs in parasitic and mutualistic symbiosis. Biol. Rev. **44**, 17—90 (1969).

SRIVASTAVA, L. M., O'BRIEN, T. P.: On the ultrastructure of cambium and its vascular derivatives. Protoplasma **61**, 257—293 (1966).

STANGE, E. L.: Plant cell differentiation. Ann. Rev. Pl. Physiol. **16**, 119—140 (1965).

STEBBINS, G. L.: Polarity gradients and the development of cell patterns. In: Trends in Plant Morphogenesis (Ed.: CUTTER), pp. 115—139. London: Longmans 1966.

STEWARD, F. C.: Physiological aspects of organization. In: Trends in Plant Morphogenesis (Ed.: CUTTER). London: Longmans 1966.

STUBBE, W.: Dreifarbenpanaschierung bei *Oenothera*. II. Wechselwirkungen zwischen Geweben mit zwei erblich verschiedenen Plastidensorten. Z. Vererbungl. **89**, 189—203 (1958).

STUTZ, E., NOLL, H.: Characterization of cytoplasmic and chloroplast polysomes in plants: evidence for three classes of ribosomal RNA in nature. Proc. nat. Acad. Sci. (Wash.) **57**, 774—781 (1967).

SUNDERLAND, N., WELLS, B.: Plastid structure and development in green callus tissues of *Oxalis dispar*. Ann. Bot **32**, 327—346 (1968).

THIELKE, C.: Die Struktur des Vegetationskegels einer sektorial panaschierten *Hemerocallis fulva*. Ber. dtsch. bot. Ges. **68**, 233—238 (1955).

THIMANN, K. V., CURRY, G. M.: Phototropism. In: Light and Life. (Ed.: McELROY GLASS), 646—672, Baltimore: John Hopkins Press 1961.

Thompson, W. W., Weier, T. E., Drever, H.: Electron microscopic studies on chloroplasts from phosphorus deficient plants. Amer. J. Botany 51, 933—938 (1964).

Turing, A. M.: The chemical basis of morphogenesis. Phil. Trans. Roy. Soc. B 237, 37—72 (1952).

Virgin, H. I., Kahn, A., Wettstein, D. von: The physiology of chlorophyll formation in relation to structural changes in chloroplasts. Photochem. Photobiol. 2, 83—91 (1963).

Whaley, W. G., Mollenhauer, H. H., Kephart, J. E.: Developmental changes in cytoplasmic organelles. Vth Internat. Congr. Electron Microscopy 2, W12 (1962).

Wooding, F. B. P., Northcote, D. H.: The development of the secondary wall of the xylem in *Acer pseudoplatanus*. J. Cell Biol. 23, 237—337 (1964).

— — The fine structure of the mature resin canal cells of *Pinus pinea*. J. Ultrastruct. Res. 13, 233—244 (1965a).

— — The fine structure and development of the companion cell of the phloem of *Acer pseudoplatanus*. J. Cell Biol. 24, 117—128 (1965b).

Yotsuyanagi, Y.: Études sur le chondriome de la levure. J. Ultrastruct. Res. 7, 141—158 (1962).